Particle Packing Characteristics

Particle Packing Characteristics

by Randall M. German

Materials Engineering Department
Rensselaer Polytechnic Institute
Troy, New York

METAL POWDER INDUSTRIES FEDERATION

Princeton, New Jersey

Library of Congress Cataloging-in-Publication Data

German, R. M.
Particle packing characteristics / by Randall M. German.
p. cm.
Includes bibliographies and index.
ISBN 0-918404-83-5 : $65.00 (est.)
1. Powder metallurgy. 2. Compacting. I. Title.
TN695.G4695 1989
671.3'7--dc20 89-50340
CIP

ISBN 0-918404-83-5
Library of Congress Catalog Card No. 89-50340

Printed in the United State of America

dedicated to my parents
Helen and Dick German

Contents

Chapter Three
Factors Affecting Particle Packing

Chapter Four
Packing of Monosized Spheres

Chapter Five
Packing of Monosized Nonspherical Particles

Chapter Six
Bimodal Mixtures of Spherical Particles

Preface

Many research and development projects have been conducted that use powders. However, insufficient attention is often given to the particle packing characteristics. I have studied this topic in great detail and am convinced that more advantage can be made from a better understanding of particle packing. My interest has been heightened by the advent of powder injection molding. In this manufacturing process, a powder is mixed with a polymeric binder and the mixture is molded into a complex shape using conventional plastic injection molding technologies. After molding, the binder is removed and the porous structure is densified by sintering. Unlike die compaction, the particle packing density after molding is close to that for the loose powder. Consequently, the overall success in fabricating high performance structures depend on the ability to work with high packing density particulate systems. A high compaction force could be applied to the powder to overcome poor packing. However, with powder injection molding the inherent packing behavior must be attended to directly. Accordingly, my attention to this subject resulted in my eventual decision to produce this book.

One reason for organizing the information on particle packing characteristics was the discovery that researchers and practitioners in many fields of technology are immersed in very similar problems. Interest in particle packing exists in studies dealing with particulates, whether they are oxide ceramics, pharmaceuticals, filter media, catalysts, minerals, nuclear fuels, metal powders, soils, cereals, casting sands, or seeds. Unfortunately, insufficient communication takes place between the various groups working in these diverse fields, resulting in replicated effort in learning the same basic principles over and over. To assess the contributions of these diverse fields, I conducted a broad survey of the literature. I was immediately overwhelmed by the quantity of information available. By no means can any book do justice to the many prior investigations. As just an example of the problem, there are well over one thousand prior citations of work on pore characterization using mercury porosimetry. Likewise, major literature compilations deal specifically with permeability, packing density, computer simulations, surface area, capillarity, compaction, strength, percolation theory, and several other topics treated here. Thus, my focus has been on distilling this diverse and large body of knowledge into a unified treatment that will serve as a basis for teaching, research, and improved industrial practices. The book contains many references (approximately 650) for the reader interested in the details. For those with the time to read the book from cover to cover, they will learn the essential elements of particle packing. This is appropriate for students newly discovering this field. Alternatively, since there is often need for quick access to the knowledge base on selected topics, the presentation is organized so that specific segments can be consulted as needed. This will be most useful to the practicing engineers and scientists that face the day-to-day problems associated with particulate systems. Each of the book's chapters contains its own set of references. The table of contents provides a quick overview of the scope and organization of the book.

Since my background is in materials science and engineering, I have a natural bias to select examples and literature that relate to this field. However, I have considered the many areas where particle packing characteristics are important and have used illustrations from several fields. The text is useful to those with practical or research interests in almost all forms of technology, including aeronautics, agriculture, biology, ceramics, chemical engineering, chemistry, civil engineering, composites, electrical engineering, foods, geology, mechanical engineering, medicine, metallurgy, nuclear engineering, paints, pharmaceuticals, physics, and polymers. Obviously, the packing of particles is fundamental to a wide variety of disciplines.

The manuscript was generated during 1987 and 1988. A sabbatical leave during the first half of 1987 played an important role in setting up the literature base and determining the scope of the book. I am grateful to Professor Guenter Petzow at the Powder Metallurgy Laboratories of the Max Planck Institute, Stuttgart, Federal Republic of Germany, for hosting my visit and providing the environment appropriate for such an effort. Subsequent work was performed as part of the industrially sponsored Advanced Powder Processing program at Rensselaer, which was ably administered by Dr. Robert Messler, Associate Director of the Center for Manufacturing Productivity and Technology Transfer. This team program provides an important link between the academic and industrial world. Indeed, many of the topics addressed in this book reflect questions posed by the research sponsors. The text was written at Rensselaer with help from several individuals. Andy Plymale and Christine Kipphut deserve special mention for their assistance in editing and proofing the manuscript. The comments of Animesh Bose, Kevin Christian, Shaji Farooq, Preston Kemp, Ronald Kurzeja, Patrick McGeehan, Shun-Tian Lin, Rainhard Laag, Barbara Lograsso, Paul Murley, Jean Murray, Robert Oddone, Richard Royer, and John Warren were very helpful. Finally, I wish to thank the sponsors of the Advanced Powder Processing program at Rensselaer for their support and interest. The members of this group are Alcoa, Allied-Signal, AMAX Specialty Metals, Boart International, Ceramics Process Systems, Cincinnati Milacron, Degussa, Digital Equipment, Fine Particle Technology, General Motors, International Business Machines, Kaiser, Kulite Tungsten, Omark Industries, Rocky Mountain Orthodontics, SCM Metals, and Teledyne Firth Sterling. Additional research support was provided by Defense Advanced Research Projects Agency, National Science Foundation, New York State Science and Technology Foundation, Metadyne, and Automated Dynamics.

I have dedicated the book to my parents, Helen and Dick German, who enjoy retirement in Murphys, California. Their tolerance and encouragement for my infatuation with science and engineering is deeply appreciated.

The Author

Randall M. German is Professor of Materials Engineering at Rensselaer Polytechnic Institute, in Troy, New York, where he teaches and conducts research in several aspects of powder metallurgy and ceramics. Much of his current research focuses on the fabrication of high performance materials and components from particulates. He has been with Rensselaer since 1980, after prior experience in industry as Director of Research (1978-1980) for the J. M. Ney Co., Bloomfield, CT; Director of Research and Development (1977-1978) for Mott Metallurgical Corp., Farmington, CT; Member of the Technical Staff (1969-1977) of Sandia National Laboratories, Livermore, CA; and Materials Scientist (1968-1969) with Battelle Columbus Laboratories, Columbus, OH. Each of these positions were involved with particulate solids. He obtained his Ph.D. degree in Materials Science from The University of California (Davis) in 1975 while employed full time at Sandia, following his B.S. degree from San Jose State University (1968) and his M.S. degree from The Ohio State University (1971).

Dr. German has conducted several research projects for industrial and government sponsors while at Rensselaer. He is the recipient of several awards, including the RPI Early Career Award (1984), SAE Teetor Engineering Educator Award (1984), Outstanding Young Men of America (1983), Honorary Professorship at Northeast University of Technology (Shenyang, PRC, 1985), ASM Geisler Young Metallurgist Award (1982),

Samsonov Prize of the International Team for the Science of Sintering (1985), and is a Patroon of Rensselaer. He serves as a reviewer and consultant for many companies, government agencies, journals, and publishers.

Dr. German's publication list includes over 200 journal articles, books, and patents, covering powder metallurgy, ceramics, and physical metallurgy. His previous books are *Powder Metallurgy Science* (1984) and *Liquid Phase Sintering* (1985). He has also edited *Processing of Metal and Ceramic Powders* (1982) with K. W. Lay.

Chapter One

Introduction to Particle Packing

Chapter Outline

Overview

An understanding of particle packing characteristics is important in many diverse fields of technology. A major concern is for the influence of the particle characteristics on the consolidation process. The desire may be to attain a high packing density with the fewest processing problems. Frequently, monosized spheres are used as models for such processes. For example, large steel ball bearings such as those shown in Figure 1.1 are employed to simulate the packing of particles. This form of modeling is quite useful in determining the key factors influencing density and coordination number. However, most practical powders are highly variable in character [1]. Real powders have distributions in size and shape, and exhibit agglomeration and cohesion affects not seen in these large sphere model packings. Figure 1.2 illustrates this point by showing scanning electron micrographs of four common

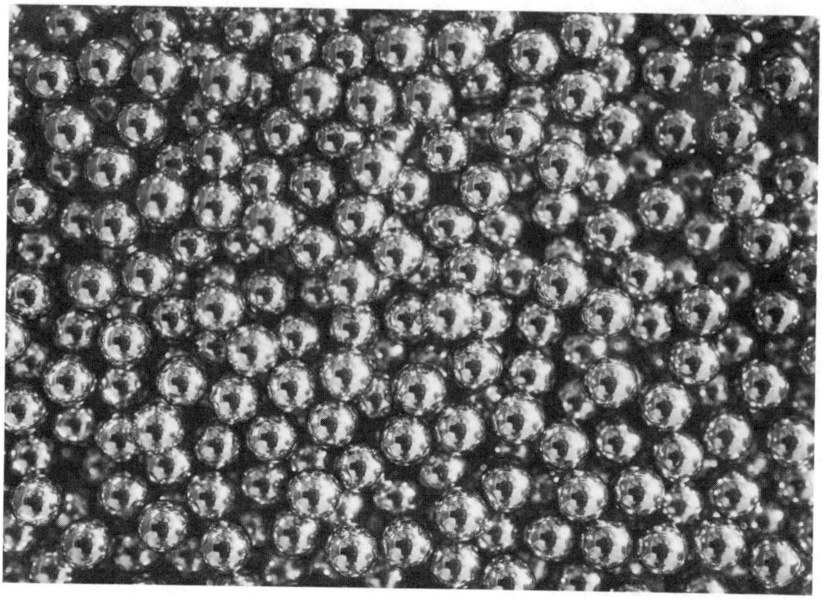

Figure 1.1 A picture of a model random packing of large spheres composed of monosized steel balls.

powders. This set of micrographs illustrates some of the variety in particle size, shape, agglomeration, and surface topography associated with particulate systems used in engineering. In contrast with Figure 1.1, these are far from the ideal monosized spheres. Thus, both large sphere modeling and actual particulate studies are needed to fully appreciate the characteristics of packed particles.

The focus in dealing with particles is on the dominant factors such as the particle size distribution, particle shape, interparticle friction, surface chemistry, and agglomeration. Each of these has an established effect on the packing characteristics. It is also possible to adjust the particles to attain high packing densities or to attain specific compact properties. For example, the use of bimodal or trimodal particle size mixtures can lead to improved packing. However, the composition and particle size ratio must be

Figure 1.2 Scanning electron micrographs demonstrating the range of particle characteristics encountered in powder systems; a) atomized aluminum, b) milled magnetite, c) reduced iron, and d) atomized palladium alloy.

simultaneously controlled for optimal properties. Furthermore, the use of fibers to form composites opens questions of the appropriate particle size, fiber diameter, fiber length, and composition for optimal behavior. The mechanical properties of the composite dictate one possible combination of characteristics, while the optimal packing characteristics require a different combination. These and other factors associated with particle packing are treated in this book. The primary concern is with forming a foundation for predicting the packing density, pore size, permeability, strength, conductivity, and other important features.

Interest in particle packing comes from very diverse fields and spans several centuries [2-5]. For example, Newton was concerned with the packing of spheres and considered the problem of the maximum coordination number. In 1694, he and Gregory disputed the maximum coordination number possible for equal sized spheres. It was thought that 13 spheres could be made to touch a central sphere in an ideal case, since the solid angle subtended by a contacting sphere is slightly less than 1/13 of the total. By 1874 Hoppe demonstrated that a maximum coordination number of 12 was possible. Many of the initial studies were concerned with ordered packings of spheres and the conditions for high packing densities. A classic early experiment reported by Hales in 1727 and Buffon in 1753, involved loading a container with dried peas and allowing them to absorb moisture. When the peas swelled and filled the container, each could be removed and the number of faces counted to estimate the coordination number. Later experiments used compressed lead shot to demonstrate that the final packing coordination number depended on both the initial packing arrangement and the mode of compression. Typical experimental coordinations were between 12 and 14. However, Smith [5] notes that theoretically there is no flat sided polyhedron that is space-filling with sharing faces that meet the requirements for surface tension equilibrium at the interfaces and junctions. The tetrakaidecahedron (14 sides) comes close to satisfying equilibrium, but often the dodecahedron (12 sides) is observed in nature. Hence, the experimental observations are not easily modeled.

The most widely considered characteristic is the density of a particle packing [4,6]. For monosized spheres the maximum packing density occurs in an ordered close-packed array with a coordination number of twelve. This packing has a fractional density of 0.7405. So far only one proof exists that this is the maximum packing density in three dimensions for a replicating structure [7], and this proof is not well accepted; the generally accepted proof is for a maximum packing density of 0.7796. There is considerable theoretical interest in dense packing structures beyond three dimensions, for example, in communication theory [3]. Thus, the field of particle packing has captured the attention of many investigators over a long time span, yet there are still simple, lingering questions.

An accurate treatment of random packing is more recent. Although random packing occurs frequently in nature, the lack of order proves a major difficulty in constructing models. Monosized steel or glass spheres have been used to create random packings, but the evaluation is often hampered by ordering at the container wall and particle repacking because of vibration. Consequently, accurate descriptions of random packing have proven elusive until recently. The random packing experiments by Scott [8], published in 1960, are recognized as being the first accurate treatment. By using various container sizes and careful loading techniques, the true random densities were obtained by extrapolating to an infinite container. Based on Scott's measurements, the loose random density is near 0.60 and the dense random value is 0.637. These values have provided a basis for testing subsequent experiments, many of which were performed using computer simulation techniques. It is recognized that only four spheres can have mutual contacts. The four sphere tetrahedral structure with variable dimensions and packing density is widely used in modeling random particle packings. However, only recently has the density obtained by numerical techniques approached the experimental value obtained by Scott using monosized spheres.

Investigations into particle packing characteristics come from a diversity of areas. As a consequence of the diversity, there is an excellent fundamental understanding of particle packing, especially for spheres. Unfortunately, this information is not well organized for use by engineers and scientists involved with powder processing. In this book, the findings from these diverse fields are combined into a statement of current understanding, with little attention to the applications behind each study. The presentation will start with a set of definitions. Subsequent chapters will treat the factors affecting packing, first of monosize spheres, then of nonspherical particles, mixed particle sizes, and mixed compositions. Also included is treatment of the capillary behavior (wetting and draining), fluid flow, surface area, pore size, and compaction of particle packings. The book summarizes these various topics to emphasize the importance of particle packing with respect to several powder processing topics.

Definitions

The first task is to provide working definitions of the terms applicable to particle packing. The following are organized in alphabetical order and are based on usual definitions [4,6,9-12]:

agglomeration - the adherence of solid particles to one another to form a cluster, typically associated with small particles

angle of repose - the natural angle of inclination from the horizontal that a powder assumes when poured into a container

apparent density - the mass of a powder in the loose condition divided by the volume, this may be normalized to the theoretical density for the material

bimodal powder - a mixture of two particle sizes which has two distinct modes in the particle size distribution

binder - a solid or liquid cementing medium that wets the particles and fills the pore space to provide strength to the particle mass

bridging - the formation of relatively large gaps or voids in a powder mass due to interparticle friction

connectivity - a measure of the interlinking contacts of one phase with similar composition particles in a multiple phase mixture of particles

coordination number - the number of neighboring particles forming contacts with a given particle

density - mass divided by volume; used synonymously with fractional packing density in this book

fractional density - the density expressed as a fraction of the theoretical density for the material *hydraulic pore diameter* - a measure of the effective pore size with respect to flow resistance and capillary phenomena

interparticle friction - the resistance to particle motion past other particles due to the character of the contacts between particles

monosized powder - particles of essentially one size

multimodal powder - a mixture of several monosized powders to create a wide and broad particle size distribution

ordered packing - a packing where the particles occupy specific sites with respect to each other such that long-range order exists, as in a crystal structure

packing - an assemblage of particles in a container

particle - the smallest discrete unit of a powder mass that can not be easily subdivided

particle shape - an index of the external profile for a particle, often expressed in terms of departure from a spherical shape

particle size - the controlling linear dimension of a particle, usually expressed in terms of the diameter of an equivalent sphere

particle size distribution - the variation in weight or population of particles shown as a function of the particle size

permeability - a measure of the ease of fluid flow through a powder bed

porosity - volume fraction of void space in a powder mass; the porosity equals one minus the fractional density

powder - a collection of matter in a finely divided state; particulate matter; discrete particles of a dry material

radial distribution function - the functional description of the number of particles, or density of particles, with respect to the radial distance from a central particle *random dense packing* - a random packing where the particles have been agitated to attain the closest packing possible without introducing long-range order or deformation; analogous to the tap density

random loose packing - the packing of particles poured into a container without agitation or vibration, representing the lowest stable packing density without introducing long-range order or deformation; analogous to the apparent density

random packing - a packing of particles without order

size segregation - the size dependent separation of particles from a homogeneous mixture, often due to agitation

sphericity - a quantitative particle shape index based on the ratio of the surface area of a sphere of equivalent volume divided by the surface area of the particle

surface area - a measure of the external area of a particle, usually expressed per unit mass or per unit volume

tap density - the density of a powder after repeated vibration, giving the maximum density attainable without compression

tesselation - any arrangement of shapes that fit together to fill space without overlap or void formation

trimodal powder - a mixture of three monosize powders which has three distinct peaks in the particle size distribution

unit cell - the smallest portion of a particle packing system that gives complete representation of the structure and the distribution of voids in the structure

void - a single volume of space not occupied by solid in a particle packing.

Key Concerns

The key characteristics of a particle mass which are especially important to processing include:

* packing density,
* strength,
* surface area,
* packing coordination,
* permeability,
* pore size, and
* connectivity.

The first major characteristic is packing density or porosity. Shrinkage in sintering, fluid storage capacity, gas permeability, deformation in compaction, and compact strength all depend on the packing density. As an example, the amount of binder added to a powder for injection molding should slightly exceed the void space between particles. There are disadvantages to either an excess or deficiency of binder: compact distortion or defect formation. The strength or stability of the powder mixture is largely dictated by the interparticle friction which arises from the particle size, shape, surface chemistry, and surface texture. In turn, the interparticle friction influences mixing, agglomeration, and size segregation tendencies. The surface area of the particles is a function of the particle size distribution and particle shape. The geometric stability, packing density, irreducible saturation, and sintering shrinkage are dependent on the number of contacts per particle as measured by the coordination number. A characteristic such as permeability is governed by the porosity and surface area. Permeability is a measure of the flow conductance through the voids between particles. Permeability controls the filtration behavior of a particle bed and the ease of fluid removal by drying. The pore size controls the trapping of fluids in the pores, as well as the wicking of fluids in and out of the pores. Finally, the structural connectivity of the particle packing is important, especially with mixed phases. This differs from the coordination number in that two or more phases are involved. For example, a continuous network of metal phase is necessary to achieve electrical conduction in a mixture of insulator and metal powders.

Compaction is an important step in working with particulates. The compaction stage not only gives shape to the powder, but also decreases the porosity and increases the coordination number. A loose powder has a relatively low fractional density as compared with its condition after high pressure compaction. The porosity for a loose powder typically ranges from 0.25 to 0.65. In the fabrication of components from powders, both compaction and sintering are used to remove this porosity. The compaction stage eliminates a great deal of the void space. Further densification relies on shrinkage of the compact during sintering. The linear shrinkage in sintering $\Delta L / L_o$ (change in compact dimension divided by the initial dimension) relates to the initial fractional solids density f_i and the final, sintered fractional solids density f_f as follows:

$$\Delta L / L_o = 1 - [f_i / f_f]^{1/3}. \tag{1.1}$$

Consider the importance of Equation 1.1 to ceramic processing. At low packing densities considerable shrinkage is needed to attain the 0.95 to 1.00 fractional density level associated with high performance materials. Figure 1.3 shows a plot of the linear shrinkage versus the initial fractional powder packing density (solids content) f_i for final fractional densities f_f of 0.95 and 1.00. As the initial solids density decreases, the amount of shrinkage increases. For a particle packing density of 0.75, the linear shrinkage needed to attain a final fractional density of 0.95 is 7.6%. By contrast, with an initial density of 0.50 the linear shrinkage is 19.3%, or nearly three times as much. Interestingly, there is relatively little difference in shrinkage for the two final density levels. The key point is that less shrinkage occurs with high initial packing densities. This contributes to easier process control. Warpage, slumping, and distortion in sintering are more prevalent at the low packing densities. A uniform packing density helps in subsequent processing by giving uniform and predictable shrinkage in sintering. Furthermore, as Figure 1.3 demonstrates, high packing densities are beneficial because of less shrinkage during sintering, making process control much easier.

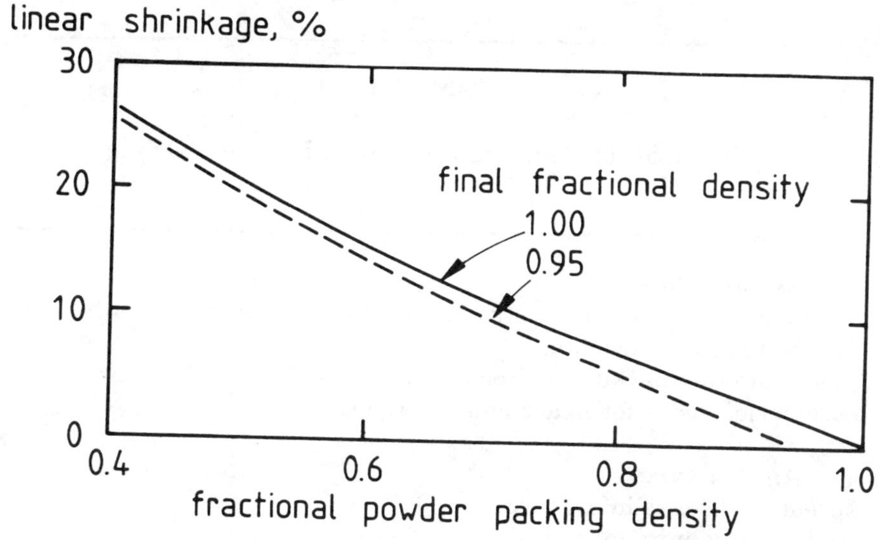

Figure 1.3 A plot of the linear shrinkage versus the initial fractional powder packing density showing the amount of uniform shrinkage needed to attain final fractional densities of 0.95 or 1.00.

Applications for Particle Packing Concepts

Interest in particle packing extends over a very broad range of subjects. Indeed, this chapter can only list the areas and provide an introductory description of a few to illustrate the diverse fields that are involved with the packing characteristics of particles. Along these lines, Table 1.1 has been constructed to introduce the breadth of interests [3,4,9,13-43]. The references provide a first basis for study on the application areas. In some cases the link between the subject and particle packing characteristics may be unclear. One example is in signal separation and encoding under the heading of electrical engineering. In this instance, the mathematics of sphere packing has been used to create a solution in multiple dimensions that has applicability to signal transmission.

Table 1.1

Examples of Concepts Relating to Particle Packing

Aeronautics

boundary-layer porous structures for turbulence control and transpiration cooling
solid state fuel packing for rockets and missiles
porous electrodes for space ionization engines

Agriculture

spread of disease in orchards
packing and storage of seeds
water seepage and drying in soils

Biology

cellular structure of plants and animals
permeation of moisture or contaminants in soil

Ceramics

density, strength, homogeneity, and defects in powder compacts
structural models of glasses
selection of feedstock for powder injection molding, slip casting, and tape casting
powder synthesis and agglomeration
shrinkage and densification during sintering
design of low thermal conductivity refractories

Chemical Engineering

filtration behavior of particulates, including reverse osmosis, biochemical separation, and fluid polishing
packing characteristics for fluidized beds or pneumatic transport
flow characteristics through packed particles for polymer orientation,
operation of air bearings, spargers, and flow restrictors
design of packed beds for catalysis and heat exchangers

Table 1.1 Continued

Chemistry

filtering, settling, and agglomeration of colloids
porosity and deposition at electrodes, electrode design
porous fuel cell and battery design
aerosol dispersion

Civil Engineering

soil foundations and asphalt mixtures for roadways
optimized concrete compositions for structures
optimized design of parking lots

Composites

selection of fibers and particles for strengthening

Electrical Engineering

porous semiconductor heat exchanger design
electrical conductivity of thick film pastes
models for signal separation and encoding, error elimination

Foods

packaging density, transport, settling of granular foods

Geology

oil and gas recovery from wells, gas permeability, fluid displacement
crystal structure of minerals
formation of structures from packed particles: sandstone, beaches, and glaciers

Mechanical Engineering

heat transfer in porous heat pipes and porous substrates

Medicine

tissue attachment, strength, and corrosion of prosthetic implants
spreading of disease

Table 1.1 Continued

Metallurgy

optimization of foundry casting sands for permeability, strength, and dimensional control
feedstock preparation for blast furnaces and ore induration
compaction, sintering, and hot isostatic compaction of particulates

Nuclear

packing density, energy concentration, radiation attenuation of fuels
design of gaseous diffusion membranes for uranium enrichment
calculation of radiation damage via particle cascades

Paint Technology

selection of particle sizes and shapes to ensure proper loading and coverage
viscosity variation with powder characteristics

Pharmaceuticals

filtration and recovery of bacterial products
compaction of pills

Physics

combustion, energy absorption, and shock wave attenuation associated with explosives
design of flame arrestors
hard sphere models for liquids, gases, melting, and thermodynamic properties
solid-state physics including structure, bonding, and coordination for amorphous metals and glasses
thin film structures including amorphous and epitaxial layers

Polymers

particle selection for electrically conductive loaded plastics
strengthening of polymer structures

The packing of particles is initially concerned with the fractional density. However, factors like the coordination number, connectivity, agglomeration, permeability, drainage, filtration, pelleting, flow, strength, compaction, and pore size are important in specific applications. The range of these attributes, especially the particle sizes, particle shapes, and particle size distribution, are very wide. As one example, the physicist is concerned with constructing a hard sphere model for amorphous metals, glass, or vapor deposited thin films. This means that the particles are individual (idealized) atoms. At the other end of the size scale, civil engineers are concerned with the packing of rocks, gravel, asphalt, sand, and concrete. Here the particles can be up to a meter in size. Thus, the particle size range is at least 10^9. At the intermediate size level are many particle systems of concern in chemical engineering. These include topics relating to filtration, catalysis, fluidized beds, pneumatic transport, and heat exchangers.

Milling is important in several fields, such as mining and ceramics. Coal mining provides a particularly good example. Granulated coal is needed for easier transport and process control. For transport and storage, a high packing density, wide particle size distribution, and spherical particle shape are desirable. However, the brittle character of coal leads to an irregular particle shape that inhibits packing and transport. As the particle size is decreased, the packing density suffers because of the higher interparticle friction. Consequently, it is desirable to have a large particle size, not only for economic reasons, but for practical handling considerations. One option is to use a wide particle size distribution to allow the smaller particles to pack between the large particles.

The deposition of thin films on substrates by sputtering or vapor deposition is a critical step in the construction of integrated electronic circuits. The deposited layers can be epitaxial or amorphous, depending on the deposition conditions, deposited material, and substrate structure. Studies on amorphous layers are often complimented by analog hard sphere particle models. For example, in the deposition of amorphous germanium on graphite, it was observed that a packing coordination number of five was most frequent [39]. The five-fold coordination structure is not crystalline, yet it fills space. Such a structure is not predicted by any crystalline concepts; however, in packing studies on compressed lead spheres it has been noted that pentagonal structures are commonly observed. Thus, five-fold symmetry exists in random particle packing arrangements. In both the amorphous layers and random packings of spheres, there are staggered structures lacking long-range order. This can be established using x-ray techniques, and the resulting radial distribution functions have features well matched by random hard sphere models formed using ball bearings. On heating, the two-dimensional layers form a structure that is partially crystalline prior to melting. Again, this has a direct analog to structures observed in random sphere packings. The packing density of packed hard spheres matches well the density of an amorphous structure. Furthermore, several of the thermodynamic, magnetic, thermal, electron, and mechanical properties can be related to characteristics of sphere packings [17,40].

One of the most common means of forming metals involves casting a melt in a mold shaped from sand. A foundry sand should have a high strength for handling ease, high permeability to allow air to be vented from the mold as the metal enters, smooth surface for a good casting finish, and high stiffness to minimize dimensional change. Empirical examinations have found that wide particle size distributions for the sand give the desired strength, surface finish, and dimensional change characteristics [41]. However, if the packing density is too high, then the permeability is unacceptable because air pockets form on the casting surface. As the proportion of small particles is increased the strength of the sand is improved, but the permeability is decreased. Here, particle packing concepts are applicable in explaining the optimized blends of sands. The smaller particles fill the voids between the larger particles, thereby improving the packing density. Furthermore, small particles contribute considerable interparticle friction to the mixture, making it stronger and stiffer. But, by filling the voids between the large particles, the space available for air passage is decreased and the permeability suffers. The optimized particle size distributions rely on a balance between these three packing characteristics.

Another example of routine use of particle packing concepts is in oil recovery from porous rocks or sands. In many cases the output from an oil well is low and never complete. This can be explained by the pendular bonds between particles and a property called the irreducible saturation. In essence, the oil is stabilized by capillary wetting of the particle contacts and resists flow to the well. The amount of trapped oil depends on the particle size, packing density, and coordination number. Some drainage will occur over time, but not all of the oil can be recovered by drainage. The output can be increased by using miscible displacement techniques [42,43]. One example is to drill a well pattern in an oil field where one well is located at each corner of a square. These are recovery wells. Then a fifth well is put in the center of the square. This well is pressurized with a fluid to displace the remaining oil from the porous structure. However, viscous fingering can occur if the displacing fluid is significantly lower in viscosity than the oil, or if there are differences in permeability or wetting. One consequence of viscous fingering is that the displacing fluid will form narrow channels to the recovery wells and largely sweep around the residual oil. In this case the recovery of oil will be quite low. There are significant economic gains possible from understanding the conditions that prevent viscous fingering in packed particle structures.

These examples involving the granulation of coal, thin film structures in microelectronic devices, foundry practice in metal casting, and oil recovery via displacement are representative of the diverse interests in particle packing. This field of study extends to many areas of science and technology. For this reason, the balance of the book largely focuses on particle packing characteristics with little direct attention to applications. The reader faces the remaining challenge of applying these concepts to the individual application areas. The fundamentals developed here should greatly accelerate this progression.

The questions addressed in this book relate to attaining a high packing density powder and identification of the roles of particle size distribution, particle shape, surface texture, particle agglomeration, and container size. Likewise, problems associated with compaction, strength, coordination number, connectivity, permeability, wetting, and pore draining are covered as they relate to packed particles. This presentation of particle characteristics sets the stage for the subsequent study of subjects directly related to applications and processing schemes.

In this book the various size and particle composition differences will not be emphasized. Rather, a general treatment of particle packing that transcends each individual field is presented. The benefit from this approach is that general principles emerge that are of value to each area, so that cross-fertilization can occur. The challenge to the reader will be to translate these general principles into specific solutions to problems relating to his or her field of interest and study.

References

1. W. J. Huppmann and K. Dalal, *Metallographic Atlas of Powder Metallurgy*, Verlag Schmid, Freiburg, Federal Republic of Germany, 1986.
2. K. Ridgway and K. J. Tarbuck, "The Random Packing of Spheres," *British Chemical Engineering*, 1967, vol. 12, pp. 384-388.
3. N. J. A. Sloane, "The Packing of Spheres," *Scientific American*, 1984, vol. 250, no.1, pp. 116-125.
4. D. J. Cumberland and R. J. Crawford, *The Packing of Particles*, Elsevier Science, Amsterdam, The Netherlands, 1987.
5. C. S. Smith, "Further Notes on the Shape of Metal Grains: Space-Filling Polyhedra with Unlimited Sharing of Corners and Faces," *Acta Metallurgica*, 1953, vol. 1, pp. 295-300.
6. C. A. Rogers, *Packing and Covering*, Cambridge University Press, London, UK, 1964.
7. A. H. Boerdijk, "Some Remarks Concerning Close-Packing of Equal Spheres," *Philips Research Reports*, 1952, vol. 7, pp. 303-313.
8. G. D. Scott, "Packing of Equal Spheres," *Nature*, 1960, vol. 188, pp. 908-909.
9. R. M. German, *Powder Metallurgy Science*, Metal Powder Industries Federation, Princeton, NJ, 1984.
10. K. Shinohara, "Rheological Property of Particulate Solids," *Handbook of Powder Science and Technology*, M. E. Fayed and L. Otten (eds.), Van Nostrand Reinhold, New York, NY, 1984, pp. 129-169.
11. W. A. Gray, *The Packing of Solid Particles*, Chapman and Hall, London, UK, 1968.
12. R. L. Brown and P. G. W. Hawksley, "Packing of Regular (Spherical) and Irregular Particles," *Nature*, 1945, vol. 156, pp. 421-422.
13. E. C. Childs, *An Introduction to the Physical Basis of Soil Water Phenomena*, Wiley Interscience, New York, NY, 1970.

14. D. M. Bigg, "Electrical Properties of Metal-Filled Polymer Composites," *Metal-Filled Polymers*, S. K. Bhattacharya (ed.), Marcel Dekker, New York, NY, 1986, pp. 165-226.
15. J. D. Bernal, "The Structure of Liquids," *Scientific American*, 1960, vol. 203, no.2, pp. 124-134.
16. J. L. Finney, "Modelling the Structures of Amorphous Metals and Alloys," *Nature*, 1977, vol. 266, pp. 309-314.
17. G. S. Cargill, "Dense Random Packing of Hard Spheres as a Structural Model for Noncrystalline Metallic Solids," *Journal of Applied Physics*, 1970, vol. 41, pp. 2248-2250.
18. R. Castells, J. Meda, J. Caprari and M. Damia, "Particle Packing Analysis of Coatings Above the Critical Pigment Volume Concentration," *Journal of Coatings Technology*, 1983, vol. 55, no.505, pp. 53-59.
19. R. J. Robertson, F. Guillon and J. P. Harrison, "Properties of Sintered Submicron Copper and Silver Powders and Their Relation to Low Temperature Heat Exchangers," *Canadian Journal of Physics*, 1983, vol. 61, pp. 164-176.
20. L. C. Graton and H. J. Fraser, "Systematic Packing of Spheres With Particular Relation to Porosity and Permeability," *Journal of Geology*, 1935, vol. 43, pp. 785-909.
21. A. Rosato, K. J. Strandburg, F. Prinz and R. H. Swendsen, "Why the Brazil Nuts Are on Top: Size Segregation of Particulate Matter by Shaking," *Physical Review Letters*, 1987, vol. 58, pp. 1038-1040.
22. R. P. Kusy and D. T. Turner, "Electrical Resistivity of a Polymeric Insulator Containing Segregated Metallic Particles," *Nature*, 1971, vol. 229, pp. 58-59.
23. T. J. Coutts, "Percolation Conduction in Mixed Insulator-Conductor Systems," *Thin Solid Films*, 1976, vol. 38, pp. 313-321.
24. F. O. Anderegg, "Grading Aggregates II - The Application of Mathematical Formulas to Mortars," *Industrial and Engineering Chemistry*, 1931, vol. 23, pp. 1058-1064.
25. H. T. Horsfield, "The Strength of Asphalt Mixtures," *Journal of the Society of Chemical Industry*, 1934, vol. 53, pp. 107T-115T.
26. J. V. Milewski, "Packing Concepts in the Use of Filler and Reinforcement Combinations," *Handbook of Reinforcements for Plastics*, J. V. Milewski and H. S. Katz (eds.), Van Nostrand Reinhold, New York, NY, 1987, pp. 14-33.
27. P. C. Carman, "Fundamental Principles of Industrial

Filtration," *Transactions of the Institution of Chemical Engineers*, 1938, vol. 16, pp. 168-188.
28. M. K. Bo, D. C. Freshwater and B. Scarlett, "The Effect of Particle Size Distribution on the Permeability of Filter Cakes," *Transactions of the Institution of Chemical Engineers*, 1965, vol. 43, pp. T228-T232.
29. S. Ergun and A. A. Orning, "Fluid Flow Through Randomly Packed Columns and Fluidized Beds," *Industrial and Engineering Chemistry*, 1949, vol. 41, pp. 1179-1184.
30. L. J. Pedicini, "Packing Characteristics of Typical Foundry Sands," *Transactions of the American Foundrymen's Society*, 1958, vol. 66, pp. 421-424.
31. T. Katan and E. A. Grens, "The Effect of Liquid Configuration in Porous Gas Electrodes," *Journal of the Electrochemical Society*, 1971, vol. 118, pp. 1881-1886.
32. J. W. Marvin, "The Shape of Compressed Lead Shot and its Relation to Cell Shape," *American Journal of Botany*, 1939, vol. 26, pp. 280-288.
33. W. C. Bell, R. D. Dillender, H. R. Lominac and E. G. Manning, "Vibratory Compacting of Metal and Ceramic Powders," *Journal of the American Ceramic Society*, 1955, vol. 38, pp. 396-404.
34. D. S. Boudreaux and J. M. Gregor, "Structure Simulation of Transition-Metal-Metalloid Glasses," *Journal of Applied Physics*, 1977, vol. 48, pp. 152-158.
35. T. M. Hare, "Statistics of Early Sintering and Rearrangement by Computer Simulation," *Sintering Processes*, G. C. Kuczynski (ed.), Plenum Press, New York, NY, 1980, pp. 77-93.
36. F. Sigulinski and I. Stamenkovic, "Packing and Consolidation of Ceramic Particles of Very Different Size," *Journal of Powder and Bulk Solids Technology*, 1977, vol. 3, pp. 61-63.
37. G. P. Korfiatis and C. M. Manikopoulos, "Maximum Dry Density for a Continuous Grain Size Distribution," *Civil Engineering for Practicing and Design Engineers*, 1984, vol. 3, pp. 1139-1158.
38. R. J. Parsick, S. C. Jones and L. P. Hatch, "Stability of Randomly Packed Beds of Fuel Spheres," *Nuclear Applications*, 1966, vol. 2, pp. 221-225.
39. R. Grigorovici and R. Manaila, "Structural Model for Amorphous Germanium Layers," *Thin Solid Films*, 1967, vol. 1, pp. 343-352.
40. L. V. Woodcock, "Glass Transition in the Hard-Sphere Model,"

Journal of the Chemical Society Faraday Transactions II, 1976, vol. 72, pp. 1667-1672.

41. J. Parisi, O. C. Nutter and C. Michalowski, "Sieve Ratios and Processing for Strong Molding Sands," *Transactions of the American Foundrymen's Society*, 1958, vol. 66, pp. 278-283.
42. B. Habermann, "The Efficiency of Miscible Displacement as a Function of Mobility Ratio," *Transactions of AIME, Petroleum Transactions*, 1960, vol. 219, pp. 264-272.
43. P. G. Saffman and G. Taylor, "The Penetration of a Fluid into a Porous Medium or Hele-Shaw Cell Containing a More Viscous Liquid," *Proceedings of the Royal Society of London*, 1958, vol. A245, pp. 312-329.

Chapter Two

Structures in One and Two Dimensions

Chapter Outline

Introductory Concepts

There are two very different types of packing structures, random and ordered. A random packing is constructed by a sequence of events that are not correlated with one another. The result of such a random assembly procedure is a structure without long-range repetition. Random structures have a lower packing density than attainable with the high packing density ordered structures. An ordered structure occurs when objects are placed systematically into periodic positions, as are the bricks in a wall or atoms in a crystal structure. Another way of viewing an ordered packing is to recognize that each particle is located by a simple mathematical translation of integer steps from every other particle.

Most packing problems are considered in three dimensions, yet there are reasons to consider other levels of packings. Indeed, problems with dimensions greater than three have been treated in pure mathematics and have proven useful in designing systems for such fields as digital communications [1,2].

The spherical particle shape has received the greatest attention in three-dimensional packings. This is because only one size parameter, the diameter, is needed to specify the dimension of a sphere. Other than regular polyhedral shapes (for example, a cube), most other shapes require multiple parameters to specify their size and shape. Also, many systems encountered in science and technology approximate a spherical shape.

There is an equivalent two-dimensional packing problem sometimes referred to as covering [1]. An object is covered by a set of shapes such that each element of the object's surface underlies at least one point of the covering shapes. The fractional coverage is the two-dimensional analog to the fractional density of particles in a volume. For some noncircular shapes, an ordered two-dimensional structure is capable of total coverage, giving a fractional density of 1.00. For example, squares, rectangles, equilateral triangles, and hexagons can totally cover a surface with no overlap. However, disks can not give total coverage without overlap.

The analogous packing problem in one dimension is termed "parking" and is analyzed in terms of segments placed on a line. The terminology arises since this can be viewed as the problem of parking automobiles along a curb. For an ordered structure, the parking density of line segments will equal 100%, because of perfect end-to-end alignment of the segments. This is similar to a string of pearls without any gaps. With a random placement of segments, the coverage is less than complete. In three dimensions the problem is termed packing, while in one and two dimensions the problem can also be referred to as parking and covering, respectively. The similarity in these three concepts leads to a generalized terminology of packing as used in this book.

In three dimensions, a sphere is defined as the locus of a point at a constant distance. The separation distance is the sphere radius, and the boundary of the sphere is a surface. Thus, for a three-dimensional sphere, the boundary is two-dimensional. In general, a "sphere" with n dimensions is defined by a boundary with n-1 dimensions [3]. The one-dimensional packing is then determined by two points along the line, the distance between the two points being equal to the size of the segment (that is, the sphere diameter). This structure is sometimes referred to as a 0-sphere, where the zero designates the dimensions of the boundary. Figure 2.1 provides a contrast between one-dimensional ordered and random packings, using the idea of pearls on a string to illustrate the difference. As shown in Figure 2.1a, an ordered packing of segments can attain a fractional density f of 1.00. Alternatively, a random packing of monosized segments attains a fractional density of approximately 0.75.

For a two-dimensional structure, the object is a disk (or circle), and the packing occurs on a plane. The maximum packing density for a two-dimensional ordered structure of monosized disks

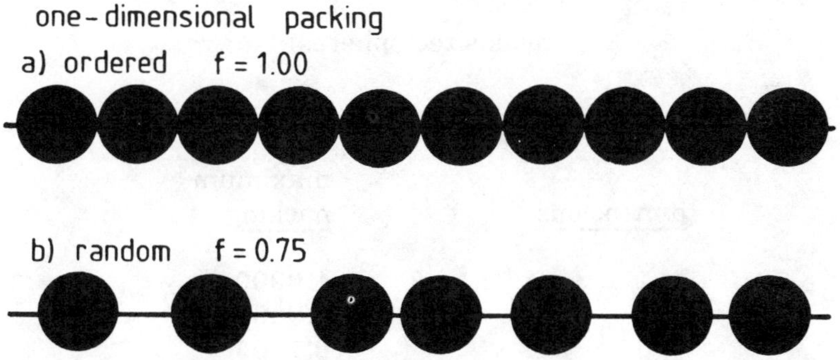

Figure 2.1 One-dimensional packing of segments along a line; a) ordered packing with full coverage, and b) random packing with incomplete coverage.

is a hexagonal array, giving a fractional density of 0.9069. When the disks are placed at random, the density decreases to approximately 0.82.

The three-dimensional case corresponds to spheres in a volume, and the maximum fractional density of 0.7405 occurs with a twelve-fold packing coordination. It is mathematically possible to extend packing concepts to more than three dimensions. At higher dimensions, the mathematical treatment must go into areas that are less directly verifiable. Beyond eight dimensions most of the maximum density packing structures are speculative. However, as shown in Table 2.1, the trend is for the maximum attainable packing density to decrease as the dimensionality of the problem increases, being 1.000 for one-dimensional, 0.9069 for two-dimensional, and 0.7405 for three-dimensional packings. For a twelve-dimensional packing, a fractional density of 0.05 is high. Rogers [1] suggests that the maximum packing density for a structure of n dimensions can be approximated by a simple set of calculations. Interestingly, the maximal packing predicted for a three-

Table 2.1

Maximum Packing Density Versus Dimensions [1]

(monosized spheres)

dimensions	maximum packing
1	1.0000
2	0.9069
3	0.7405
4	0.6168
5	0.4652
6	0.3729
7	0.2952
8	0.2536

dimensional array of spheres using his concepts is 0.7797, which is higher than the realized value of 0.7405. The problems with dimensionality higher than three is of little concern in the balance of this book.

In this chapter, one-dimensional and two-dimensional packings are given preferential discussion. One reason for covering these structures first is that they are easy to illustrate in graphical form. Secondly, the basic concepts of ordered versus random packing are readily contrasted with these structures. Additionally, many of the concepts developed in greater detail with three-dimensional packings are initially established for the simpler structures. These include the effects of composition and size ratio in bimodal mixtures, and the effects of size distributions on the packing density. Beyond the important introductory concepts, the one-dimensional and two-dimensional packings have practical implications in such areas as bacterial cells, surface coatings, parking of automobiles, filter structures, energy cascades, amorphous metals, packing in composites, and retinal pigmentation.

Packing of Monosized Segments in One Dimension

As noted earlier, one-dimensional packing of monosized segments is analogous to parking cars along a curb. In the ordered structure, the assembly of the packing is systematic, and the fractional packing density is 1.00. As illustrated by the string of pearls (or disks) in Figure 2.1a, the coordination number is exactly two with one contact at each end. Conceptually, the first segment is placed at one end of the string, the second is placed just in contact with the first, and the process is repeated to fill the string. Each added segment creates one new contact which is shared by two segments, giving a net coordination increase of two contacts per segment. Total coverage occurs with this end-to-end ordered structure.

The less obvious situation occurs with random parking of monosized objects, as illustrated in Figure 2.1b. The procedure for attaining the final random structure is shown in Figure 2.2, where the structure along the line is represented at four stages of parking. For conceptualization, think of each segment as

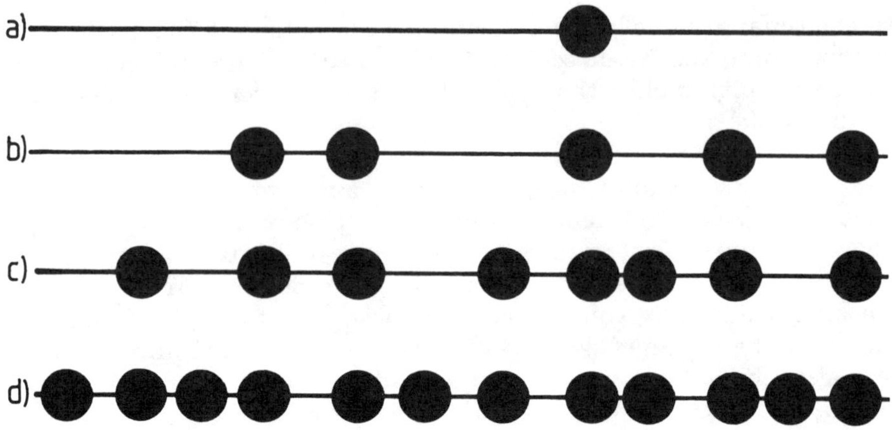

Figure 2.2 The one-dimensional parking sequence for forming a random packing along a line.

representing a car with the first placed at random along the test line as shown in Figure 2.2a. Likewise, the next segment is placed either to the left or right of the first, again at random as long as no overlap occurs. If a selected site would cause overlap with an existing segment, then the attempt at parking is rejected and a new site is randomly selected. Thus, each time a site is selected with no effort to maximize the use of space. This process continues for the next few segments, as illustrated in Figure 2.2b, where six segments have randomly arrived and parked on the line. Subsequently, the number of spaces that can accommodate another segment decreases (the number of failed parking attempts increases dramatically during the later stage). This random process continues with gradual filling of the line and consequential reduction in size of the remaining line (parking spaces) as illustrated in Figure 2.2c. Should a segment arrive at its randomly selected spot and find it occupied, then a new random site is selected. Eventually the line has no remaining spaces large enough to accommodate another segment, and is saturated as shown in Figure 2.2d. In spite of being saturated, the packing density is less than that obtainable with an ordered structure. This is the parking analog to the packing of monosized segments in one dimension. The maximal random packing density for this case has been estimated by computer simulation, experiment, and theoretical analysis [3-6]. The generally agreed upon fractional density is 0.7476, although some rough measurements ranged as high as 0.78.

If the random parking situation is modified so that an arriving segment (car) can move slightly in order to find an available spot, then the packing density increases. Depending on the algorithm, various higher densities are attained. For example, if the arriving segment (car) is able to move slightly forward or backward (up to a half segment length) to find a space after its arrival, then the fractional density improves to 0.809 [4]. Motion to a new spot can increase the local parking density to 1.00 because of the resulting end-to-end structure. In later chapters, a distinction will be made between loose and dense random packing structures. The dense structure is induced by vibration of a loosely packed structure, thereby improving the use of space. Similarly, the modified parking problem with slight motion to find space is comparable to vibration induced packing of a three-dimensional structure.

Mixed Segment Sizes in One Dimension

The random placement of segments is important in packing problems, since random events are more typical in natural processes. Although randomly placed segments produce a packing density below that for an ordered structure, if two segment sizes are used to randomly pack a line, then it is possible that the smaller segments can preferentially fill the voids left in packing of larger segments. For a bimodal one-dimensional packing, the fractional packing density depends on the size ratio between the segments and the compositional ratio. As shown in Figure 2.3, segments that are very different in size will give more of a packing benefit than those of similar size. The size ratio is denoted by D_L/D_S, which is the ratio of the large to small sizes. The composition is given in terms of the fraction of small segments X_S, and the packing density is given as f. If the segments are very different in size, then the smaller segments will increase the packing efficiency by filling many of the voids between the larger segments. However, the additional factor of composition becomes important: if there are too few of the small segments, then there is little packing benefit. Alternatively, if there are too many of the small segments, then the

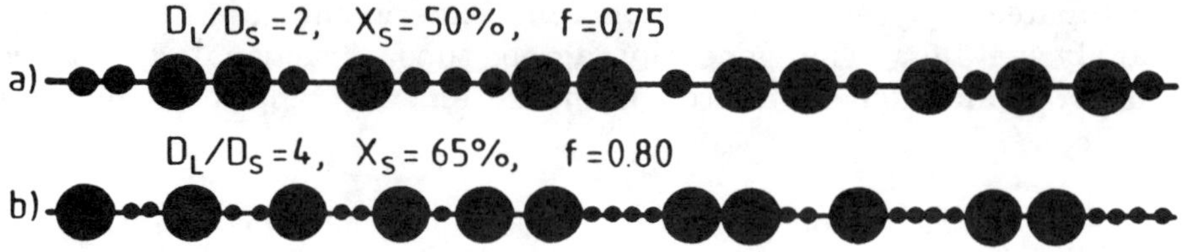

Figure 2.3 Two binary random one-dimensional packings showing the improved packing density possible with a large size ratio and an abundance of small segments.

packing density is controlled by their inherent void space.

In a bimodal packing of segments, packing density is improved if smaller segments fill the voids between large segments. If the large segments give a random packing density of 0.7476, then a fractional voidage of 0.2524 remains for the small segments to fill. Assuming the small segments can pack into the voids just as well as the large segments initially packed along the line gives a gain in packing density equal to 0.2524 times 0.7476. The combined packing density is 0.7476 + (0.2524)(0.7476), which equals a fractional packing density of 0.9363. This would occur for a very large size ratio and the exact composition of small segments that just fills the voids between large segments. For example, if the size ratio between the large and small segments is 10:1, then for every large segment the optimal composition requires 2.52 small segments. In this case, the optimal composition will consist of approximately 72% small segments by population. For compositions away from optimal, the packing density decreases. Figure 2.4 shows the effect of composition on the fractional packing density for a bimodal mixture of segments with size ratios of 1.96 4.80, and 9.57 [4]. For the size ratio of 4.80, the optimal concentration of small segments is expected to be near 55%.

Following these concepts, it is possible that a distribution in segment sizes would be beneficial in attaining a high packing density from randomly placed segments. The character of the optimal distribution presents a difficult problem. In computer simulations of random one-dimensional packings, it has been determined that wide

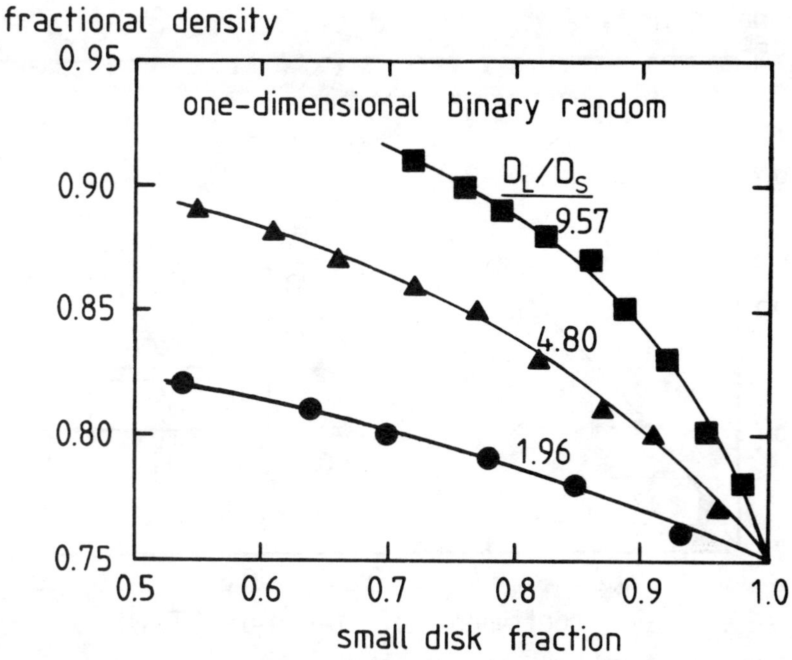

Figure 2.4 The effect of composition on the packing density for one-dimensional structures composed of segments with size ratios of 1.96, 4.80, and 9.57 [4].

variations in the distribution width are indeed beneficial. The packing density increases as the normalized coefficient of variation increases. The coefficient of variation for a distribution is defined as the standard deviation divided by the mean size. Goldman et al. [4] examined several types of distributions and found that the packing density not only increased with a large coefficient of variation, but that a high population of large segments was most favorable. For a given packing density the small segments break up the line length more rapidly than the large segments. That is, the small segments preempt space disproportionately to their packing contribution. As a consequence, a distribution with a high population of small segments decreases the site probability for the larger segments that give high coverage. Figure 2.5 shows that with a uniform distribution, the packing density increases as the coefficient of variation increases. The greater the range between the smallest and largest segment

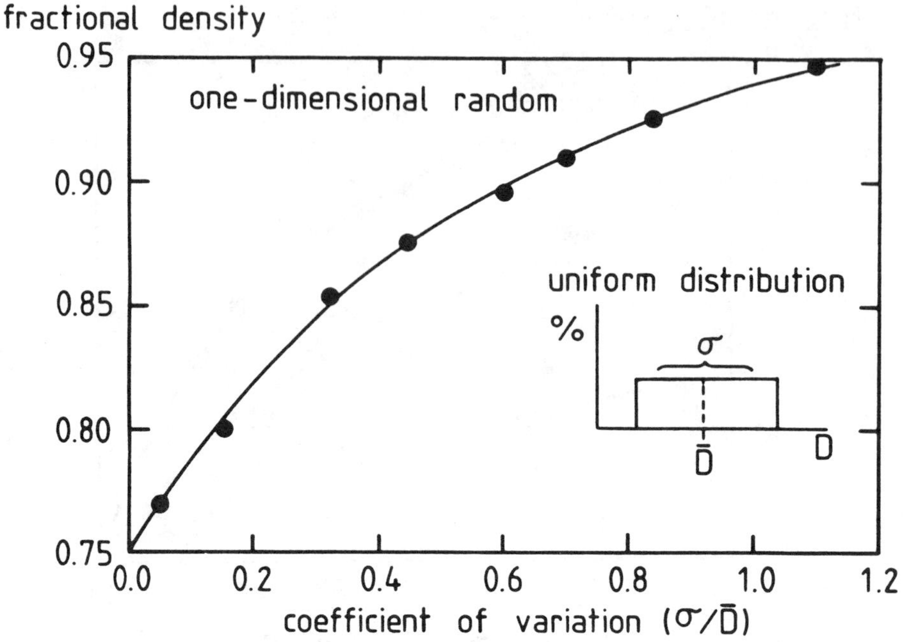

Figure 2.5 The effect of the coefficient of variation in the segment size distribution on the packing density for one-dimensional packings [4].

sizes, the higher the packing density. This is equally true for skewed distributions with a high population of large size segments. Later, we will see that the packing of spheres in three dimensions is sensitive to distribution breadth and shape.

Packing of Monosized Disks in Two Dimensions

Ordered Packing

In two dimensions, a disk or circle can be packed into either ordered or random structures. Because of the shape, the packing structure is not totally covering for the underlying structure. Indeed, significant overlap of the disks is required to attain total

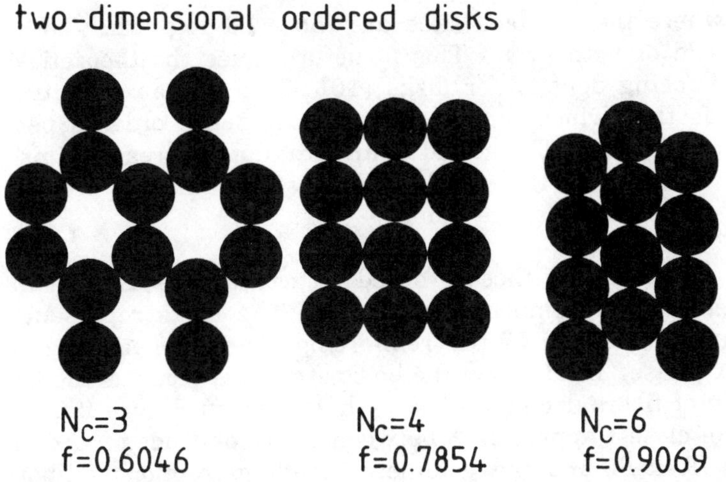

Figure 2.6 A sketch of the three ordered structures of disks packed in two dimensions with coordination numbers of three, four, and six.

coverage, with an estimated excess in disk area of 21% over the covered area necessary to produce a complete covering [1,7]. For this reason, a disk (or sphere in three dimensions) is termed a low packing efficiency object.

There are three ordered packings of disks that can be repeated to fill space, as shown in Figure 2.6. These are best characterized by the number of contact points for each disk, which is the coordination number N_c, and the fractional density f. The lowest density structure has a coordination number of three and the highest density structure has a coordination number of six. Note that the packing density increases with the coordination number as shown in Figure 2.6. There is no repetitive unit with five-fold symmetry. However, a statistical model of two-dimensional packing gives an estimated density of 0.78 for the five-fold geometry [8,9]. Likewise, this model predicts a density of 0.42 for a coordination of two.

The six-fold geometry is termed close-packed. It can be constructed by packing disks on every other lattice site in a square grid, where the occupied sites are those having either both coordinates odd or both even. This structure gives the theoretical maximum packing density for disks [10]. This planar structure is the same as that which makes up the most dense ordered packing in three dimensions. The actual minimum unit for creating this geometry is a cluster of three disks. This is the unit cell in two-dimensional space that is space filling.

Variants on these ordered structures are used in modeling the behavior of composite materials. The disks represent a two-dimensional section of the reinforcing fibers penetrating into the matrix phase. Dilations of the basic structure occur as the fractional density of fibers decreases below that given in Figure 2.6. The distance of closest approach δ between fibers depends on the diameter of the fiber D and the volumetric loading fraction of fibers a as follows:

$$\delta = D\,[(a/f)^{0.5} - 1] \qquad (2.1)$$

where f is the packing density for the structural coordination as given in Figure 2.6.

Random Packing

The photographs in Figure 2.7 show the difference between an ordered and random packing for monosized disks. The ordered structure has a coordination number of six, and shows a very uniform arrangement. In contrast, the random structure is variable from point to point in coordination number, packing density, and void size.

The random packing density of disks in two dimensions is largely dependent on the procedure used in assembling the packing. Indeed, two random conditions are possible, loose and dense. Most studies use some force in maintaining contact between the disks, giving a maximum packing density for the random structure, the dense random packing. This can be a unidirectional force, such as gravity, or a central attraction force. The central attraction force is most typical in computer simulations, while experimental measurements use a unidirectional force (tilted container with the disks sliding under gravity into a random packing). Random packing

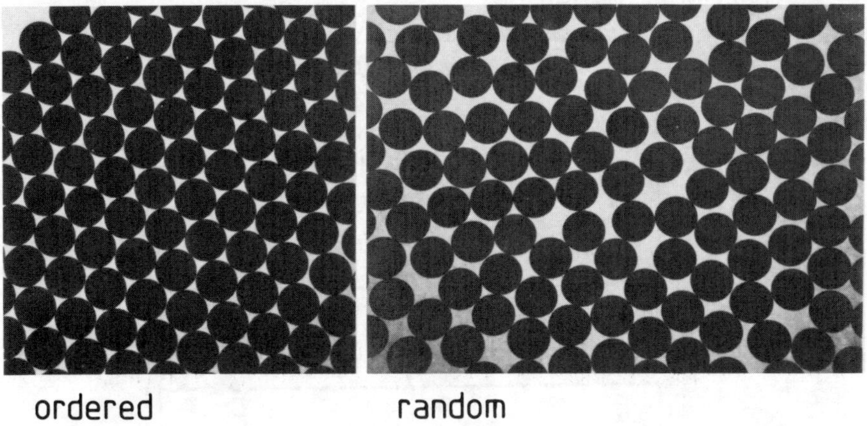

Figure 2.7 Photographs contrasting the structures of monosized disks in a close-packed ordered array and random array.

without such a force will allow low density regions (near 0.73) with zero or one contact per disk, the random loose packing. Alternatively, packings formed with a force will have at least two contacts per disk and a correspondingly higher density. Vibration has a slight influence on the packing density, but this influence is less than observed in three-dimensional packings. With small packing clusters, cluster size is also an effect [11]. Generally, at least 1000 disks are needed to ensure behavior representative of a true random packing without disruption from the container [12].

The actual packing density for a random mixture can vary over a considerable range, yet the transition between random and ordered packing occurs when the fractional density exceeds 0.82. Figure 2.8 illustrates the behavior change associated with the random packing limit [11]. For these experiments the disks were placed on a stretched flexible membrane. As the stretch was relaxed, the packing characteristics were measured. The upper plot gives the fractional packing density versus the relative contraction during assembly. The lower plot shows the mode coordination number as a function of the fractional density for the same experiment. A shift in slope occurs for both plots with a change from a

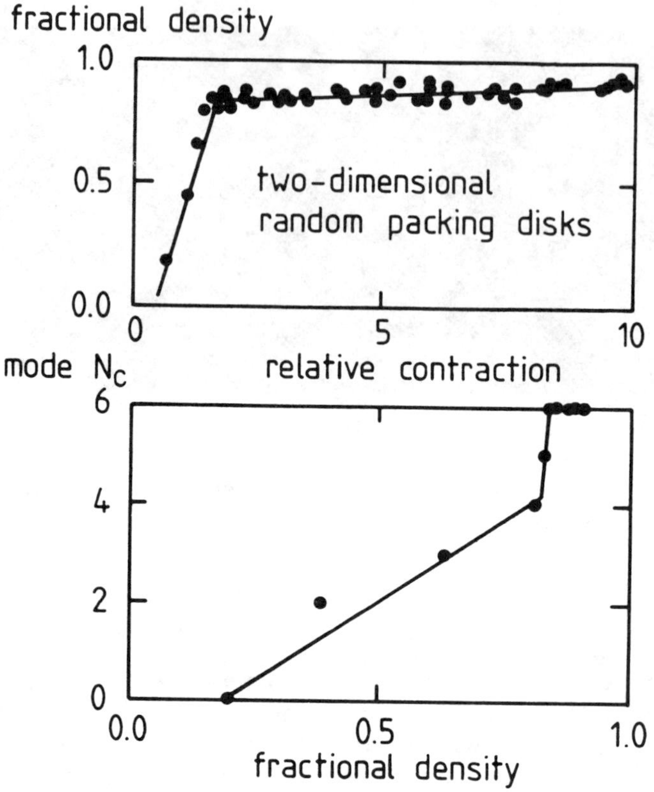

Figure 2.8 The effect of contraction on the fractional density and coordination number variation for a random two-dimensional packing of monosized disks [11]. A behavior shift occurs at the onset of ordering.

random mixture to the onset of ordering. The packing coordination number makes a dramatic change from four to six over a narrow range in packing density between 0.80 and 0.84. There have been several measurements of the random packing density for disks [6,9, 11-23]. The fractional density reports vary from a low value of 0.78 to a high value of 0.89, with a mean value of 0.82 and a standard deviation of 0.02. Additionally, the most frequently reported values are near 0.82. As noted above, some of this difference is due to the force state used to assemble the packing. Gamba [17] argues from probability considerations that the most probable packing of disks on a plane is 0.8225, which agrees well with the various computer and laboratory results.

The coordination number in a random packing is variable from point to point. Some regions with a coordination of six will exist while other regions will have values as low as two. On the one hand, the statistical model of Nikolenko and Kovalchenko [9] overestimates the occurrence of regions with a coordination number of six. Alternatively, Quickenden and Tan [11] report roughly an equal mixture of coordinations of four and five in their experimental structures. However, as noted earlier, the five-fold coordination is not capable of generating a space filling structure. Figure 2.9 plots the distribution in measured coordination numbers from two studies [11,22]. Considering the measurement error, these two distributions are very similar. The average coordination number has been reported as 4.66 and 4.31, with a mode value of four. At densities greater than approximately 0.82 the structure shows a degree of order which is indicated in Figure 2.8 by a rapid rise in coordination

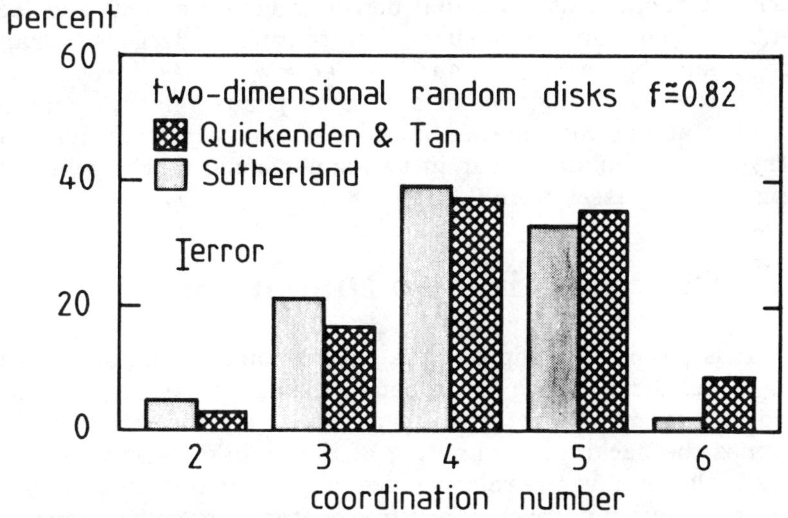

Figure 2.9 The distributions in coordination number for random two-dimensional packings of disks as observed by Quickenden and Tan [11] and Sutherland [22].

number. Accordingly, the structure is better characterized as regions of order separated by random or defective regions. Consequently, a random packing with a density over approximately 0.82 can be treated as a mixture of random and ordered regions. For such a case, the packing coordination can be estimated based on the fractional density. Assuming a linear combination of regions with coordination numbers of either four or six, the fractional density f is as follows:

$$f = X_4\, 0.7854 + (1 - X_4)\, 0.9069 \tag{2.2}$$

where X_4 is the fraction of the structure with a packing coordination of four (the balance assumed to have a coordination of six). If the structure is composed of just these two coordinations, then the density of 0.82 for the random structure leads to a predicted mean coordination number of 4.6.

Several characteristics of surfaces correlate with a random-to-ordered transition at fractional densities between 0.80 and 0.88 [11,14,24]. The vapor pressure of a coating undergoes a rapid change when the molecular surface density passes through this range. Likewise, various models of amorphous (glassy) coatings indicate the atomic coordination number is approximately five. The first crystalline domains occur in two-dimensional liquids when the packing density passes from 0.76 to 0.85.

Mixed Disk Sizes in Two Dimensions

It is possible to improve the packing or covering of disks by placing small disks in the voids between large disks. Again, both ordered and random structures are possible. The size ratio of disks determines the packing coordination at the highest packing density. Table 2.2 shows how the relative size of surrounding disks influences the coordination number. For three contacts, as noted above, the surrounding disks should be 6.464 times the size of the central disk. Alternatively, for ten contacts, the surrounding disks are smaller than the central disk, with a diameter ratio of 0.447.

Consider the first case of a close-packed lattice of monosized disks with a fractional density of 0.9069. As Table 2.2 shows, a

Table 2.2

Diameter Ratio of Surrounding Disks
and Corresponding Coordination Numbers

(diameter ratio = diameter of surrounding disk divided by diameter of central disk)

diameter ratio	coordination number
6.464	3
2.414	4
1.426	5
1.000	6
0.766	7
0.620	8
0.520	9
0.447	10

disk with a diameter of 1/6.464 (or 0.1547) times that of the lattice disks will just fit into the void and touch the three surrounding disks. To place one small disk into every void requires twice as many small disks as large disks (33% of the disks are large and 67% are small). In this case the packing density increases to 0.9503.

For the most efficient packing improvement, defined as the gain in density per disk, the added disk should just touch the three neighboring disks. Using the generalized geometry shown in Figure 2.10, the three existing disks have radii of R_1, R_2, and R_3. The radius of the disk that just touches these three disks is noted as R_4 and can be calculated as follows:

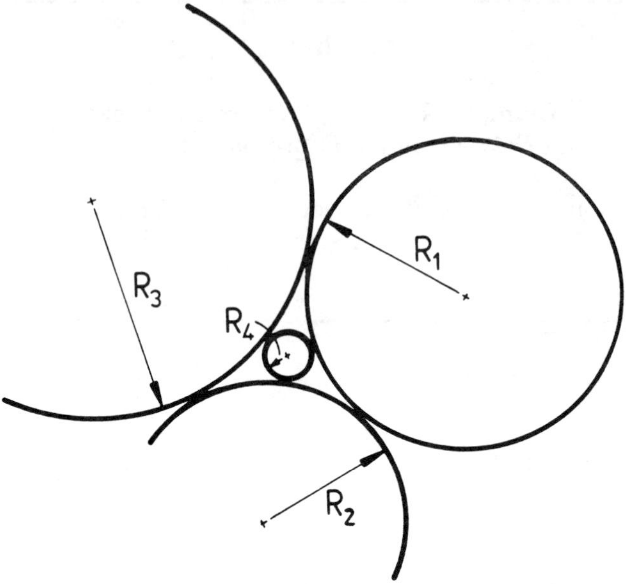

Figure 2.10 The void at a three disk junction which is just filled by a touching disk of radius R_4.

$$1/R_4 = 1/R_1 + 1/R_2 + 1/R_3 + 2[1/(R_1R_2) + 1/(R_1R_3) + 1/(R_2R_3)]^{1/2} . \qquad (2.3)$$

If the three large disks are equal in size, this equation predicts R_4 will have a radius of 1/6.464 or 0.1547 of the large disk radius. The procedure for filling with smaller disks can be continued with still smaller disks fitted into each of the voids remaining after the first level of filling. Accordingly, the number of disks increases by a factor of three as each new size class is added. Gilbert [25] provides a table giving nine levels of ordered nesting with the appropriate sizes and compositions. With this sequence it is possible to combine disk sizes to attain a coverage of 0.9895. From this basis a generalized equation is possible for predicting the fractional coverage as a function of the number of size classes. This leads to consideration of a continuous distribution in disk sizes and possible packing densities. The continuous population distribution will be skewed toward the smaller disks for those cases giving a high packing density. Consequently, a generalized size distribution can be calculated to optimize the packing density [25,26]. This size distribution is essentially independent of the initial packing conditions. Likewise, there are formulae for predicting the coordination number versus the size ratio for ordered binary mixtures of disks [13,27]. An interesting variant is a spiral packing created with a size ratio of 1.25.

A more general case is that of disks packed into the voids of an existing structure where the disk sizes are very different. In this case more than one disk is allowed to fit into each void. If each disk size has associated with it a fractional density f, then by filling the voids with smaller disks the packing density of the mixture f_{mix} is improved to give,

$$f_{mix} = 1 - (1 - f)^k \tag{2.4}$$

where k is the number of levels of sizes. This says that the voids existing after positioning the first level of disks, equal to $1 - f$, are filled to increase the packing density by an additional amount f. For an ordered bimodal mixture with an infinite ratio of large to small disk sizes, the peak packing density will converge to 0.9913 at the optimal composition. This differs from the efficient bimodal case where the large to small size ratio is 6.464, which gives a maximum packing density of 0.9503. Note that the fractional density approaches unity as the number of size levels increases. However, in practice putting the small disks into the proper sites presents a mechanical problem that limits the actual coverage. Consequently, the experimentally realized densities are less than those calculated using such idealized models.

When disks of slightly differing sizes are mixed, there is an increased tendency toward randomness [28]. This is especially true when the disk sizes differ by at least 25% [29]. Figure 2.11 shows a photograph of mixed disks of three sizes with a diameter ratio of 1.4:1.1:1.0 (formed from a mixture of quarters, nickels, and pennies). The packing density is not significantly improved by the use of a small size ratio. In such cases, the large disks have the

Figure 2.11 A photograph of mixed disks illustrating the high degree of randomness associated with small variations in disk sizes; the ratio of sizes is 1.4:1.1:1.0.

most disrupting influence on the structure and cause randomness at small concentrations. Alternatively, the small disks do not necessarily contribute to randomness, yet they decrease density by creating larger voids in the packing.

The structure of binary disk mixtures has important implications with respect to amorphous materials and random packing of particles. Consequently, several studies have been conducted to view the two-dimensional structure using disks or spheres with a size ratio close to unity [12,13,16,20,28-30]. The two dominant factors are the size ratio (diameter of the large disk divided by the diameter of the small disk) and composition of the mixture. The fractional packing density of a random assembly is increased by mixing large and small disks. The degree of improvement depends on the size ratio: If the sizes of the disks in a mixture are not very different, then the packing density shows little improvement over

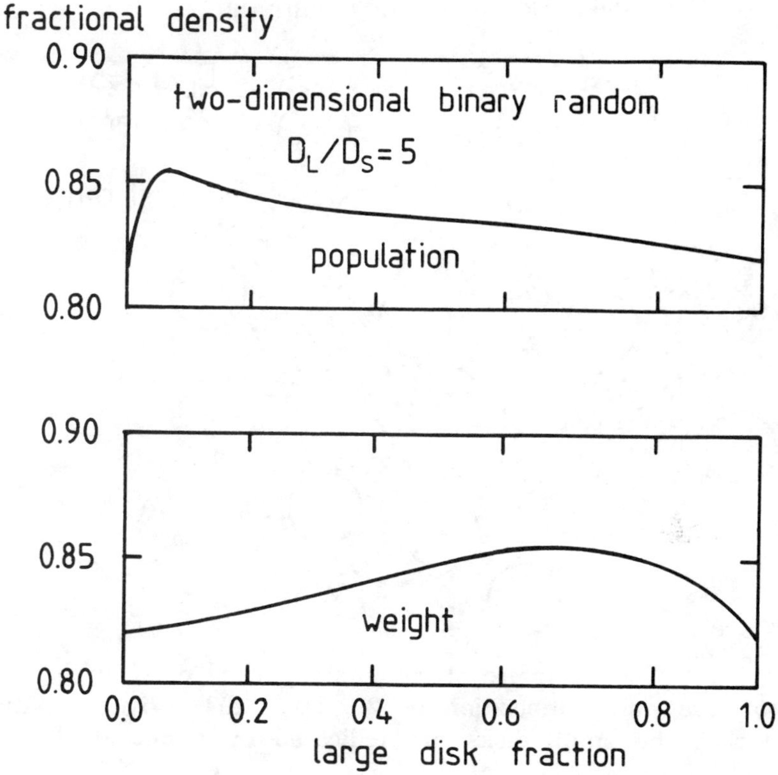

Figure 2.12 The fractional density versus composition for binary random mixtures of disks with a size ratio of five. The upper plot is based on disk population and the lower plot is based on weight [12].

the random case. Figure 2.12 plots the density versus composition on both a number and weight basis for a size ratio of five [12]. Here the weight is based on the number of disks and their area. Figure 2.13 shows one example of this structure at a composition of 2% large disks by population. For clarity, only the centers of the small disks are shown. The density is greater for a mixture containing a large number of small disks, but a high weight of large disks. The maximum density occurs at the point where the large disks are just separated from contact with one another. However, even with a size ratio of five the packing benefit is relatively small.

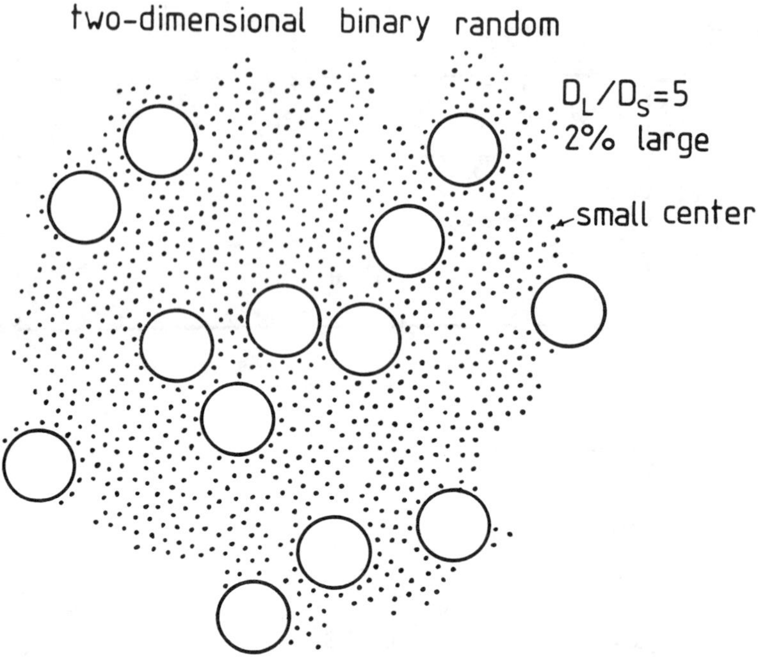

Figure 2.13 The structure of randomly packed disks with a size ratio of five and composition of 2% large disks by population. For clarity, the small disks are indicated by a dot at the disk center [12].

In a mixture of large and small disks, three structures can form, depending on the density. A random packing exists at low densities. At high densities, the structure is ordered, or at least exhibits regions of order surrounded by regions of disorder. Depending on the composition and size ratio, segregation by size can also be seen, especially when the mixture is vibrated. At intermediate densities with small size ratios the structure is termed hexatic since there are local regions of order. Figure 2.14 shows a diagram of such structure fields. This figure plots the density versus the composition for disks with relatively small size ratios [13]. The extent of the three regions depends on the size ratio of the mixture. The region of order exists at the highest packing densities. At the terminal compositions, order occurs at lower packing densities because the distortion induced by a different sized disk can be easily accommodated by dislocations, vacancies or other structural defects.

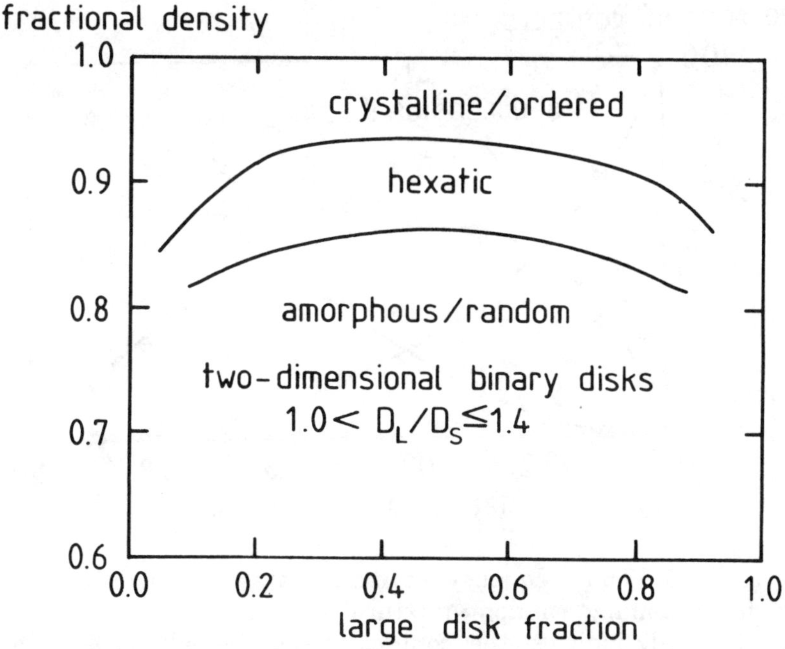

Figure 2.14 Composition and packing density affects on the observed regions of ordered, hexatic, and random structures for small disk size ratios [13].

Within the random mixtures, there are three types of contacts, designated L-L, L-S, and S-S, where L corresponds to the large disk and S corresponds to the small disk. The frequency of each contact type depends on the composition of the mixture, as illustrated in Figure 2.15 for a size ratio of 2:1 [30]. It is curious that over the entire composition range the average coordination number remains near 4.7 [20,30].

Finally, mention should be made of the connectivity concept. If disks of differing compositions, for example a conductor and an insulator, are randomly mixed, then for certain compositions it would be possible for the mixture to behave as a conductor. In two dimensions, this condition requires a large fraction of the disks be conducting. Likewise, at other compositions, the mixture would behave as an insulator if the conducting disks were not in sufficient

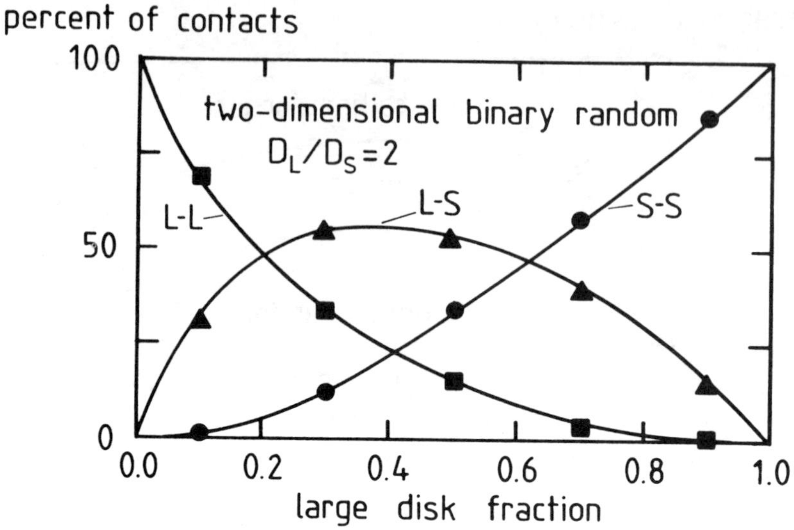

Figure 2.15 The frequency of disk contact versus the composition for a binary random mixture with a size ratio of 2; S-S, L-S, and L-L indicate the contact is small-small, large-small, or large-large [30].

contact with one another [31]. As conducting disks are added to a two-dimensional packing of insulating disks, the onset of conductivity is predicted to occur at a critical level denoted f_c where

$$f_c = 2/N_c \tag{2.5}$$

with N_c equal to the mean coordination number [32]. For an ordered packing with a coordination number of six, conduction is not expected at any fractional concentration of conducting disks below 0.33. Alternatively, for a random packing with a coordination number of 4.7, conductive mixtures would not exist at fractional concentrations of conductive disks below approximately 0.42. The conductivity problem for mixtures of differing sizes is yet to be solved.

Ordered Packing of Other Two-Dimensional Shapes

Shape is a very important factor in the packing of particles. In three dimensions a spherical shape minimizes interparticle friction, thereby aiding packing, even though the sphere is not space filling. In a two-dimensional ordered array, the disk is not as efficient as other shapes in covering a surface. Some of the high packing density shapes can be easily generated by constructing a Voronoi polygon around the ordered structures of disks. This procedure is illustrated in Figure 2.16 for coordination numbers of three, four, and six. A perpendicular bisector is drawn on the line between disk centers. The polygon formed by the intersection of these perpendiculars is termed a Voronoi polygon. The resulting tilings are equilateral triangles, squares, and hexagons. The number of sides represents the packing coordination numbers. In an ordered array, these two-dimensional shapes give perfect coverage. There is a three-dimensional analog for the Voronoi polygon that creates space filling polyhedral shapes.

Summary and Generalizations

This chapter has focused on packing problems in one and two dimensions. Brown [33] points out several analogous optimization problems built on the concepts introduced in this chapter. A classic problem is the cutting off of lengths of steel bars to satisfy a random batch of orders. The inventory of steel consists of equal length bars. The challenge is to fill the maximum number of orders with the minimum waste. Other examples include studies of travel schedules, packing, cutting, and depletion.

Some useful concepts have been generated in this chapter on one-dimensional and two-dimensional arrays. These concepts will be explored in greater detail with respect to three-dimensional structures in the balance of this book. The differences in packing between ordered and random structures have been illustrated. It is clear that close-packed order always gives a higher packing density than disorder. There is greater difficulty in predicting the density for a random packing since the procedure used in constructing the structure influences the density. Unlike an ordered packing, the

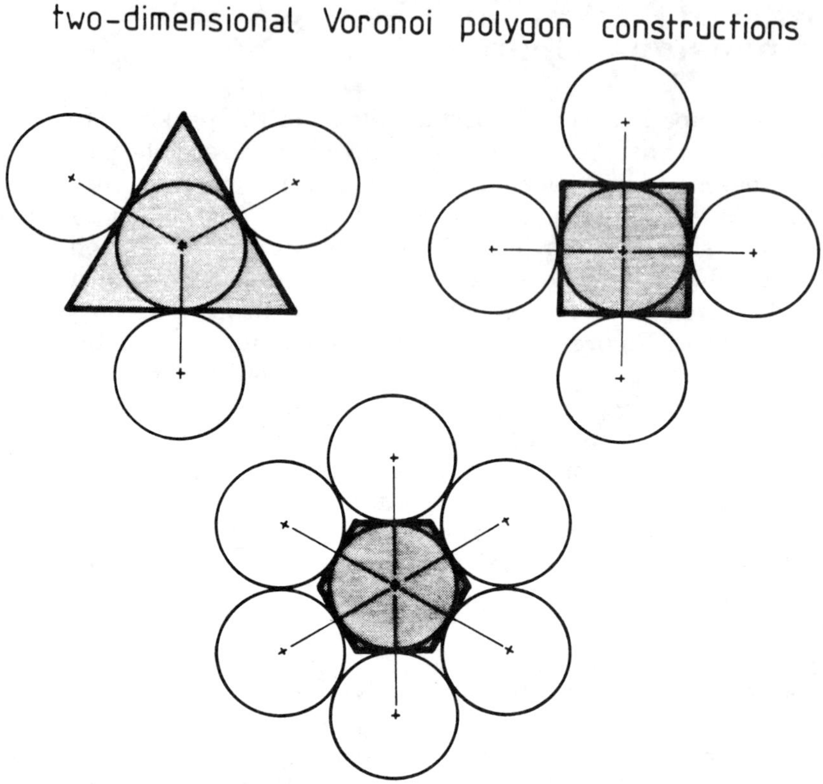

Figure 2.16 Two-dimensional Voronoi polygon constructions around disk packings with coordination numbers of three, four, and six.

coordination number is distributed in a random structure. The distribution in the coordination number within the array actually provides one of the most sensitive measures of randomness.

In this chapter, the effect of mixing different sizes has been illustrated. A large difference in sizes can improve packing, both for the ordered and random situations. The small objects fit into the interstitial voids between the large objects. Interestingly, the mean coordination number shows little variance with composition in binary mixtures. Optimization in packing occurs at an intermediate composition where the small objects fill all of the voids between the large objects without dilating the structure. An excess of either the large or small fraction will decrease the packing density. Usually, optimization occurs with a majority of small objects, when measured on a population basis. From such concepts, it is possible to create continuous size distributions that give high packing efficiencies.

Although spheres (and disks) are not efficient in filling space, they constitute the basis for most of the packing studies performed to date. There are regular polygonal shapes that can give higher packing densities when placed in an ordered array. However, in many instances real structures are random and the sphere or disk shape has sufficiently low friction to give the highest packing density. One major difference between two-dimensional and three-dimensional packings is the effect of contraction. In a two-dimensional packing of disks the structure can be ordered by a contraction force. In three dimensions, spheres resist ordering from an external force. Indeed, extremely high pressures are needed to densify spheres beyond the random packing density; this is because a shape change is produced by the pressurization.

Finally, it is interesting to note the similarity between close packing structures as the dimensionality is increased. In one dimension, the highest packing density corresponds to segments (spheres) in point contact along a straight line. This gives a density of 1.000 with a coordination number of two. In two dimensions, the highest density corresponds to three of these one-dimensional structures aligned in a plane to form an equilateral triangle. This corresponds to a close-packed plane with six contacts per disk. The fractional packing density is now decreased to 0.9069. For the three-dimensional case, the maximum density occurs with three of the two-dimensional planes stacked one upon another. The coordination number is twelve and the fractional density is 0.7405. This progression is not paralleled by a similar systematic behavior for the fractional density in the random packing cases: 0.75 for one-dimensional, 0.82 for two-dimensional, and 0.64 for three-dimensional packings.

References

1. C. A. Rogers, *Packing and Covering*, Cambridge University Press, London, UK, 1964.
2. N. J. A. Sloane, "The Packing of Spheres," *Scientific American*, 1984, vol. 250, no. 1, pp. 116-125.
3. I. J. Smalley, "Packing of Equal 0-Spheres," *Nature*, 1962, vol. 194, p.1271.
4. A. S. Goldman, H. D. Lewis and W. M. Visscher, "Random Packing of Particles: Simulation With a One Dimensional Parking Model," *Technometrics*, 1974, vol. 16, pp. 301-309.
5. J. S. Griffith, "Packing of Equal 0-Spheres," *Nature*, 1962, vol. 196, pp. 764-765.
6. P. E. Ney, "A Random Interval Filling Problem," *Annals of Mathematical Statistics*, 1962, vol. 33, pp. 702-718.
7. J. F. C. Kingman, "Book Review on *Packing and Covering*," *Nature*, 1965, vol. 205, p.738.
8. A. N. Nikolenko and M. S. Kovalchenko, "Analysis of the Random Packing of Identical Particles. V. Morphological Properties of Transversely Isotropic Fiber Composites," *Soviet Powder Metallurgy and Metal Ceramics*, 1987, vol. 26, pp. 106-111.
9. A. N. Nikolenko and M. S. Kovalchenko, "Analysis of the Random Packing of Identical Particles II. Structural Characteristics of the Packing of Disks on a Plane Surface," *Soviet Powder Metallurgy and Metal Ceramics*, 1985, vol. 24, pp. 912-914.
10. C. A. Rogers, "The Packing of Equal Spheres," *Proceedings of the London Mathematical Society*, 1958, vol. 8, ser.3, pp. 609-620.
11. T. I. Quickenden and G. K. Tan, "Random Packing in Two Dimensions and the Structure of Monolayers," *Journal of Colloid and Interface Science*, 1974, vol. 48, pp. 382-393.
12. H. H. Kausch, D. G. Fesko and N. W. Tschoegl, "The Random Packing of Circles in a Plane," *Journal of Colloid and Interface Science*, 1971, vol. 37, pp. 603-611.
13. M. Rubinstein and D. R. Nelson, "Order and Deterministic Chaos in Hard-Disk Arrays," *Physical Review B*, 1982, vol. 26, pp. 6254-6275.
14. D. Turnbull and R. L. Cormia, "A Dynamic Hard Sphere Model," *Journal of Applied Physics*, 1960, vol. 31, pp. 674-678.

15. R. K. Bordia, "A Theoretical Analysis of Random Packing Densities of Mono-Sized Spheres in Two and Three Dimensions," *Scripta Metallurgica*, 1984, vol. 18, pp. 725-730.
16. G. F. Round and R. Newton, "Random Packing of Equal Spheres on a Plane Surface," *Nature*, 1963, vol. 198, pp. 747-749.
17. A. Gamba, "Random Packing of Equal Spheres," *Nature*, 1975, vol. 256, pp. 521-522.
18. W. M. Visscher and M. Bolsterli, "Random Packing of Equal and Unequal Spheres in Two and Three Dimensions," *Nature*, 1972, vol. 239, pp. 504-507.
19. J. G. Berryman, "Random Close Packing of Hard Spheres and Disks," *Physical Review A*, 1983, vol. 27, pp. 1053-1061.
20. D. Bideau and J. P. Troadec, "Compacity and Mean Coordination Number of Dense Packings of Hard Discs," *Journal of Physics C: Solid State Physics*, 1984, vol. 17, pp. L731-L735.
21. V. C. Aguilera-Navarro, M. Fortes, M. DeLlano and A. Plastino, "Pade Approximants and the Random Close Packing of Hard Spheres and Disks," *Journal of Chemical Physics*, 1982, vol. 76, pp. 749-751.
22. D. N. Sutherland, "Random Packing of Circles in a Plane," *Journal of Colloid and Interface Science*, 1977, vol. 60, pp. 96-102.
23. M. Shahinpoor, "Statistical Mechanical Considerations on the Random Packing of Granular Materials," *Powder Technology*, 1980, vol. 25, pp. 163-176.
24. F. H. Stillinger, E. A. DiMarzio and R. L. Kornegay, "Systematic Approach to Explanation of the Rigid Disk Phase Transition," *Journal of Chemical Physics*, 1964, vol. 40, pp. 1564-1576.
25. E. N. Gilbert, "Randomly Packed and Solidly Packed Spheres," *Canadian Journal of Mathematics*, 1964, vol. 16, pp. 286-298.
26. H. H. Kausch-Blecken von Schmeling and N. W. Tschoegl, "Osculatory Packing of Finite Areas with Circles," *Nature*, 1970, vol. 225, pp. 1119-1122.
27. D. J. Cumberland and R. J. Crawford, *The Packing of Particles*, Elsevier Science, Amsterdam, The Netherlands, 1987.
28. D. R. Nelson, M. Rubinstein and F. Spaepen, "Order in Two-Dimensional Binary Random Arrays," *Philosophical Magazine A*, 1982, vol. 46, pp. 105-126.
29. A. S. Nowick and S. R. Mader, "A Hard-Sphere Model to Simulate Alloy Thin Films," *IBM Journal of Research and*

Development, 1965, vol. 9, pp. 358-374.
30. J. A. Dodds, "Simplest Statistical Geometric Model of the Simplest Version of the Multicomponent Random Packing Problem," *Nature*, 1975, vol. 256, pp. 187-189.
31. B. J. Last and D. J. Thouless, "Percolation Theory and Electrical Conductivity," *Physical Review Letters*, 1971, vol. 27, pp. 1719-1721.
32. T. J. Coutts and B. Hopewell, "Monte-Carlo Studies of Thin Film Growth," *Thin Solid Films*, 1972, vol. 9, pp. 37-55.
33. A. R. Brown, *Optimum Packing and Depletion*, Macdonald, London, UK, 1971.

Chapter Three

Factors Affecting Particle Packing

Chapter Outline

Introduction

Generally the density of a material has no significant influence on its fractional packing density [1]. Particles of an equal size and shape will pack to the same fractional density in spite of differing theoretical densities. However, several other factors do cause differences in fractional packing densities. The key factors treated here are particle size, particle shape, agitation, particle size distribution, surface texture, agglomeration, container size,

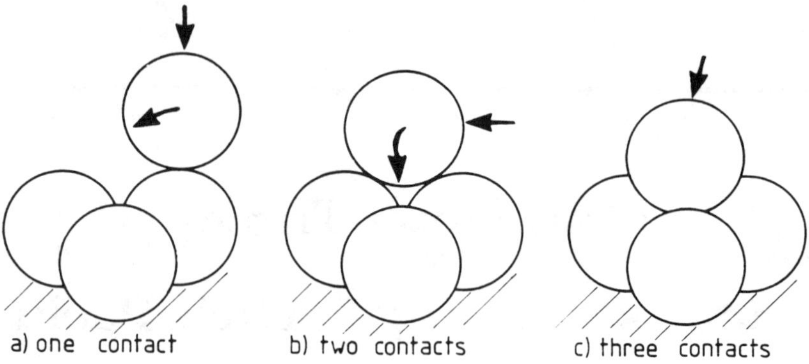

Figure 3.1 An added particle attains a stable packing position by first impacting on one existing particle, rolling to contact two particles, and finally rolling into a valley between three existing particles.

segregation, bridging, surface active agents, internal powder structure, and cohesion. These concepts are introduced prior to discussion of various packing geometries, models, and the properties of powder packings, since the assessment of these factors is necessary in each case.

Stable Positions

For gravitational stability a particle must have contact with at least three other particles. Figure 3.1 illustrates the addition of a new particle to an existing layer of particles. The initial point contact is unstable in comparison with the lower energy double contact. In turn, a lower height and energy can be achieved by further rolling the sphere into the valley between the three stationary spheres. Within a powder mass, an individual particle requires at least four contacts within a powder mass to ensure stability. These four contacts cannot lie on a single equator or single hemisphere of the particle [2]. This is analogous to the tetrahedral bonding observed in the diamond crystal structure. In a random packing of spheres, the particle coordination generally reaches an average value of six to seven contacts per particle.

As the packing density of a powder decreases, conceptually it reaches a point where the compact is no longer stable. Under perfect conditions stability might persist at very low packing densities. However, a random structure becomes unstable if there are fewer than 4.75 average contacts per particle [3]. This corresponds to an approximate fractional packing density of 0.35 for monosized spheres. Gotoh et al. [4] have computer simulated the packing of monosized spheres at low packing densities. Figure 3.2 shows their results for the normalized mean nearest neighbor spacing r/R (where r = distance, R = sphere radius) versus volume fraction of solid. The curve extrapolates to zero mean separation at a fractional packing density between 0.5 and 0.6, similar to experimental densities. Based on various correlations between packing density and coordination number, a fractional density of 0.35 for monosized spheres corresponds to approximately 3.5 to 5.0 contacts per sphere. This range agrees with the mean value of 4.75 contacts per sphere mentioned above. Packed particles with fewer contacts will be unstable.

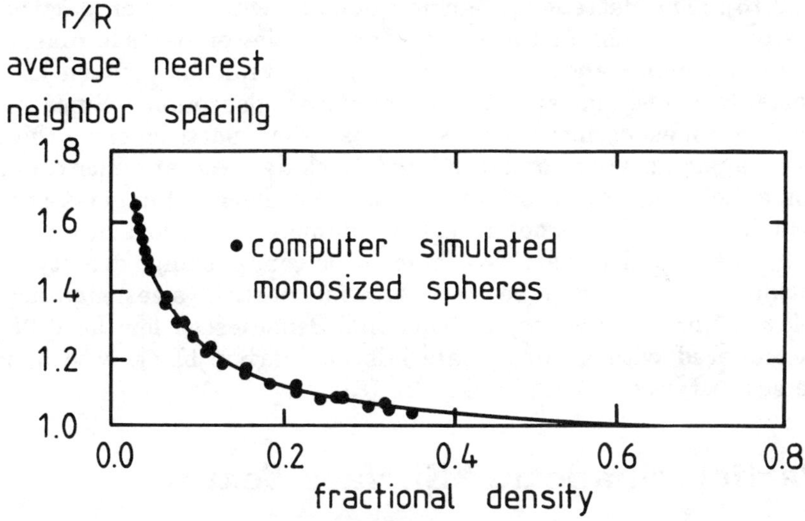

Figure 3.2 Computer simulated average nearest neighbor spacing for monosized spheres versus the fractional packing density [4]. The packing is unstable at densities less than 0.35.

In theory, very low packing densities are possible with ordered packings of spheres. The requirements are that the spheres touch in a continuous, chain-like network to produce a structurally stable arrangement. Melmore [5] describes two such packings with coordination numbers of three and fractional densities of 0.056. These are the lowest packing densities that have been described for monosized spheres; however, they are not random structures.

Particle Size

Once a stable packing is established, considerable variability is still possible in the structure. For packings composed of large particles, the particles size is not important to the density. However, when the mean particle size is below approximately 100 μm there is more interparticle friction, and particle bridging is more likely to occur. For example, tungsten carbide particles of 4.6 μm size pack to a 0.55 fractional density, while for those of 1.8 μm size the density is 0.31, and for 0.6 μm size particles the fractional density is 0.12 [6]. The decreasing packing density with smaller particles is due to an increase in the surface area, a lower particle mass, and a greater significance of the short-range, weak forces such as electrostatic fields, moisture, and surface adsorption. Since interparticle cohesion increases with a smaller particle size, there is more agglomeration and inhibited packing. As an illustration of this effect, Figure 3.3 shows the packing density for glass spheres as a function of the mean particle diameter on a logarithmic scale [7]. The smaller particles give a lower packing density. The packing densities can be very low for particle sizes significantly below 1 μm. For example, fractional densities as low as 0.05 are encountered with natural materials like carbon black with a particle size between 0.1 and 0.01 μm [8].

Particle Shape and Surface Texture

Another form of interparticle friction arises from irregularities on the particle surface. The greater the surface roughness or the more irregular the particle shape, then the lower the packing density. For example, in one study a 15-fold increase in surface roughness lowered the fractional packing density from 0.65 to 0.20 [9]. The data for the particle shape effect on packing density are

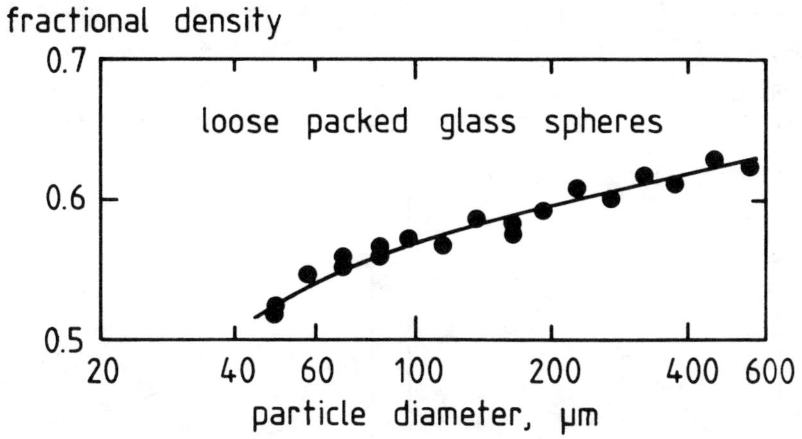

Figure 3.3 A plot of the fractional packing density for loose packed glass spheres versus the particle diameter [7].

scattered, yet some general patterns are apparent. Figure 3.4 provides a schematic of the general particle shape and surface roughness effects on the fractional packing density [10]. On the left, the packing density is shown as a function of the relative sphericity, which is defined as the surface area of a sphere of equivalent volume divided by the actual surface area of the particle. The closer the particle shape is to being spherical, the larger the relative sphericity. On this basis both particle shape and surface texture are included in the relative sphericity. As shown in Figure 3.4, the density improves as the particles approach a spherical shape. The right half of this figure shows the effect of the relative surface roughness. The relative roughness is a measure of the texture on the powder surface for an otherwise spherical shape [11]. In this regard, the effect of surface roughness is similar to the particle shape effect. This is due to bridging of the particles. The use of vibration or lubricants can help attain a high packing density, but problems may arise with agglomeration or size segregation. Water or various oils can reduce the interparticle friction [12]; however, treatments that increase the surface stickiness of a powder will degrade the packing density.

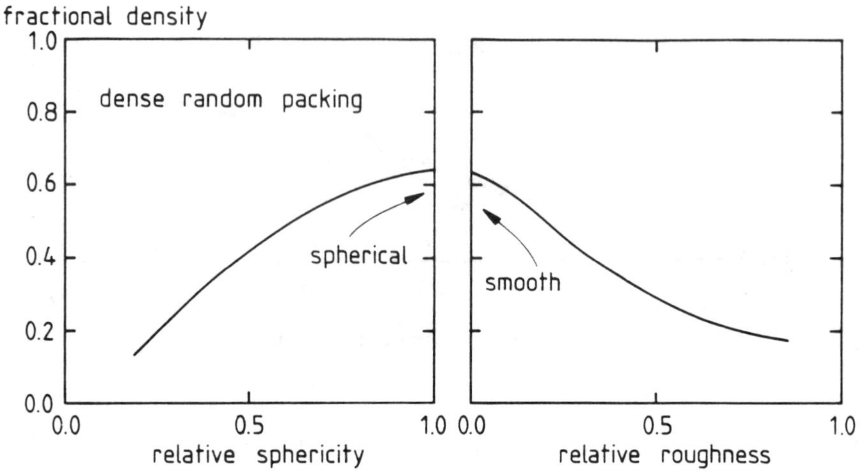

Figure 3.4 Schematic plots of the effect of particle shape and surface texture on the dense random fractional packing density [10]. The highest density is associated with smooth spherical particles.

For powders of the same size but different shapes, the packing density will decrease as the shape departs from equiaxed (spherical) [13-16]. This is easily seen in the packing of fibers compared to spheres. The length to diameter ratio provides a measure of the departure from an equiaxed shape. As the shape becomes more fibrous, with a larger ratio of length to diameter, the packing density is reduced. In powder mixing, an irregular particle shape will interfere with mixing, but will also help maintain a homogeneous mixture by interfering with demixing. Density can be improved by mixing different sizes of particles. This packing benefit is independent of shape, but the starting densities are lower with irregular particle shapes [17]. However, with certain shapes under vibration, a high packing density may be achieved by orienting the irregular particles. Such a high packing density occurs most typically with equiaxed particles. For example, cubic particles can be vibrated to place the cubes in contact along their faces. Such close proximity would give a packing density higher than that attainable with spheres. With a mixture of spherical and irregular powders, the packing density of the spheres is not significantly degraded by the irregular powder until the mixture contains approximately 10% of the irregular powder [13]. Thus, the addition of irregular particles may not harm packing density for spheres, yet may improve compact strength by providing more interparticle friction.

Agglomeration

Small particles cause difficulty in attaining a high packing density because of agglomeration due to cohesion. The attractive forces between particles become larger as the surface area

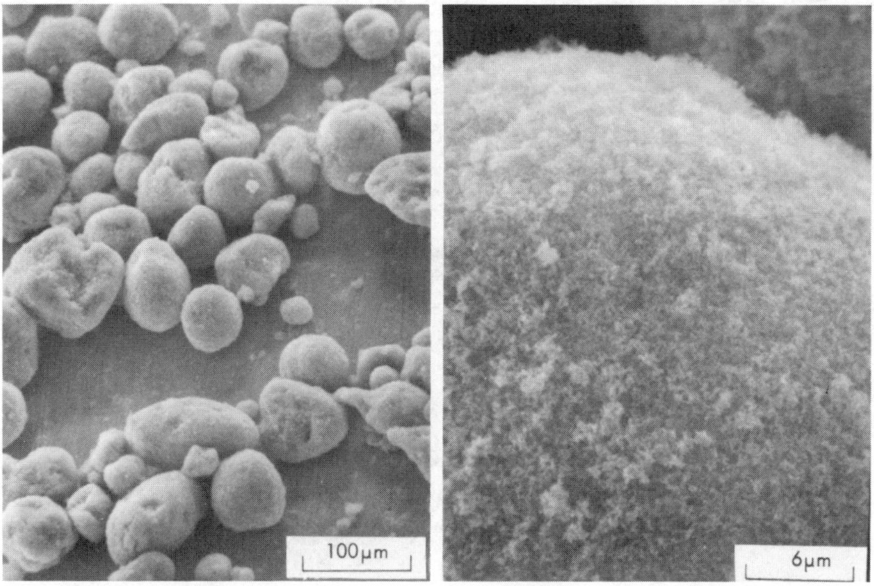

Figure 3.5 Two scanning electron micrographs of agglomerated iron oxide particles.

increases and the particle mass decreases [18]. In addition, small particles have a greater tendency for vapor condensation at the particle contacts. Agglomeration can be induced in a powder by inhomogeneously distributing a wetting liquid. Most commonly, a condensed vapor will form pendular bonds between particles. These bonds strengthen the particle cluster, but inhibit dense packing [19-21]. This is a particular problem with submicron sized powders exposed to humid air. Figure 3.5 shows two scanning electron micrographs of an extensively agglomerated iron oxide powder composed of submicrometer sized particles. Agglomeration makes mixing more difficult. An alternative is to create conditions that give rise to repulsive forces between particles by using thin coatings of polar molecules. Surface repulsive forces contribute to high packing densities by reducing the interparticle friction found with powders possessing cohesive forces.

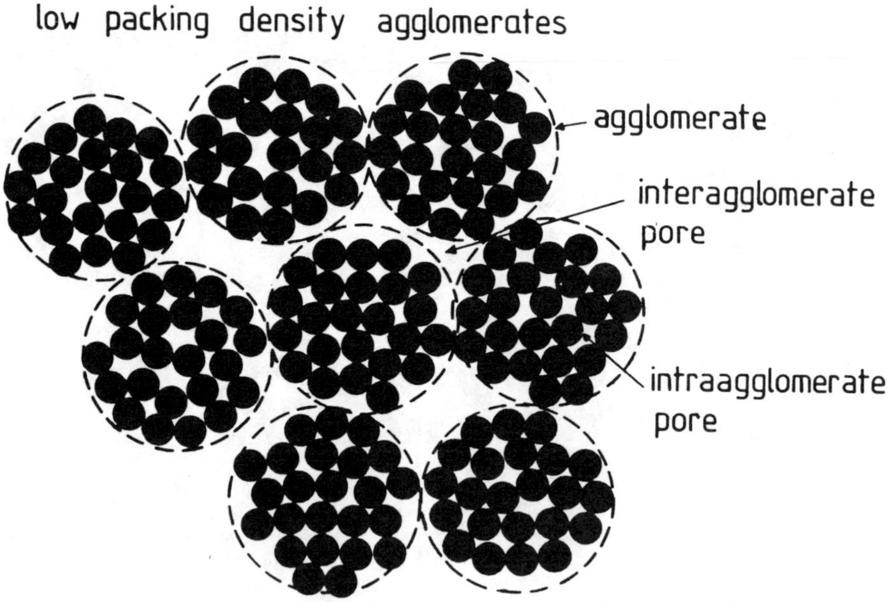

Figure 3.6 A sketch showing the large pores associated with the interagglomerate regions in an agglomerated spherical powder; the overall packing density is reduced by agglomeration.

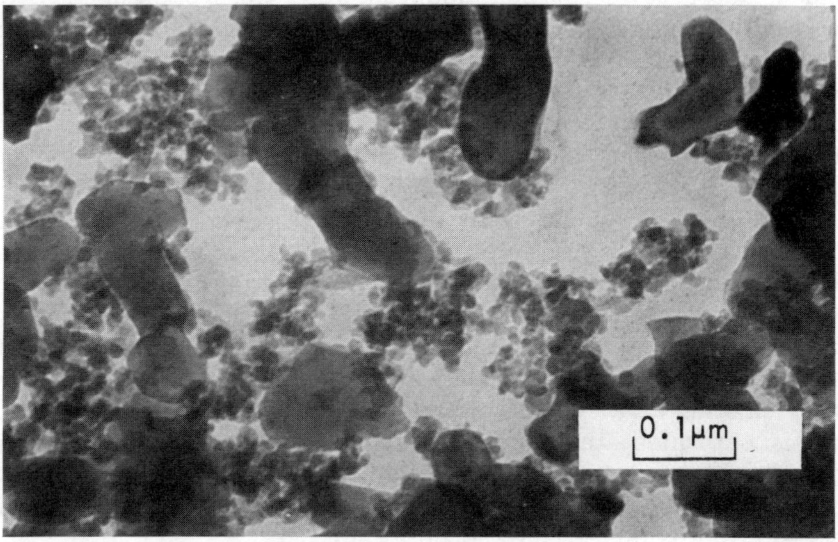

Figure 3.7 A transmission electron micrograph of an agglomerated submicrometer sized alumina powder.

Particles agglomerate into clusters of high coordination number separated by high porosity regions, as sketched in Figure 3.6. Although this is an idealized situation, many examples of agglomeration are evident in submicron powders. Figure 3.7 is a transmission electron micrograph of a fine alumina powder showing considerable agglomeration. Agglomeration occurs mostly with smaller particles, because of a high surface area and the action of one of the weak forces. The common weak forces are van der Waals attraction, electrostatic charges, chemical bonding, capillary liquid forces or magnetic forces [22,23]. The van der Waals force is significant for particles below 0.05 μm in size [9]. Agglomeration can also occur during mixing due to cold welding at the particle contacts.

Most typically, particles agglomerate due to adsorbed surface films, especially during periods of agitation. Atmospheric vapors condense and establish a surface concentration dependent on their partial pressure. Water is the most typical vapor to condense on a powder. The amount of water adsorption depends on the relative humidity and the surface curvature [22,24]. Four cases of adsorption are possible [25]. First, at low vapor pressures the powder surface is uniformly coated with a thin layer of adsorbed vapor. Second, as the partial pressure increases, a critical level is reached where the vapor condenses to form capillary bridges localized at the particle contacts. These bridges are termed pendular bonds and are shown schematically in Figure 3.8a. Third, as the vapor pressure increases, the funicular state occurs. In the funicular state the pendular bonds merge, but the pores are less than totally filled by liquid. In this state the pores are smooth and surrounded by liquid. A connected path exists in the both the vapor and liquid phases as shown in Figure 3.8b, with the vapor phase existing as a cylindrical shape. Finally, when the pore structure is saturated, the pores are filled with liquid. This case is shown in Figure 3.8c. For water vapor and spherical powders, the pendular bond state is anticipated at relative humidity levels between approximately 65 and 80% [26].

A measure of the pore space filling is given by the saturation, defined as the volume of pore space filled by fluid divided by the total pore space. The saturation increases with the vapor partial pressure. The agglomerate strength has a nonlinear dependence on the saturation. Figure 3.9 shows a plot of the tensile strength versus saturation for limestone powder. The strength of the pendular bond increases with the liquid-vapor surface energy and depends inversely on the square of the particle diameter. At less than total saturation, the strength of the agglomerates

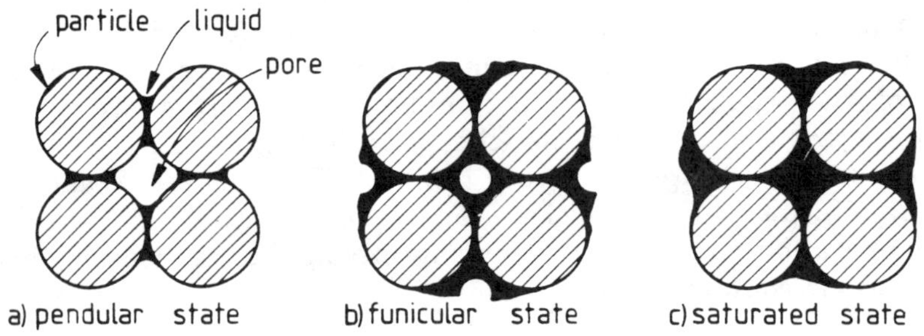

Figure 3.8 The states associated with agglomeration of powders due to a wetting liquid: (a) pendular, (b) funicular, and (c) saturated.

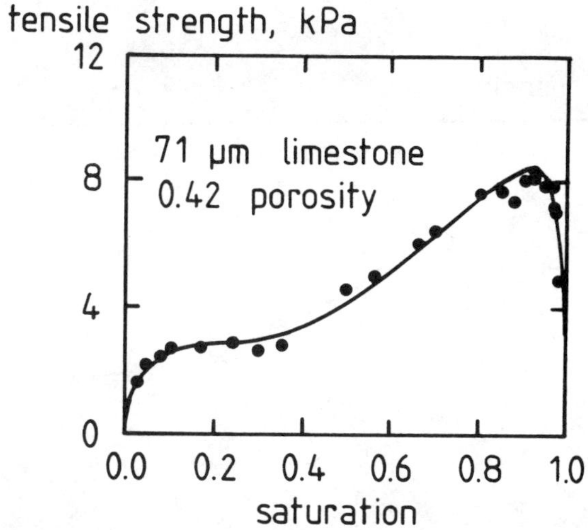

Figure 3.9 The effect of agglomeration on the tensile strength of limestone powder as a function of the moisture saturation [25].

increases with the amount of liquid and the surface energy of the liquid. The absence of internal liquid-vapor surfaces at 100% saturation causes the strength to suddenly decrease at this point. When the pores are filled with liquid, only the outer surfaces provide an agglomeration force. Since the external curvature dictates the compressive stress, the agglomerate strength decreases as the agglomerate size increases. As a consequence, large agglomerates have a lower packing density than small agglomerates [21,27]. For an irregular particle shape, the density initially decreases with the addition of a wetting liquid [14].

A negative consequence of agglomeration is a decrease in the packing density. This is a particular problem with small particles. Clustering of particles, especially those of similar size, is an unexplained characteristic of many structures [28,29]. As the particle size decreases and the number of particles in each agglomerate increases, the agglomerates become stronger and exhibit a lower maximum packing density [19,21,30]. This arises because the pores between agglomerates grow in size as the degree of

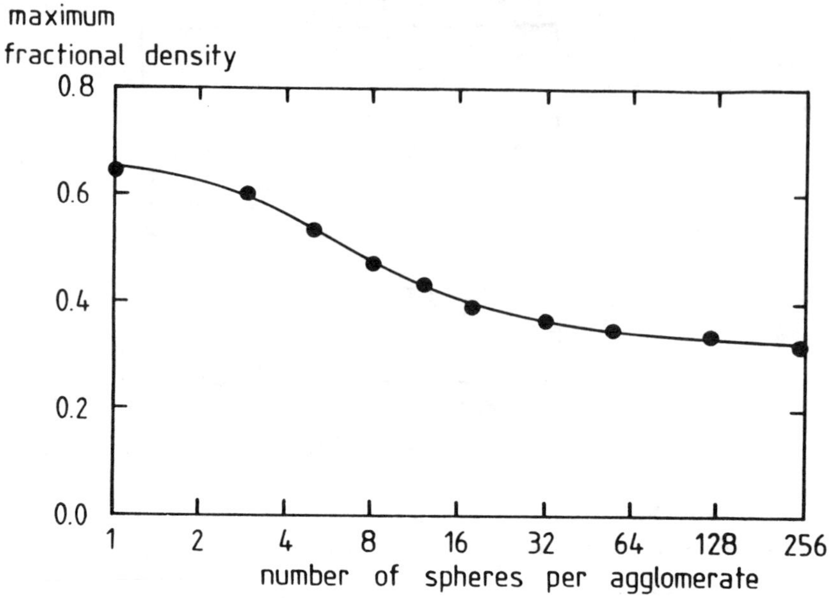

Figure 3.10 The calculated role of particle agglomeration on the maximum packing density for monosized spheres [21].

agglomeration increases. The results of calculations of the maximum packing density versus the number of spheres in an individual agglomerate are plotted in Figure 3.10 [21]. This plot shows the negative effect of agglomeration on packing density. The predicted lower limit packing for monosized spheres due to agglomeration is 0.37 [31]. Obviously, with other particle shapes and particle size distributions, the packing density can be even lower. Overall, the problem with agglomeration becomes more acute as the particle size decreases below approximately 1 μm [32].

Besides decreasing packing density, agglomeration also creates problems with mixing, settling from suspensions, compaction, and sintering, especially for those systems with wide particle size distributions [20,32-36]. The sensitivity to the particle size distribution arises because the small particles are the primary cause of agglomeration. In a wide particle size distribution, the smallest particles can exert a strong effect on the larger particles. Alternatively, agglomeration can be selectively used to minimize size segregation in powder handling.

Surface Active Agents

Small quantities of surface active agents are often added to particles to alter packing or mixing characteristics. Some common additives are polyvinyl alcohol, stearic acid, sodium oleate, glycerine, and oleic acid [6]. These additives reduce interparticle friction by lubricating surfaces via short range repulsive forces. Generally, the flow and packing of particles are improved by the presence of the appropriate surface active agent. The level of improvement is dependent on the molecular size of the additive, its polar character, the layers of coverage, the particle surface condition, the particle size, and the temperature. Polar molecular coatings with short range interactions aid in keeping particles from agglomerating [19, 32,37-40]. This is most important with the submicrometer particles that are dispersed in a fluid for processing, such as in slip casting. The particle dispersion is maintained by use of short range repulsive forces, possibly induced by surface charges as measured by the zeta potential. The viscosity and surface energy of the additive are not as important as the polarity and wetting ability. Agglomeration is less of a problem with a narrow particle size distribution and larger particle sizes. In contrast, sticky particle surfaces will have a high level of agglomeration, leading to a lower particle packing density.

Intermediate chain-length polymeric molecules are used to lubricate powder compaction, where they are used in high concentration. Alternatively, at low concentrations small organic molecules are optimal because of their polar character. Generally, a backbone chain length of approximately ten carbon atoms is optimal as a surface additive for inorganic particles. Even so, for large particles the relative packing benefit is low. For example, in one experiment with large copper particles (~60 μm) having a clean surface (no additive or oxide), the fractional packing density was 0.53 [41]. When the powder was surface oxidized to reduce interparticle friction the packing density improved to 0.56. When the powder was coated with one monolayer of lauric acid the packing density was 0.57, a relatively small improvement. As the particle size decreases, packing is limited by the high interparticle friction and the natural tendency to agglomerate. Here surface active agents have their greatest benefit [42,43].

Internal Powder Porosity

Many powders contain internal porosity which is sometimes totally isolated from the powder surface. Figure 3.11 shows sketches of cross-sectioned particles with differing internal pore structures: a) fully dense, b) entrapped pore, and c) open pore. The closed porosity will not interact with a penetrating fluid. In contrast, the sponge-like structure has vapor phase communication from the external surface to the inner pores. Figure 3.12 contains two optical photographs of cross-sectioned and polished iron powders. These two pictures show all three forms of particles; dense, enclosed pores, and open pores.

The packing laws for particles ignore the internal porosity within the particles; the focus is on the interparticle pores with the assumption of dense particles. If a particle contains internal pores, then the intraparticle porosity will degrade the packing density. Thus, in mixed particle size systems, the predicted mixture density

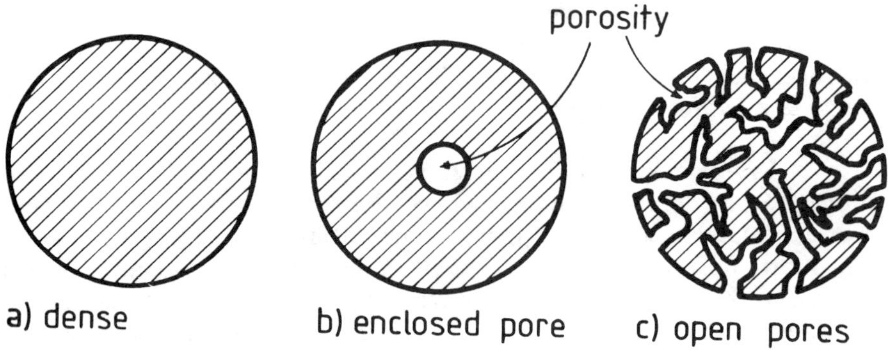

Figure 3.11 Sketches of three forms of particles with varying internal pore structures; a) dense (no pores), b) closed internal pore, and b) open internal pore.

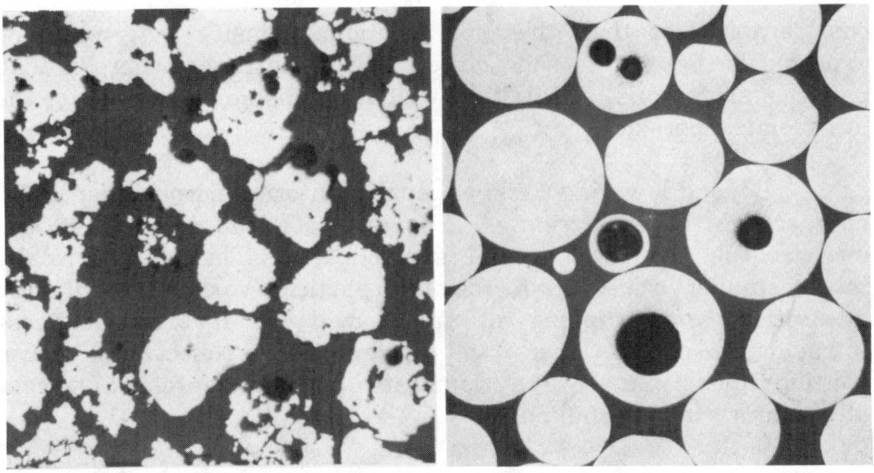

Figure 3.12 Optical micrographs of polished cross sections through iron powders showing dense, and porous particles with both closed and open pores.

must be corrected for the internal porosity. Let f_s be the fractional solids content for the powder, giving the fractional intraparticle porosity as $1 - f_s$. Then, with f_e as the predicted external fractional packing density (interparticle porosity is $1 - f_e$), the overall packing fractional density f is,

$$f = f_e f_s \,. \tag{3.1}$$

This equation says that the overall packing density is the product of the fractional densities of the particles times the particle packing density. Generally in this treatment, the focus will be on the interparticle porosity. However, when porous particles are involved in the packing, the actual solids content will be reduced by the intraparticle porosity. For very porous particles, this correction can be substantial. Consider the packing of hollow particles having an egg shell structure. For such an oval shape, Allen [44] predicts a packing density of 0.55. An egg shell constitutes only a small fraction of the actual volume of an egg, probably near 5% of the total. Hence, a random packing of drained egg shells (not cracked) would have a fractional solids density of approximately 0.0275. This is considerably less than the 0.55 fractional density that would be expected if these were solid objects. In those cases where not all particles are porous, a statistical weighting should be applied for the intraparticle porosity.

Beyond lowering packing density, internal porosity can influence several other attributes. For example, the open pores will increase the measured specific surface area. These are typically much smaller pores than the interparticle voids, consequently adsorption and trapping of fluids will be favored in these intraparticle pores. These will preferentially collect and retain wetting fluids. In a sense, agglomerated particles represent one case of powders with internal pores.

Container Wall

The container used to hold a powder will induce a local area of order at the container wall in an otherwise random packing [45-67]. The effect is more pronounced for flat, smooth containers, giving local regions of oscillating high and low porosity in the first few particle layers near the wall. Even in flexible containers there is ordering near the wall. Besides packing density, the container wall influences measurements because of the low density regions near the wall. The packing coordination is higher in the first few layers of particles near a container wall. Figures 3.13 and 3.14 illustrate the magnitude of the container wall effect in terms of the fractional packing density [10,49]. Figure 3.13 plots the local packing density versus distance from the wall in sphere diameters. This decaying oscillation in packing density with distance has been confirmed in several studies [50-53]. The wall effect persists for several particle diameters into the packing. This same influence is evident at the interface between a crystallized solid and its liquid; damped density oscillations can be seen in the liquid bordering on the solidification interface. Figure 3.14 shows the integral packing density versus distance for spheres packed in a cylinder. The integral density decreases from a low value at the wall to a constant value of approximately 0.64 within the first two particle diameters. The oscillations are not as apparent in the integral density, as can be seen by contrasting Figures 3.13 and 3.14. Depending on the particle size, particle shape, and container shape, it takes from one to ten particle diameters from the wall to establish truly random

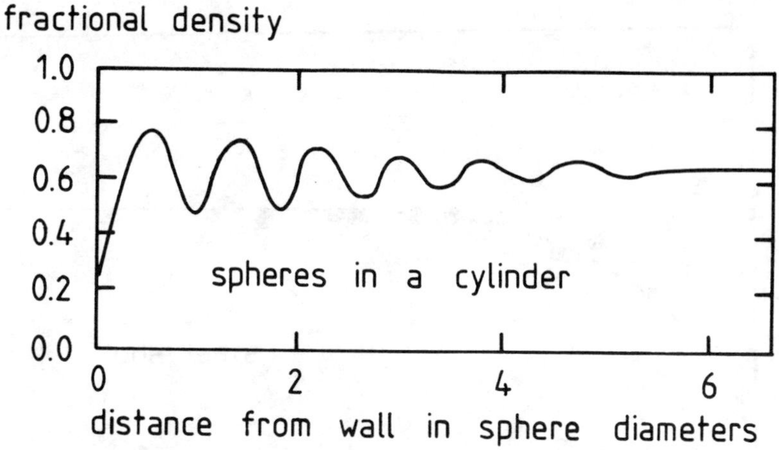

Figure 3.13 The local fractional density versus the distance from the container wall for monosized spheres packed in a cylindrical container [10].

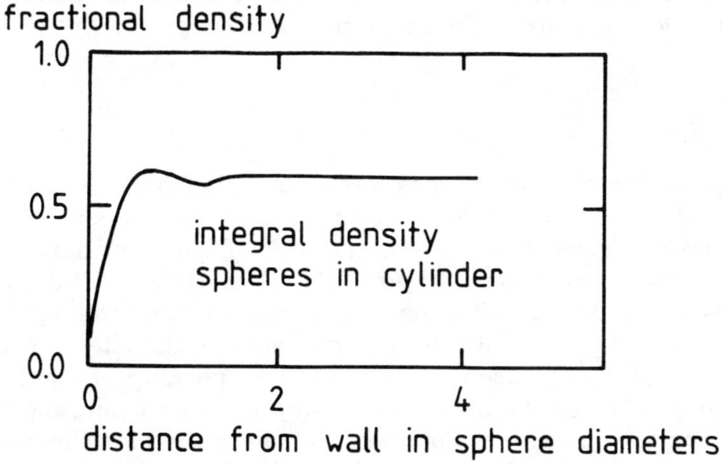

Figure 3.14 The integral fractional packing density for spheres in a cylindrical container [49]. There is less sensitivity of the integral density in comparison with the local density shown in Figure 3.13.

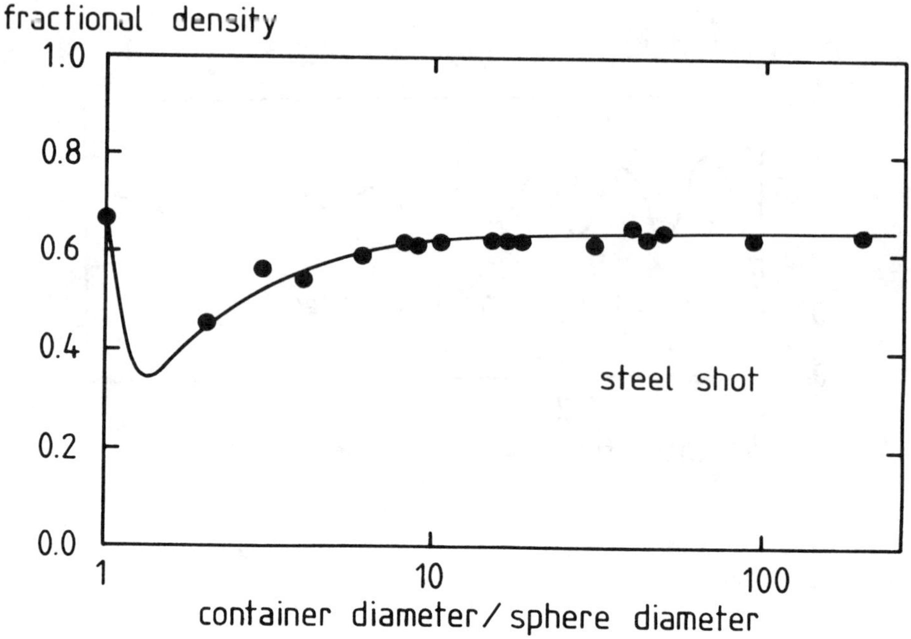

Figure 3.15 The data of McGeary [54] for the container size effect on packing density for steel shot.

packing [17,50-58]. The effect is larger at higher packing fractions. In each case, as the randomness increases, the packing density approaches an asymptotic value as the ratio of container diameter to particle diameter increases. Figure 3.15 plots McGeary's data [54] for the variation in fractional packing density versus the logarithm of the ratio of the container diameter to sphere diameter. In these experiments McGeary used a dense random packing of steel shot. The density is low for small ratios, but approaches the limiting value of 0.64 for ratios over ten. For the case where the particles just fill the cylinder (container diameter/sphere diameter = 1), an ordered packing exists with a fractional density of 0.67.

The effect of the container wall on the packing density has been expressed in various mathematical forms [9,17,50,55-62]. For a given container, the container size influence on density increases with the particle size and with the surface area of the container. This same behavior can be seen when a close-packed structure is separated. The resulting unfilled depressions on the surface of the close-packed layer result in a slight density decrease. For spherical particles the results by Ayer and Soppet [55] fit the following equation:

$$f = 0.635 - 0.216 \exp(-0.313\ D_c/D) \tag{3.2}$$

where D is the sphere diameter, D_c is the container diameter, and f is the factional packing density. They found the overall packing density to be within 1.5% of the calculated value. Alternatively, Haughey and Beveridge [56] express the container size effect in the form of a polynomial,

$$f = 0.61 - 0.07\ (D/D_c) - 0.54\ (D/D_c)^2\ . \tag{3.3}$$

In both cases the container wall effect has less influence on the total packing density as the container becomes larger. Note these two equations predict fractional packing densities that differ slightly from the accepted value of 0.637 for an infinite container.

The experiments by Scott [59] are a classic demonstration of container influences on packing. Figures 3.16 and 3.17 show the fractional packing density versus the inverse of the powder bed height and container diameter for monosized spheres in cylindrical containers. Both loose and dense random packings are included in these plots. The extrapolated density for the random dense packing is 0.637. Scott also showed that the packing density varied with the cube root of the number of particles in the packing. Since the container volume varies with the number of particles, this effectively shows that the packing density scales with the container size. Larger containers have less of an effect on the packing density; however, the container size role is dependent on the particle size; larger containers are needed for larger particle sizes to give an equivalent effect [1,9,46,60,61,63]. The true fractional density can be estimated based on the number of particles needed to span the container n;

$$f_m = f_o\ [(n - 1)/n]^3 \tag{3.4}$$

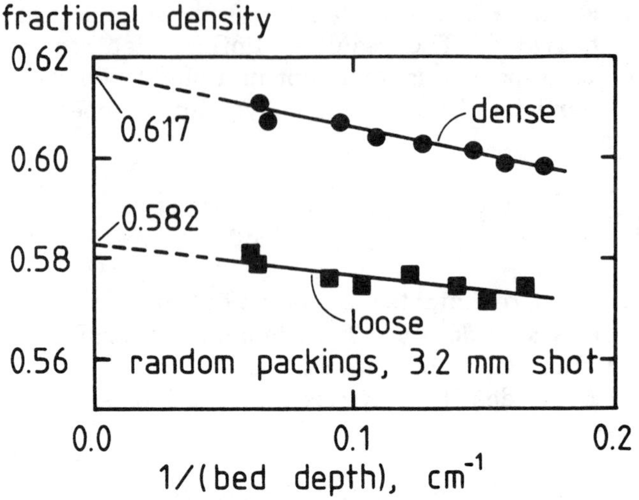

Figure 3.16 The effect of the powder bed depth on the packing density for loose and dense random packings of 3.2 mm steel shot in a 63 mm diameter cylinder [59].

where f_m is the measured fractional density and f_o is the actual random packing density. Calculations based on the ratio of the sizes can give a good estimate of the effect. Such estimates indicate that between 10^7 and 10^9 particles are needed to reduce the experimental density error to 0.1%.

For nonspherical powders there is inherently more randomness to the packing; thus, the wall effect decays over a shorter distance from the container wall. Equations for rough spheres and cylinders are analogous to those shown above for hard, monosized spheres. Mixed particle sizes aid in minimizing the porosity variations at the container wall. Accordingly, as the width of the particle size distribution increases, there is less influence from the container wall [1]. Any departure from a smooth, continuous plane for the container wall will induce more random packing in the neighboring powder. Thus, sharply curved walls or complex surfaces give random packing closer to the wall. The nonrandom packing due to a container wall occurs for both convex and concave surfaces; hence, it is also seen near internal posts or

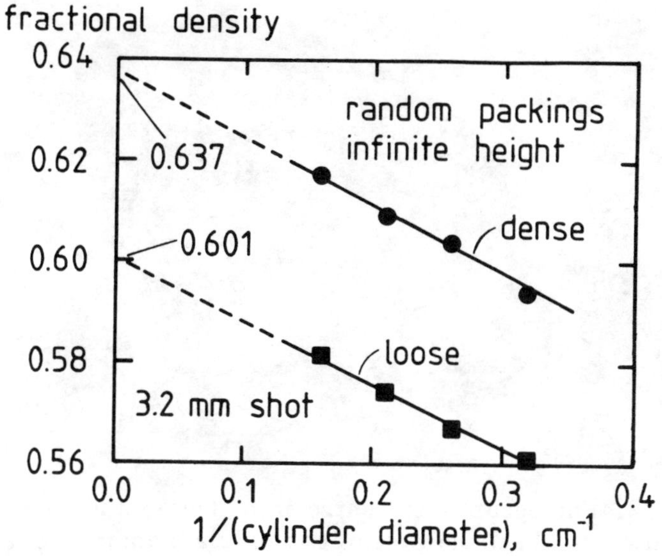

Figure 3.17 Fractional packing density for random loose and random dense packings of steel spheres, showing the extrapolated limiting densities with no container wall effects [59].

core rods in the packing as well as on the external surface [67,57]. When measuring packing density it is necessary to correct for the wall effect to ensure accurate results.

Segregation

Another concern in studying powder packing characteristics is segregation in mixed powders. This can lead to uneven packing densities and to distortion in compaction and sintering. There are three causes of segregation: differences in particle size, density, and shape [8,68-70]. Of these three, size segregation is dominant. For example, a powder poured into a container will segregate by size if the small particles can pass through the holes between the large particles. This leads to banding, especially at low filling rates [8]. Figure 3.18 shows two photographs of size segregation involving mixed coarse iron and fine alumina powders. On the left is a photograph of the segregation evident when the mixed powders

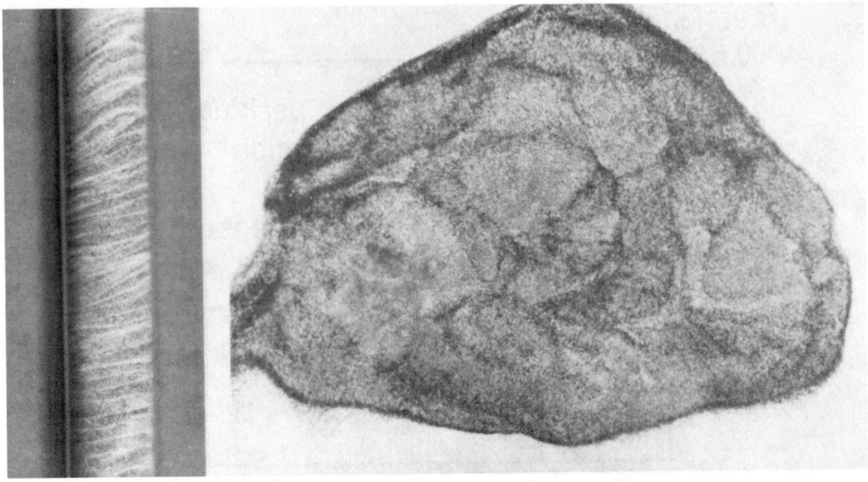

Figure 3.18 Photographs of coarse iron (black) and fine alumina (white) powders that show segregation via banding in a transparent cylindrical container and in a loose pile of powder.

were poured into a cylindrical glass container. Because the alumina was much smaller than the iron, it passed through the interparticle voids and collected into bands that are visible along the container length. On the right side of Figure 3.18 is a photograph of this same powder after having been poured into a loose pile. Again, segregation is apparent based on the inhomogeneous color and banding. Such size segregation is more of a problem with larger mean particle sizes and larger differences in particle size. One consequence of size segregation is that the overall density will decrease and there will be a point-to-point density variation [20,71]. The density at any point in the particle mass will depend on the local particle size distribution.

A common cause of size segregation is agitation. In a randomly mixed powder with differing particle sizes (or with the presence of agglomerates), the larger particles segregate to the top of the container when the particles are agitated. Such segregation is most pronounced when there is a vertical component to the agitation. The agitation opens local interstices between particles, making it more probable for the small particles to fall by gravitational action into these interstices [72]. This process enriches the

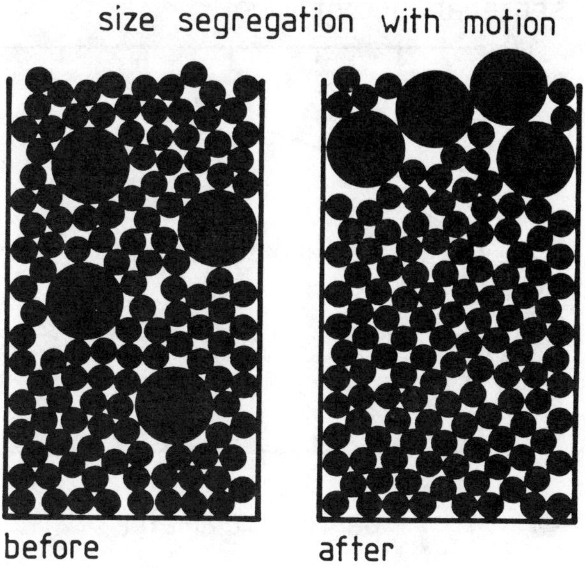

Figure 3.19 Size segregation of mixed powders occurs by percolation of the smaller particles to lower gravity positions during agitation. The net effect is a rise of the larger particles to the container top.

lower gravity positions with smaller particles over time, leading to a rise in the large particles as illustrated by the schematic diagram of Figure 3.19. Such a process can occur with particles in a size ratio as small as 1.7. When the coarse particles are over six times larger than the small particles, the interstitial pore size between the coarse particles will be sufficient for the small particles to fall through. As shown in Figure 3.20, size segregation is more of a problem at the larger particle size differences. In this figure, the abscissa corresponds to different ratios of the two sand particle sizes. The ordinate shows the degree of segregation as given in terms of the segregation coefficient C_s calculated as follows:

$$C_s = (X_t - X_b) / (X_t + X_b) \tag{3.5}$$

where X_t is the fraction of large particles in the top half of the

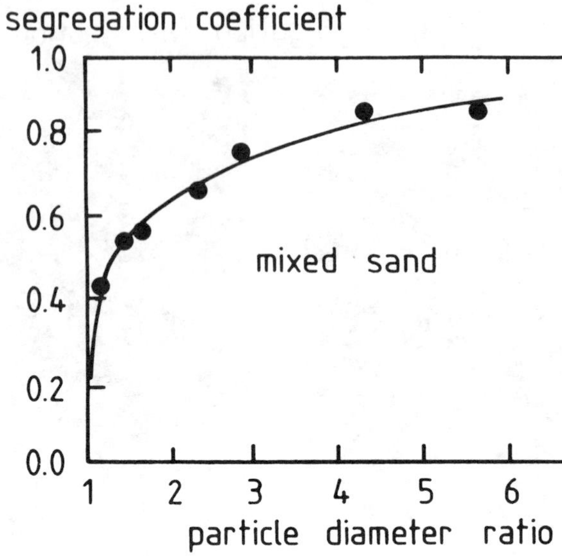

Figure 3.20 Size segregation as measured by the segregation coefficient for mixed sand of different particle size ratios [68]. Even small particle size differences can lead to size segregation.

container and X_b is the fraction of large particles in the bottom half of the container. Particle segregation requires that the agitation be large enough for the small particles to percolate into dilated interstitial cavities. During agitation the probability distribution in dilated pore sizes has an inverse relation to the pore size; small pores are created more frequently by agitation. The occurrence of large pores is unlikely, thereby inhibiting the fall of large particles. The preferential falling of small particles leads to segregation of the large particles to the top of the container.

Although segregation due to a particle size difference is prevalent, there is a weak dependency on particle density and particle shape. An irregular particle shape will inhibit size segregation. Likewise, as shown in Figure 3.21, less size segregation occurs with the smaller particle sizes because of the greater interparticle friction. Size segregation is less of a problem for particle sizes below approximately 100 μm [68,73]. Also, as noted in Figure 3.19, mean particle sizes over approximately 500 μm show no further

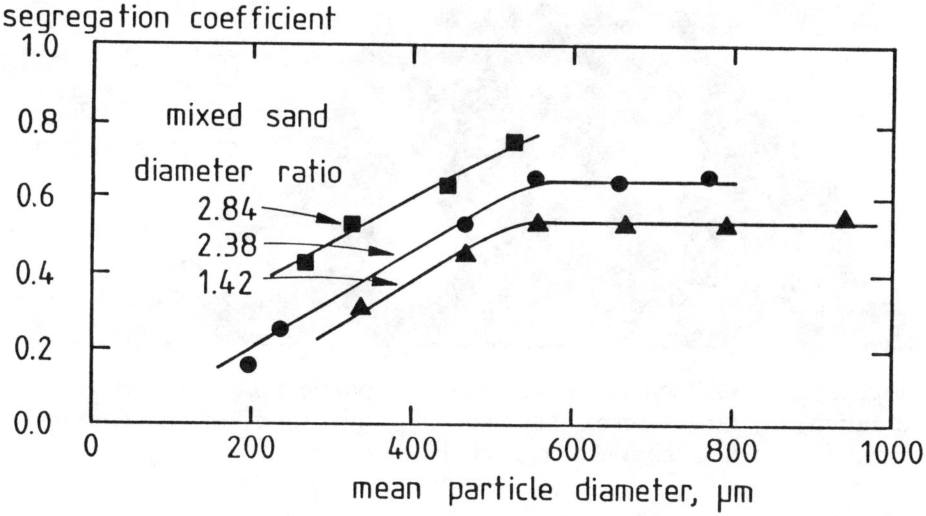

Figure 3.21 A plot of the size segregation coefficient versus the mean particle diameter for three different size ratios of sand [68].

change in segregation tendency. In practice, size segregation can be reduced by using wetting agents, though agglomerates created by uneven wetting can contribute to segregation [74]. The powder handling procedures can be adapted to minimize segregation. The best practice is to gently vibrate the particles simultaneous with loading [75].

Bridging and Vibration

The low packing density found in many powders is attributed to bridging, which is the creation of large pores by arching, such as shown in the two dimensional schematic of Figure 3.22. A nonspherical particle shape, especially an angular shape, will increase the tendency of a powder to bridge. Likewise, as the particle size decreases there is less mass to an individual particle and more surface friction. Both factors contribute to a lower packing density [9,56]. Bridging is most apparent near container walls,

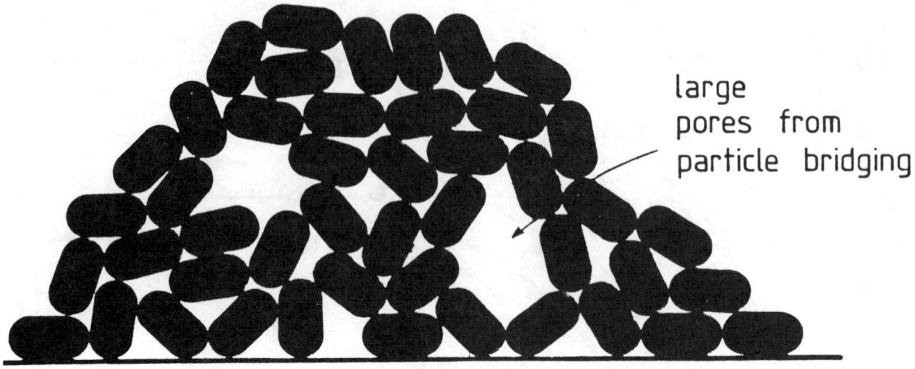

Figure 3.22 A schematic diagram of particle bridging and the creation of large pores because of high interparticle friction, irregular particle shapes, or particle cohesion.

where it gives voids larger than the particle size [76,77]. Extensive bridging has been simulated in a computer model of loose packing of monosize spheres under gravity [3]. Particles with sticky or cohesive surfaces exhibit extensive porosity because of bridging.

To eliminate the large pores associated with bridging, agitation is used to bring a powder to a high packing density. This is performed by vibrating the container without a direct external pressure on the powder. This density requires the particles to attain a close proximity to one another. The tap density is generally accepted as the maximum density attainable under vibration or agitation. The tap or vibrated density depends on the material, vibration amplitude, vibration direction, applied pressure, vibration frequency, particle density, shear, and test apparatus [48,54,69, 75,78-86]. During vibration the density varies with the number of vibrations with an exponential function [87,88]

$$(f_f - f_n) = (f_f - f_i) \exp(-K /N) \qquad (3.6)$$

where K is a constant that depends on the device, height of fall, and velocity, N is the number of vibration cycles, f_f is the final solids

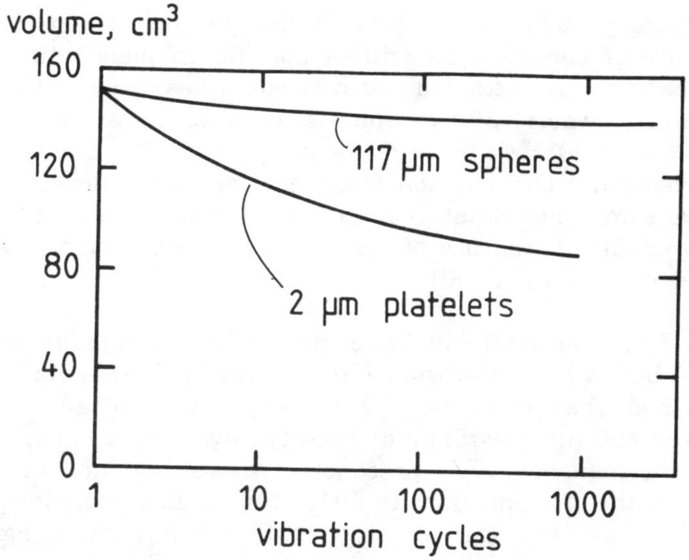

Figure 3.23 The tap volume versus the number of vibration cycles for spherical glass and platelet shaped titania particles [88].

density, f_n is the solids density after N cycles, and f_i is the initial solids density. Example data for vibrational packing are shown in Figure 3.23 for 117 μm glass spheres and 2 μm titania platelets [88]. The spherical glass beads undergo little densification during vibration. In contrast, the packing of the titania platelets is considerably improved by vibration. Generally, the more irregular the particle shape, the greater the packing benefit from vibration. Additionally, powders will reach the dense random packing limit more rapidly as the particle size increases. Likewise, a dense packing is aided by a spherical shape and a lack of interparticle cohesion.

For any powder there is an optimal condition for attaining rapid packing. The angle, type, amplitude, and frequency of vibration must all be considered; their effects vary between powders. The optimal procedure is to vibrate the powder with both a lateral and an axial component. Generally, the most rapid packing occurs with an acceleration near 6 to 7 times gravity, and with an amplitude near 1 mm. Many powder systems maximize their rate of densification with a specific frequency [9,54,82,89]. Best packing is attained with frequencies over 300 Hz for submicron powders, 200 to 300 Hz for intermediate particle sizes, and 100 to 200 Hz for large particles over 100 μm. Because of this dependence on vibration frequency, variable vibration frequencies are sometimes employed, especially with particle mixtures. A three component mixture of particles has been vibrated to a fractional density of 0.94 [90].

Under optimal conditions the packing variability within the particle bed will be less than 0.005% from point-to-point. It is not established that a variable frequency leads to any significant improvement in asymptotic density over that from a fixed vibration frequency. The presence of a wetting fluid has a small effect on the limiting density [91]. Small and irregularly shaped powders require longer times to attain the highest density by vibration. With anisotropic particles, the final structure may exhibit a preferred orientation [92]. Likewise, agglomerated powders will not attain the same level of tap density as nonagglomerated powders.

Summary

Early studies on powder packing failed to recognize the various effects noted here, such as the density variation at the container wall, as well as the role of particle size, surface roughness, particle shape, agglomeration, and vibration. Hence, the reported packing densities were highly variable. Recent studies have more closely controlled these variables and have shown consistent results on particle packing.

A stable random packing of particles requires at least four contacts on each particle and most random packings of large spherical particles exhibit an average of six contacts per particle.

Various factors will degrade the packing density of a powder, especially characteristics that increase the interparticle friction. For this reason smaller particle sizes, irregular particle shapes, rough surface textures, and cohesive films are detrimental to attaining high packing densities. Low packing densities are associated with bridging and agglomeration. Agglomeration occurs as the particle size decreases and the interparticle attractive forces become significant. This tendency can be offset by use of appropriate surface active agents. Bridging occurs with small particles and angular particle shapes under poorly agitated conditions. It is further influenced by the surface texture or stickiness of the powder.

The randomness of a packed powder is influenced by the container wall. The container wall effect on packing is less apparent with roughened container walls and irregular particles. Size segregation is a common problem with powder mixtures when they are passed through free fall or are subjected to agitation. Size segregation decreases as the particle size becomes smaller and the particle shape becomes more irregular, but agglomeration then becomes a greater concern.

An irregular particle shape creates special problems in packing and in working with powders. Although an irregular shape typically degrades the particle packing density and interferes with mixing, there are some offsetting benefits. For example, the compact strength increases with an irregular particle shape, and the greater interparticle friction is helpful in preventing segregation or demixing. Generally, the handling and packing deficiencies associated with an irregular shape are significant concerns.

References

1. M. Leva and M. Grummer, "Pressure Drop Through Packed Tubes, Part III, Prediction of Voids in Packed Tubes," *Chemical Engineering Progress*, 1947, vol.43, pp.713-718.
2. A. N. Nikolenko and M. S. Kovalchenko, "Analysis of the Random Packing of Identical Particles III. Tapping of Spherical Powders," *Soviet Powder Metallurgy and Metal Ceramics*, 1986, vol. 25, pp. 24-26.
3. L. Oger, J. P. Troadec, D. Bideau, J. A. Dodds and M. J. Powell, "Properties of Disordered Sphere Packings I. Geometric Structure: Statistical Model, Numerical Simulations and Experimental Results," *Powder Technology*, 1986, vol. 46, pp. 121-131.
4. K. Gotoh, W. S. Jodrey and E. M. Tory, "Average Nearest Neighbor Spacing in a Random Dispersion of Equal Spheres,"

Powder Technology, 1978, vol. 21, pp. 285-287.
5. S. Melmore, "The Densest and the Least Dense Packings of Equal Spheres," *Mineralogical Magazine*, 1949, vol. 28, pp. 479-485.
6. I. G. Shatalova, N. S. Gorbunov and V. I. Likhtman, "Physicochemical Principles of Vibratory Compacting," *Vibratory Compacting Principles and Methods*, H. H. Hausner, K. H. Roll and P. K. Johnson (eds.), Plenum Press, New York, NY, 1967, pp. 1-206.
7. R. J. Wakeman, "Packing Densities of Particles with Log-Normal Size Distributions," *Powder Technology*, 1975, vol. 11, pp. 297-299.
8. R. L. Brown and J. C. Richards, *Principles of Powder Mechanics*, Pergamon Press, Oxford, UK, 1970.
9. W. A. Gray, *The Packing of Solid Particles*, Chapman and Hall, London, UK, 1968.
10. K. Shinohara, "Rheological Property of Particulate Solids," *Handbook of Powder Science and Technology*, M. E. Fayed and L. Otten (eds.), Van Nostrand Reinhold, New York, NY, 1984, pp. 129-169.
11. J. L. Finney, "Random Packings and the Structure of Simple Liquids I. The Geometry of Random Close Packing," *Proceedings of the Royal Society of London A*, 1970, vol. 319, pp. 479-493.
12. W. C. Bell, R. D. Dillender, H. R. Lominac and E. G. Manning, "Vibratory Compacting of Metal and Ceramic Powders," *Journal of the American Ceramic Society*, 1955, vol. 38, pp. 396-404.
13. J. R. Parrish, "Packing of Spheres," *Nature*, 1961, vol. 190, p. 800.
14. H. E. White and S. F. Walton, "Particle Packing and Particle Shape," *Journal of the American Ceramic Society*, 1937, vol. 20, pp. 155-166.
15. J. V. Milewski, "The Combined Packing of Rods and Spheres in Reinforcing Plastics," *Industrial Engineering Chemistry Product Research and Development*, 1978, vol. 17, pp. 363-366.
16. J. V. Milewski, "Packing Concepts in the Utilization of Filler and Reinforcement Combinations," *Handbook of Fillers and Reinforcements for Plastics*, H. S. Katz and J. V. Milewski (eds.), Van Nostrand Reinhold, New York, NY, 1978, pp. 66-78.
17. K. Ridgway and K. J. Tarbuck, "Particulate Mixture Bulk

Densities," *Chemical and Process Engineering*, 1968, vol. 49, no. 2, pp. 103-105.
18. K. E. Meiners, *Process Handbook, Complex Metal Powder Shapes by Injection Molding*, Battelle Columbus Division, Columbus, OH, 1985.
19. M. D. Sacks, "Rheological Science in Ceramic Processing," *Science of Ceramic Chemical Processing*, L. L. Hench and D. R. Ulrich (eds.), Wiley, New York, NY, 1986, pp. 522-538.
20. G. L. Messing and G. Y. Onoda, "Inhomogeneity-Packing Density Relations in Binary Powders - Experimental Studies," *Journal of the American Ceramic Society*, 1978, vol. 61, pp. 363-366.
21. R. F. Fedors and R. F. Landel, "Effect of Surface Adsorption and Agglomeration on the Packing of Particles," *Powder Technology*, 1979, vol. 23, pp. 219-223.
22. M. C. Coelho and N. Harnby, "Moisture Bonding in Powders," *Powder Technology*, 1978, vol. 20, pp. 201-205.
23. J. V. Milewski, "Packing Concepts in the Use of Filler and Reinforcement Combinations," *Handbook of Reinforcements for Plastics*, J. V. Milewski and H. S. Katz (eds.), Van Nostrand Reinhold, New York, NY, 1987, pp. 14-33.
24. M. Floman and J. L. Sherehfesky, "Liquid-Vapor Equilibrium in Microscopic Capillaries. II. Non-Aqueous Systems," *Journal of Physical Chemistry*, 1955, vol. 59, pp. 607-610.
25. H. Schubert, "Capillary Forces - Modeling and Application in Particulate Technology," *Powder Technology*, 1984, vol. 37, pp. 105-116.
26. M. C. Coelho and N. Harnby, "The Effect of Humidity on the Form of Water Retention in a Powder," *Powder Technology*, 1978, vol. 20, pp. 197-200.
27. J. C. Macrae and W. A. Gray, "Significance of the Properties of Materials in the Packing of Real Spherical Particles," *British Journal of Applied Physics*, 1961, vol. 12, pp. 164-172.
28. B. N. Boots, "The Arrangement of Cells in a Random Network," *Metallography*, 1982, vol. 15, pp. 53-62.
29. A. N. Nikolenko and M. S. Kovalchenko, "Analysis of the Random Packing of Identical Particles IV. Zonal Segregation in Powder Bodies," *Soviet Powder Metallurgy and Metal Ceramics*, 1986, vol. 25, pp. 96-99.
30. D. M. Newitt and J. M. Conway-Jones, "A Contribution to the Theory and Practice of Granulation," *Transactions of the Institution of Chemical Engineers*, 1958, vol. 36, pp. 422-442.

31. R. F. Fedors, "A Relationship Between Maximum Packing of Particles and Particle Size," *Powder Technology*, 1979, vol. 22, pp. 71-76.
32. M. Strauss, T. Ring, A. Bleier and H. K. Bowen, "The Effect of Particle Size Distribution on the Coagulation of Ceramic Suspensions," *Particulate and Multiphase Processes*, vol. 3, T. Ariman and T. N. Veziroglu (eds.), Hemisphere, New York, NY, 1987, pp. 57-67.
33. M. T. Kamel, E. M. Tory and W. S. Jodrey, "The Distribution of kth Nearest Neighbors and its Application to Cluster Settling in Dispersions of Equal Spheres," *Powder Technology*, 1979, vol. 24, pp. 19-34.
34. K. D. Reeve, "Non-Uniform Shrinkage in Sintering," *Ceramic Bulletin*, 1963, vol. 42, p. 452.
35. D. E. Niesz, R. B. Bennett and M. J. Snyder, "Strength Characterization of Powder Aggregates," *Ceramic Bulletin*, 1972, vol. 51, pp. 677-680.
36. H. F. Huisman, "Pigment Volume Concentration and an Interpretation of the Oil Absorption of Pigments," *Journal of Coatings Technology*, 1984, vol. 56, no. 712, pp. 65-79.
37. T. P. Fisher and C. Fisher, "The Basic Principles of Electrophoretic Paint Deposition and its Application to the Development of Coating Systems for Metal Powders on Metal Substrates for Subsequent Diffusion Treatment," *Surface Technology*, 1981, vol. 12, pp. 107-125.
38. H. Itoh, R. L. Pober, M. V. Parish and H. K. Bowen, "Dispersion and Particle Packing of Magnetic Oxide Powders," *Ceramics International*, 1986, vol. 12, pp. 85-93.
39. M. J. Vold, "A Numerical Approach to the Problem of Sediment Volume," *Journal of Colloid Science*, 1959, vol. 14, pp. 168-174.
40. W. Drost-Hansen, C. V. Braun, R. Hochstim and G. W. Crowther, "High Precision Dilatometry on Aqueous Suspensions: Volume Contraction upon Settling," *Particulate and Multiphase Processes*, vol. 3, T. Ariman and T. N. Veziroglu (eds.), Hemisphere, New York, NY, 1987, pp. 111-124.
41. D. Yarnton, "Effect of Lubricant Chain Length on Flow and Packing Characteristics of Atomized Copper Powder," *International Journal of Powder Metallurgy and Powder Technology*, 1977, vol. 13, pp. 189-196.
42. R. E. Cowan, "Slip Casting," *Ceramic Fabrication Processes*, F.

F. Y. Wang (ed.), Academic Press, New York, NY, 1976, pp. 153-171.

43. E. A. Barringer, R. Brook and H. K. Bowen, "The Sintering of Monodisperse Titania," *Sintering and Heterogeneous Catalysis*, G. C. Kuczynski, A. E. Miller and G. A. Sargent (eds.), Plenum Press, New York, NY, 1984, pp. 1-21.
44. J. R. L. Allen, "Notes Towards a Theory of Concentration of Solids in Natural Sands," *Geological Magazine*, 1969, vol. 106, pp. 309-321.
45. H. Susskind and W. Becker, "Random Packing of Spheres in Non-Rigid Containers," *Nature*, 1966, vol. 212, pp. 1564-1565.
46. J. D. Bernal and J. L. Finney, "Random Packing of Spheres in Non-Rigid Containers," *Nature*, 1967, vol. 214, pp. 265-266.
47. J. M. Culson, "The Flow of Fluids Through Granular Beds: Effect of Particle Shape and Voids in Streamline Flow," *Transactions of the Institution of Chemical Engineers*, 1949, vol. 27, pp. 237-257.
48. D. J. Cumberland and R. J. Crawford, *The Packing of Particles*, Elsevier Science, Amsterdam, The Netherlands, 1987.
49. P. Le Goff, D. Leclerc and J. Dodds, "The Structure of Packed Beds: Continuity of Research in Nancy and Some New Results," *Powder Technology*, 1985, vol. 42, pp. 47-53.
50. K. Ridgway and K. J. Tarbuck, "Voidage Fluctuations in Randomly-Packed Beds of Spheres Adjacent to a Containing Wall," *Chemical Engineering Science*, 1968, vol. 23, pp. 1147-1155.
51. A. Bonissent and B. Mutaftschiev, "A Computer Built Random Model for Simulation of the Crystal-Melt Interface," *Philosophical Magazine*, 1977, vol. 35, pp. 65-73.
52. K. K. Pillai, "Voidage Variation at the Wall of a Packed Bed of Spheres," *Chemical Engineering Science*, 1977, vol. 32, pp. 59-61.
53. M. C. Thadani and F. N. Peebles, "Variation of Local Void Fraction in Randomly Packed Beds of Equal Spheres," *Industrial and Engineering Chemistry Process Design and Development*, 1966, vol. 5, pp. 265-268.
54. R. K. McGeary, "Mechanical Packing of Spherical Particles," *Journal of the American Ceramic Society*, 1961, vol. 44, pp. 513-522.
55. J. Ayer and F. E. Soppet, "Vibratory Compaction: I, Compaction of Spherical Shapes," *Journal of the American Ceramic Society*, 1965, vol. 48, pp. 180-183.

56. D. P. Haughey and G. S. G. Beveridge, "Structural Properties of Packed Beds - A Review," *The Canadian Journal of Chemical Engineering*, 1969, vol. 47, pp. 130-140.
57. R. F. Benenati and C. B. Brosilow, "Void Fraction Distribution in Beds of Spheres," *American Institute of Chemical Engineers Journal*, 1962, vol. 8, pp. 359-361.
58. K. Gotoh, W. S. Jodrey and E. M. Tory, "Variation in the Local Packing Density Near the Wall of a Randomly Packed Bed of Equal Spheres," *Powder Technology*, 1978, vol. 20, pp. 257-260.
59. G. D. Scott, "Packing of Equal Spheres," *Nature*, 1960, vol. 188, pp. 908-909.
60. R. L. Brown and P. G. W. Hawksley, "Effect of Container Walls on Packing Density of Particles," *Nature*, 1946, vol. 157, pp. 584-585.
61. R. L. Brown and P. G. W. Hawksley, "Letter to the Editor," *Nature*, 1946, vol. 157, p. 585.
62. O. K. Rice, "On the Statistical Mechanics of Liquids, and the Gas of Hard Elastic Spheres," *Journal of Chemical Physics*, 1944, vol. 12, pp. 1-18.
63. L. C. Verman and S. Banerjee, "Effect of Container Walls on Packing Density of Particles," *Nature*, 1946, vol. 157, pp. 584-585.
64. L. H. S. Roblee, R. M. Baird and J. W. Tierney, "Radial Porosity Variations in Packed Beds," *American Institute of Chemical Engineers Journal*, 1958, vol. 4, pp. 460-464.
65. J. M. Campbell and R. L. Huntington, "Heat Transfer and Pressure Drop in Fixed Beds of Spherical and Cylindrical Solids," *Petroleum Refiner*, 1951, vol. 30, no. 12, pp. 127-133.
66. J. S. Goodling, R. I. Vachon, W. S. Stelpflug, S. J. Ying and M. S. Khader, "Radial Porosity Distribution in Cylindrical Beds Packed with Spheres," *Powder Technology*, 1983, vol. 35, pp. 23-39.
67. R. P. Seelig, "Introduction to Seminar - Review of Literature on Pressing of Metal Powders," *Transactions of AIME*, 1947, vol. 171, pp. 506-518.
68. J. C. Williams, "The Segregation of Particulate Materials. A Review," *Powder Technology*, 1976, vol. 15, pp. 245-251.
69. J. C. Williams and G. Shields, "The Segregation of Granules in a Vibrated Bed," *Powder Technology*, 1967, vol. 1, pp. 134-142.
70. J. Bridgwater, W. S. Foo and D. J. Stephens, "Particle Mixing

and Segregation in Failure Zones - Theory and Experiment," *Powder Technology*, 1985, vol. 41, pp. 147-158.
71. N. Ouchiyama and T. Tanaka, "Porosity of a Mass of Solid Particles Having a Range of Sizes," *Industrial and Engineering Chemistry Fundamentals*, 1981, vol. 20, pp. 66-71.
72. A. Rosato, K. J. Strandburg, F. Prinz and R. H. Swendsen, "Why the Brazil Nuts Are on Top: Size Segregation of Particulate Matter by Shaking," *Physical Review Letters*, 1987, vol. 58, pp. 1038-1040.
73. D. S. Parsons, "Particle Segregation in Fine Powders by Tapping as Simulation of Jostling During Transportation," *Powder Technology*, 1976, vol. 13, pp. 269-277.
74. J. Bridgewater and I. Tranter, "The Effect of Interstitial Liquid on Inter-Particle Percolation," *Journal of Powder and Bulk Solids Technology*, 1978, vol. 2, no. 4, pp. 9-14.
75. C. D. Ripple, R. V. James and J. Rubin, "Radial Particle Size Segregation During Packing of Particulates into Cylindrical Containers," *Powder Technology*, 1973, vol. 8, pp. 165-175.
76. K. Ridgway and K. J. Tarbuck, "The Random Packing of Spheres," *British Chemical Engineering*, 1967, vol. 12, pp. 384-388.
77. P. C. Carman, "Fundamental Principles of Industrial Filtration," *Transactions of the Institution of Chemical Engineers*, 1938, vol. 16, pp. 168-188.
78. G. D. Scott, A. M. Charlesworth and M. K. Mak, "On the Random Packing of Spheres," *Journal of Chemical Physics*, 1964, vol. 40, pp. 611-612.
79. T. G. Owe Berg, R. L. McDonald and R. J. Trainor, "The Packing of Spheres," *Powder Technology*, 1969, vol. 3, pp. 183-188.
80. Y. A. Libermand and N. V. Mikhailov, "Close Packing of Quartz Sand of Discontinuous Particle Size Composition Under Vibration," *Colloid Journal of the USSR*, 1967, vol. 27, pp. 273-274.
81. J. C. Macrae, P. C. Finlayson and W. A. Gray, "Vibration Packing of Dry Granular Solids," *Nature*, 1957, vol. 179, pp. 1365-1366.
82. E. V. Degtyareva and Y. M. Grinberg, "Vibrovacuum Packing of Powder Masses," *Refractories*, 1982, vol. 23, pp. 331-339.
83. A. K. Singhal and P. M. Dranchuk, "Influence of Particle Diameter and Vibrational Amplitude Upon Compaction of Uniform Spheres," *Canadian Journal of Chemical Engineering*,

1973, vol. 51, pp. 503-506.
84. R. J. Parsick, S. C. Jones and L. P. Hatch, "Stability of Randomly Packed Beds of Fuel Spheres," *Nuclear Applications*, 1966, vol. 2, pp. 221-225.
85. K. Kawakita and K. H. Ludde, "Some Considerations on Powder Compression Equations," *Powder Technology*, 1970, vol. 4, pp. 61-68.
86. P. E. Evans and R. S. Millman, "The Vibratory Packing of Powders," *Powder Metallurgy*, 1964, vol. 7, pp. 50-63.
87. G. D. Scott and D. M. Kilgour, "The Density of Random Close Packing of Spheres," *British Journal of Applied Physics (Journal of Physics D)*, 1969, vol. 2, ser. 2, pp. 863-866.
88. M. Takahashi and S. Suzuki, "Numerical Analysis of Tapping Behavior of Ceramic Powders," *Ceramic Bulletin*, 1986, vol. 65, pp. 1587-1590.
89. J. Van Brakel and P. M. Heertjes, "Homogeneous Particle Packings and Porous Masses. Some Experimental Techniques," *Powder Technology*, 1974, vol. 9, pp. 263-271.
90. S. I. Furmanov, G. M. Khutortsov and N. V. Mikhailov, "Vibro-Packing of Free Flowing Materials from Fractions with a Wide Particle Size Range," *Colloid Journal of the USSR*, 1969, vol. 31, pp. 612-616.
91. R. Sempere, "Sur la Porosite Reele Minimale des Assemblages de Spheres Identiques," *Powder Technology*, 1969, vol. 3, pp. 1-8.
92. J. R. F. Arthur and T. Dunstan, "Radiological Techniques Developed to Describe Particle Packing," *Powder Technology*, 1969, vol. 3, pp. 195-207.

Chapter Four

Packing of Monosized Spheres

Chapter Outline

Introduction

There are three main packing types of interest with monosized spheres: ordered, random loose, and random dense. An ordered packing has periodic structure that repeats over a specific, small distance. For example, the atoms in a crystal and cannon balls in a stack both represent packings with an ordered, periodic structure. Figure 4.1 illustrates the difference in packing structures between ordered and random arrangements using monosized steel spheres. Random loose packing is represented by particles placed in a container without agitation. Often this is reported as the apparent density. In contrast, random dense packings are attained by agitating or vibrating a random loose packing to the

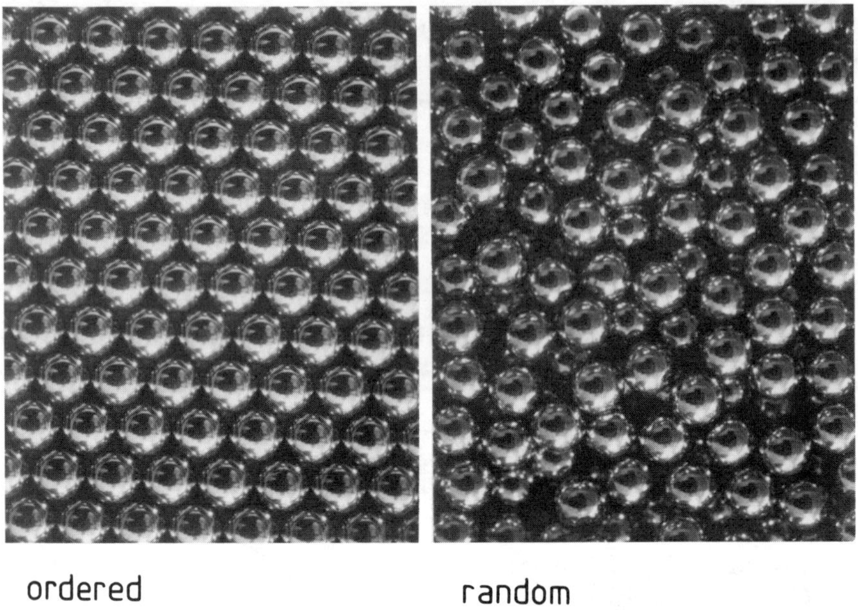

Figure 4.1 Photographs of steel sphere packings showing an ordered and random arrangement.

maximum density. This corresponds to the tap density. It is the random dense packing which most typically correlates with the packing density of a powder in processes such as powder injection molding and hot isostatic compaction. Likewise, random dense sphere packings give the best simulation of the amorphous structure of glasses and liquids.

Coordination Number

The determination of the packing coordination of monosized spheres is one of the classic problems encountered in science. Monosized spheres can be packed in an orderly close array with a coordination number of twelve and a fractional density of 0.7405. Curiously, Hudson [1] shows that a maximum packing coordination of 11.4 is predicted based on surface coverage for a central sphere. Generally, for geometric stability the packing coordination can not go below four [2]. A random packing assembled under the action of gravity requires each sphere that joins the packing rest on at least three contact points. Thus, each sphere adds three new contacts to the packing. Since each contact point is shared by two spheres, the net increase in packing coordination number is six contacts per sphere. Accordingly, an average coordination number of six is expected for a random yet stable packing of monosized spheres. This value is generally substantiated by experiments using large spheres, including packings of mixed sphere sizes.

In experimental investigations of particle packing, it is often unclear how close in proximity two spheres must be to constitute an actual contact. If the close contacts (within 10% of the sphere diameter in separation) are included in the count, then the coordination number is near 9.5 [3-5]. Thus, the reported coordination number depends on the measurement technique. Indeed, this

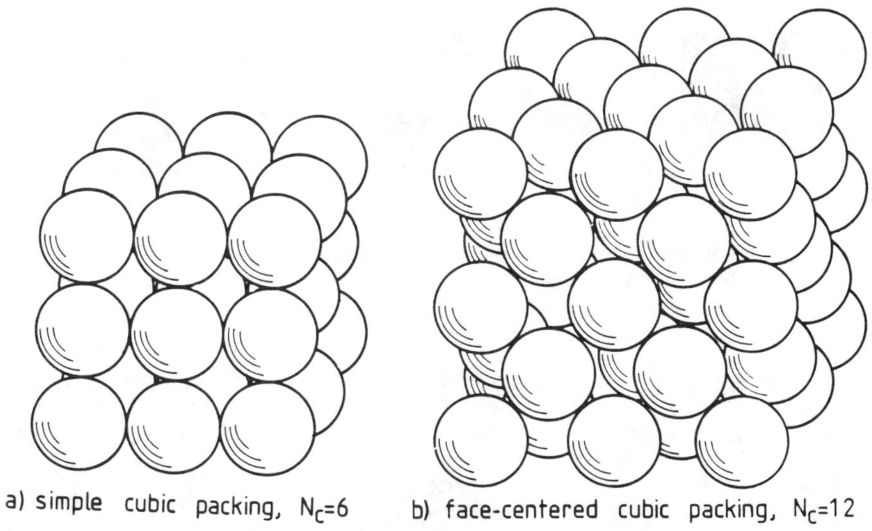

Figure 4.2 The structure of two ordered packings of monosized spheres; a) simple cubic packing with a coordination number of six and b) face-centered cubic packing with a coordination number of twelve.

is an error source in many of the experimental studies. For a random loose packed array of monosized spheres the packing coordination has been reported to be as high as 9.5 and as low as 6.0 [3,6-10]. The latter value is probably correct based on the more recent precision measurements and computer simulations [11].

Ordered Packing

In analyzing the packing of monosized spheres, first attention will be given to ordered packings. These are well known because of their similarities to hard sphere models of crystalline atomic structure. For example, Figure 4.2 shows the geometries of regular ordered packings with coordination numbers of six and twelve. These particle arrays are the simple cubic and face-centered cubic packings. In an early analysis of particle packing Graton and Fraser [12] introduced several simple unit cells to characterize packing and pore structures. Figure 4.3 shows their

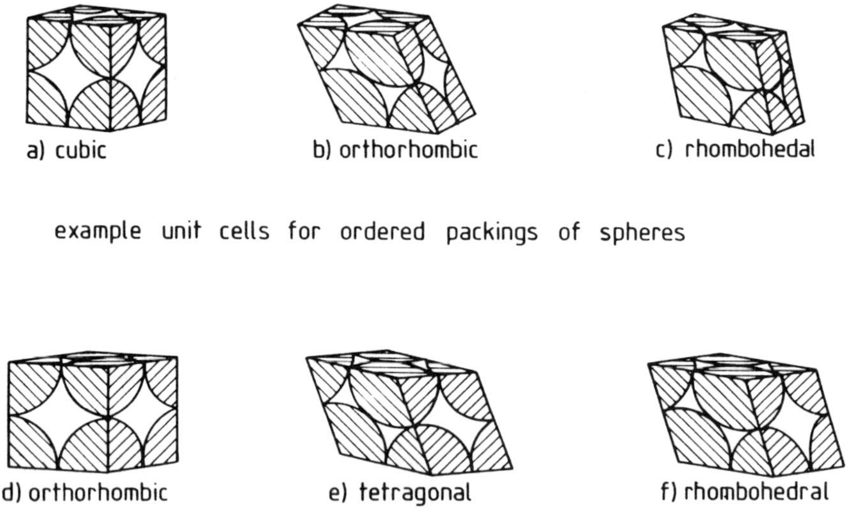

Figure 4.3 The unit cells of the six packing geometries introduced by Graton and Fraser [12] for construction of random packings; cubic, orthorhombic (two forms), rhombohedral (two forms) and tetragonal geometries.

six basic packings, corresponding to cubic, orthorhombic (two forms), rhombohedral (two forms), and tetragonal geometries. Graton and Fraser noted that some variations on these structures are possible by twinning, but that twinning does not change the packing density or coordination number. Building on this concept, Smalley [13] claims that nine basic units are needed for assembling all possible particle packings. Other researchers [14-16] place the number of geometric units at a higher value, but basically agree that all random packings can be generated by assemblies of simple ordered geometries.

Various models linking the coordination number and packing density are possible for ordered packings at integer coordination numbers. Table 4.1 collects the recognized solutions for standard even and odd coordinations while Figure 4.4 presents a bar chart of coordination number versus fractional density [3,7,8, 12,17-26]. There are two recognized packings involving a coordination of eight. As the coordination number increases, the spacing

Table 4.1

The Densities and Coordination Numbers for Ordered Packings [3,7,8,12,17-26]

packing type	coordination number	packing density	porosity
face-centered cubic	12	0.7405	0.2595
	11	0.7120	0.2880
tetragonal-sphenoidal	10	0.7081	0.3019
	9	0.6134	0.3866
body-centered cubic	8	0.6802	0.3198
orthorhombic	8	0.6046	0.3954
	7	0.5612	0.4388
simple cubic	6	0.5236	0.4764
	5	0.4031	0.5969
diamond	4	0.3401	0.6599

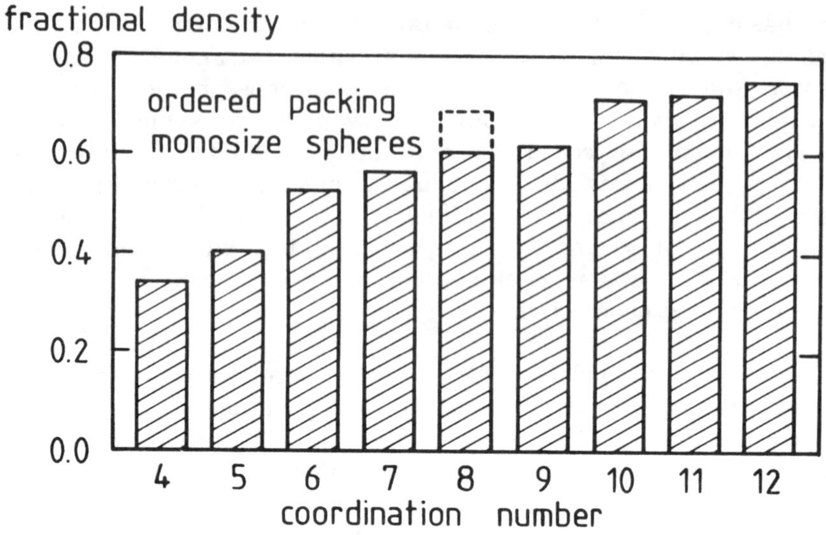

Figure 4.4 The fractional packing density versus the coordination number for ordered packings of monosized spheres.

between layers of spheres decreases. From Figure 4.4 it is clear that the coordination number varies with the fractional solids content. Although the proof is difficult, it is mathematically accepted that the maximum possible packing density for equal size spheres is given by a coordination number of 12 [2,27,28]. This packing structure has special characteristics because of the 48 symmetry groups [28,29]. Furthermore, the close-packed structure is a mixture of tetrahedral and octahedral units with two tetrahedral units for every octahedral unit. The packing density of the tetrahedral unit is 0.7797, while the density of the octahedral unit is 0.7209. Although the tetrahedral unit has a locally high packing density, it can not be repeated to fill space.

Some novel ordered packings involving spirals, cones, and twinned structures have been proposed in the past [30-32]. There have been attempts to predict the lowest packing density possible in a three-dimensional ordered array of monosized spheres [2,33-35]. The lowest possible fractional density for monosized spheres is estimated to range from 0.056 to 0.042. This corresponds to a chain-like structure of spheres with a coordination number of three.

Several mathematical approximations have been proposed to link the packing density and the coordination number for monosized spheres. Most of these are based on correlations using data

Table 4.2

Mathematical Expressions for the Packing Density and Average Coordination Number for Monosized Spheres

f = fractional density
N_c = coordination number

Equation 1:

$$f = -0.072 + 0.1193\,N_c - 0.00431\,N_c^2$$ [8,38]

Equation 2:

$$N_c = \pi/(1 - f)$$ [3]

Equation 3:

$$N_c = 26.49 - 10.73/f$$ [3]

Equation 4:

$$N_c = 2\,exp(2.4\,f)$$ [22,36]

Equation 5:

$$f = ((N_c - 1)/N_c)^3$$ [39]

Equation 6:

$$f = 0.4696 + 0.02539\,N_c$$ [40]

Equation 7:

$$N_c = 16\,f - 2$$ [41]

Equation 8:

$$N_c = 19.7\,f - 2.5$$ [42]

for ordered packings. There is no exact relation between packing density and coordination number, although a strong correlation exists. Various simple models have been applied to data such as shown in Table 4.1 to approximate the relation of coordination number N_c and fractional density f. Some of these expressions are collected in Table 4.2 [3,8,22,26,36-42]. The relations range from the simple to complex. Not included in Table 4.2 are several polynomials that involve undefined coefficients or infinite series. Figure 4.5 plots these relations and includes for reference the data for ordered packings shown in Table 4.1. It can be seen that Equations 3 and 6 from Table 4.2 have the wrong curvatures. Equation 5 underestimates the behavior of ordered packings. Equation 1 has too high a coordination number at the high densities. Alternatively, Equations 7 and 8 are straight lines that have reasonable behavior, but lack curvature. Thus, Equations 2 and 4 seem most reasonable based on the ordered packing density results, although random packings tend to give a lower coordination number at all density levels.

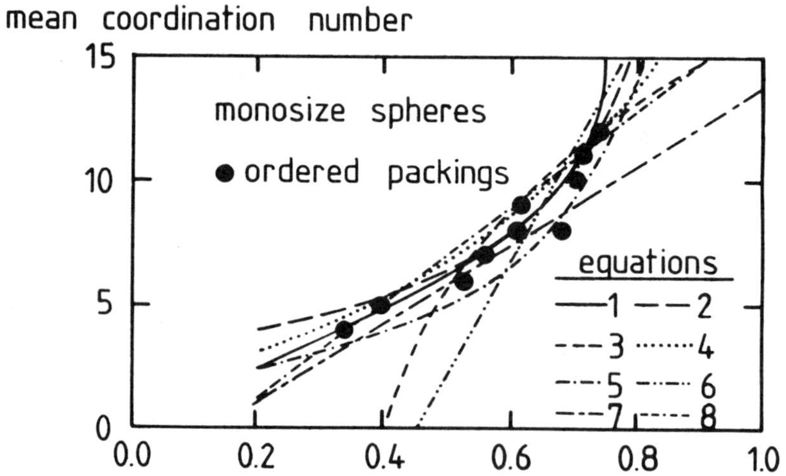

Figure 4.5 The functional relations between coordination number and fractional density from Table 4.2 compared with the ordered packing data from Table 4.1.

When the spheres are compressed they will form new contacts only if particle rearrangement can occur. Ordered packings of monosized spheres at a high coordination number do not show an increase in coordination with compression; the final geometric shape is usually a dodecahedron (12 sided polyhedra). In contrast, compressed random spheres deform to polyhedra with 14 sides.

Random Packing

There are three recognized random packing densities for monosized spheres. Loose and dense random packing concepts have already been introduced. The third is defined by the density at the onset of fluidization, which ranges from fractional densities of 0.53 to 0.54 [7]. Of the three densities, the loose and dense random packings are important to most uses of particulate materials. Various reports have placed the fractional densities at 0.610 to 0.667 for the random dense packing, while values between 0.560 and 0.625 are cited for the random loose density. One difficulty in dealing with random structures is that several different particle arrangements give similar densities and other attributes [43]. As more energy is put into assembling a particle packing by cascading or impacting particles, the fractional density increases to the random dense packing limit [44,45]. Likewise, the more rapidly a packing is assembled the lower the packing density. This is the basis for forming glass structures in rapidly solidified materials. For every material there is a critical particle entry energy level needed to ensure dense packing.

In scientific studies on the packing of spheres there is considerable interest in the radial distribution function. The character of this function can be related to the structure of liquids and glasses. The radial distribution function is the average number of sphere centers per unit area in a spherical shell about a central sphere. Figure 4.6 plots the radial distribution function for both random loose and random dense packings of monosized spheres [8,10,46]. The number of sphere centers per unit area is shown as a function of the radial distance from the sphere surface as measured in sphere diameters. Mason [10] measured these distributions using 25 clusters of 330 spheres each embedded in packings of 1000 spheres. There are subtle differences between the two packing types. A peak in both distributions occurs at the first nearest neighbor distance. A

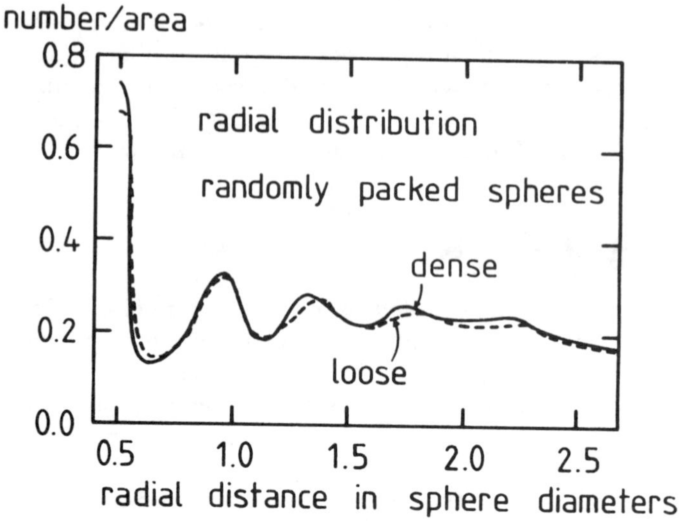

Figure 4.6 The radial distribution of monosized spheres randomly packed in both the loose and dense structures. The number of particles per unit spherical area is shown versus the normalized radius.

splitting of this first peak has been noted in some packings and is thought to represent partial or short range ordering [47,48]. Minima exist at approximately 1.4, 2.2, 3.1, and 4.1 sphere diameters. In the first shell of neighbors, there is a definite angular correlation between neighbors to avoid overlap; however, by the third shell of nearest neighbors this angular character has been lost [49]. As expected for a random packing, there is a uniform probability distribution as the distance increases. As noted in Figure 4.6, the radial distribution functions for the random dense and loose packings of monosized spheres are similar [10,46,50-56]. More of a difference is apparent in the coordination number distributions as shown in Figure 4.7 [57]. Bernal et al. [51] claim that the coordination number distribution is a more sensitive characteristic of particle packing than is the radial distribution function. Random packings are characterized by clusters of high packing coordination and small pores embedded in a matrix of low packing density and low packing coordination. Such clustering is noted in soap bubbles and polycrystalline solids, although it is generally assumed that random packing

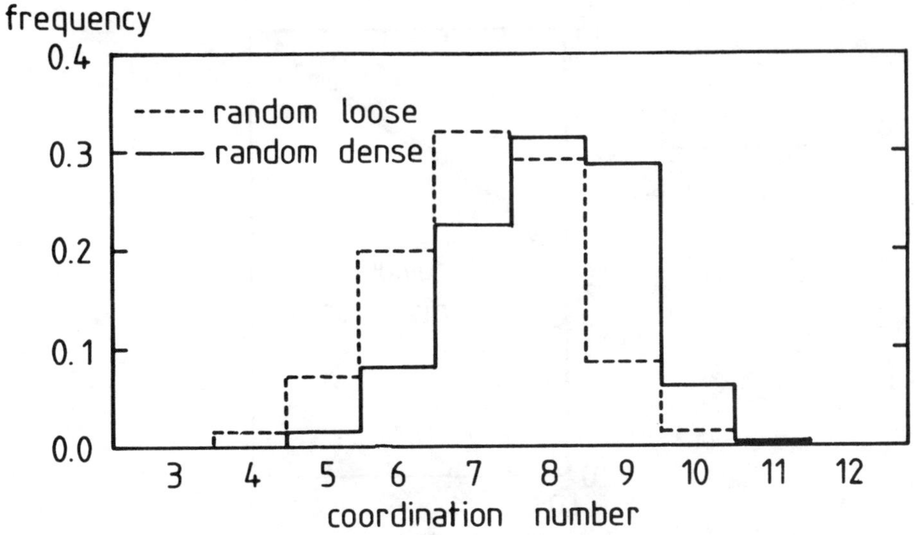

Figure 4.7 Histograms showing the distributions in packing coordinations among spheres in random loose and dense structures [57].

has no correlation in particle placement beyond the first nearest neighbor region. The cumulative radial distribution function provides another measure of the packing coordination number behavior in a random packing. Figure 4.8 shows the cumulative radial distribution function for a random dense packing [53]. This figure plots the cumulative number of neighboring spheres versus the radial distance in sphere diameters. The linear approximation to this curve is as follows:

$$G(r) = 7.3 + 15.5\,(r/D - 1) \tag{4.1}$$

where $G(r)$ is the cumulative number of neighbors per unit radial distance r from the sphere surface and D is the particle diameter [6]. This model assumes a first neighbor coordination number of 7.3. This function has proven useful in modeling particle sintering as well as compaction behavior.

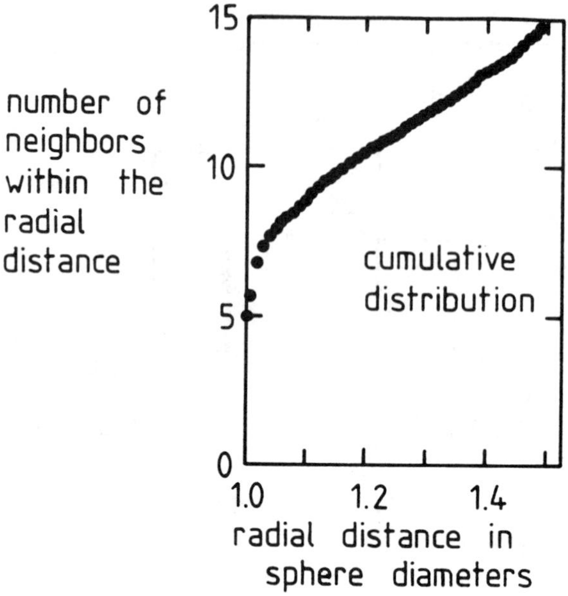

Figure 4.8 The cumulative radial distribution function for monosized randomly packed spheres [53].

An early model by Smith et al. [58] assumed that random packing was composed of a mixture of close packed and simple cubic regions with packing coordinations of six and twelve, respectively. On this basis the fraction of the structure with a packing coordination of twelve was calculated from the fractional density as follows:

$$X_{12} = 4.6\,f - 2.4 \tag{4.2}$$

where X_{12} is the fraction of the structure with a packing coordination of twelve and f is the fractional packing density. It is then assumed that the balance of the structure has a packing coordination of six. Using this concept, the average packing coordination number was calculated as a function of the packing density as

$$\bar{N}_c = 26.5 - 10.7/f \qquad (4.3)$$

where $\bar{N}_c$ is taken as the average packing coordination number. Tests of this model have found that the predicted average packing coordinations for hard spheres are higher than experimentally measured. Most recent models of random packing assume that the structure can be decomposed into a few unit cells, typically represented as distorted tetrahedral units [15,59-62]. Various other theories exist for predicting the average coordination number or the distribution in coordination numbers for random packings [3,7,20,21,39, 58,63-67]. A binomial distribution is a good representation of the coordination number variation in a random particle mass. Besides models based on hard sphere assemblages, there has also been considerable attention to computer simulation of random packings. It is established that the numerical procedure for adding spheres to the packing influences the limiting density. Widom [68] showed that random sequential selection of sites gives a monosize sphere packing limit between 0.35 and 0.38. These values are far below those attained in practice. Filling schemes that force each sphere to nest into a low energy stable position from one direction give fractional

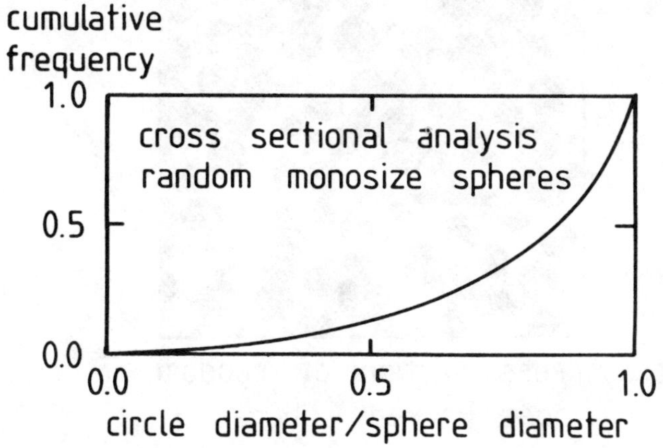

Figure 4.9 The cumulative frequency distribution of circle sizes expected from a random cross section through a packing of monosized spheres [55].

densities closer to reality [69]. But these packings are anisotropic because of the need for a gravitational force.

Many observations of particle packing are made on random cross sections through the packing. It is reassuring that the size distribution of the cross-sectioned circles in the simulations agree with theory for the size distribution for monosized spheres in random cross sections [70]. Figure 4.9 plots the cumulative distribution of circle sizes seen in two dimensions for random cross sections through monosized spheres. In such two-dimensional cross sections there is an apparent slight clustering of the particles as can be seen in Figure 4.10. The local clusters indicate a distribution in pore sizes. Figure 4.11 shows the probability distribution for pore size as determined in a computer simulation [55]. In a two-dimensional section both the pore size and particle size will appear smaller than actual. Even so, the presence of large pores is apparent in Figure 4.10.

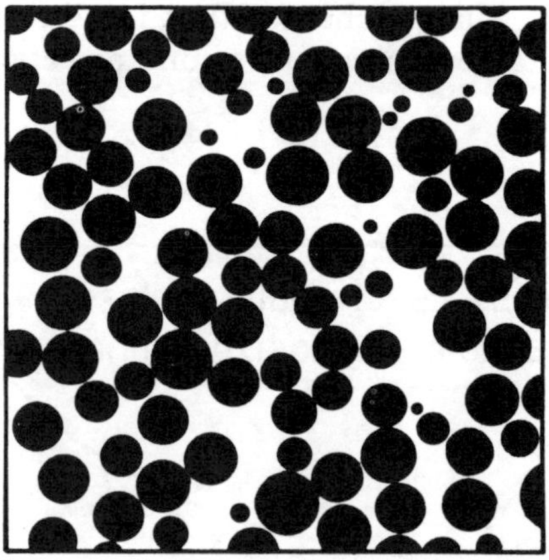

Figure 4.10 An example of the particle clustering and circle size distribution apparent in a random cross section of monosized spheres [55].

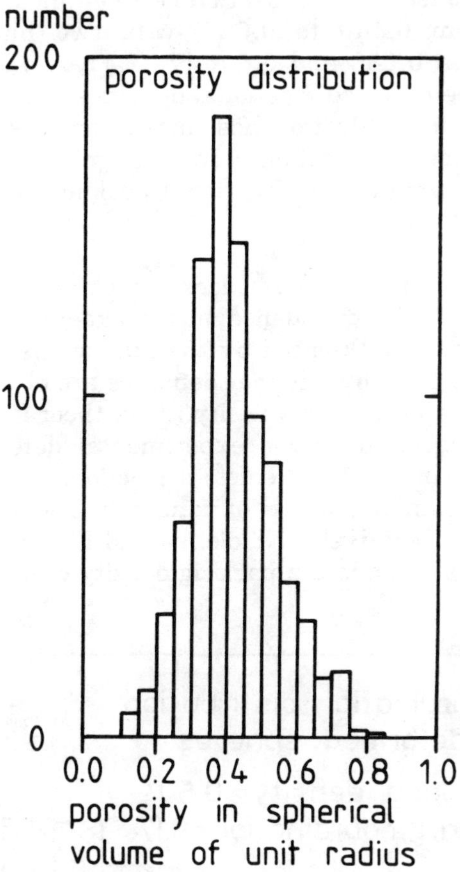

Figure 4.11 A histogram of the porosity distribution within unit sphere volumes in a random packing [55].

Random Loose Packing

In practice the random loose density is best represented by the apparent density. Several investigations have reported the fractional density for random loose packings of monosized spheres. This density is very sensitive to experimental techniques. As a consequence, there is considerable variation between measurements. The values range from 0.560 to 0.625, with two-thirds of the reported values between 0.580 and 0.606 [7,21,41,44,45,55,57,70-82]. The average of the reported fractional densities is 0.593 with a standard deviation of 0.014. Most of these measurements have been corrected for wall effects on packing. In the limit of an infinite container, Scott [75] estimated the fractional packing density to be 0.601.

A problem with the random loose packing is that the observed density has a high dependence on the experimental procedure. The measurement is influenced by the kinetic energy of each particle as it enters the bed; thus, higher densities are observed with particles of higher mass or higher velocity. The theoretical predictions are as widely scattered as the experimental determinations [50,71,72,76,83,84]. Gamba [76] predicts a fractional packing density of 0.596 from a geometric model of randomly oriented spheres in three dimensions. Alternatively, Nikolenko and Kovalchenko [21] use a statistical model to calculate a packing density of 0.604. Their

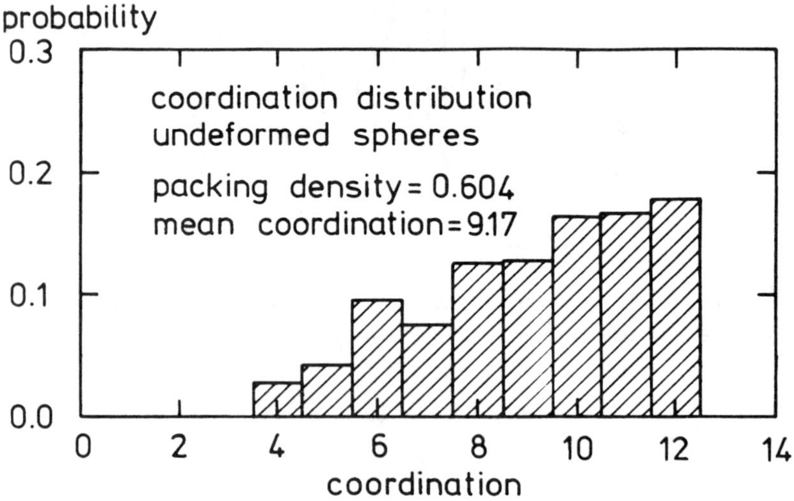

Figure 4.12 The coordination number distribution for the statistical model of randomly packed spheres of Nikolenko and Kovalchenko [21].

calculated coordination number distribution is shown in Figure 4.12. The mean fractional density of 0.604 is reasonable. However, the predicted mean coordination number of 9.17 contacts per sphere is higher than most other measurements or computer calculations. A value near six is more typical for the average coordination number.

Computer simulations of random loose packing are largely inaccurate representations because of the assumptions needed concerning the assembly technique. These assumptions are random placement, with a central unidirectional force, with either a dilated or minimum energy configuration [11,16,47,48,55,68,85-87]. The computer predicted random packing densities range from 0.35 to 0.64, with mean coordination numbers ranging up to nine. Both parameters tend to be high in comparison with experimental values. The coordination number distribution from a computer simulated packing is shown in Figure 4.13 along with the experimental

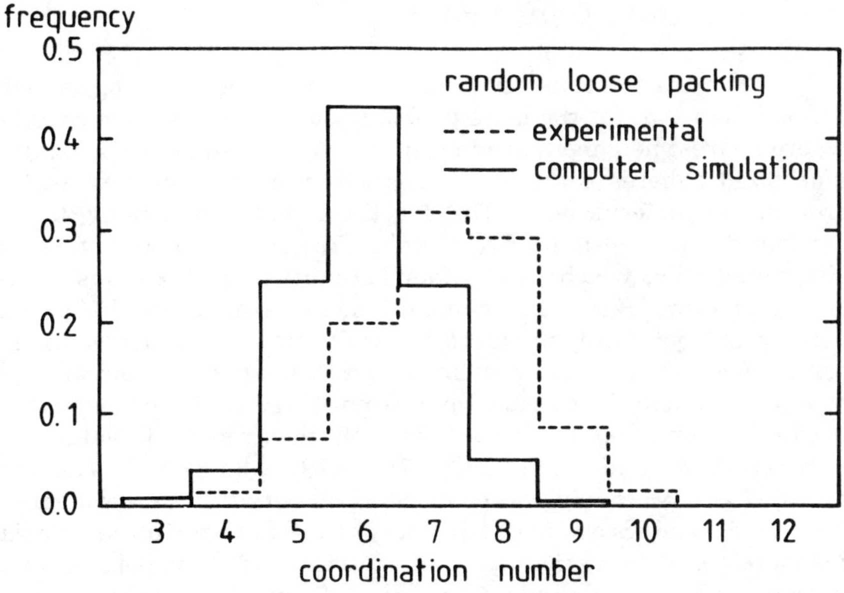

Figure 4.13 A comparison of coordination number distributions for experimental [57] and computer simulated [55] random loose packings.

packing shown previously in Figure 4.6 [55]. Most of the computer generated distributions give a mean coordination near six, in agreement with other reports [53,69,70,88], and the distribution in contacts is nearly binomial. However, as is evident in Figure 4.13, the experimental distribution is shifted to a higher coordination number. If the close contacts are included (less than 10% of a sphere diameter from contacting), then the computer generated coordination number is near eight [87]. Because of these problems, the results of computer simulations must be treated with caution.

The most probable value for the fractional packing density of a random loose array of monosized spheres is between 0.59 and 0.60, with approximately six contacts per sphere. There is an inherent variation point-to-point in the packing density and coordination number. The mean and mode coordination numbers are near six and the distribution is approximately binomial.

Random Dense Packing

The fractional density for a random dense packing is better defined than that for the loose packing, since it is a limiting condition where nearest neighbors are in contact. A random dense packing for monosized spheres corresponds to their maximum density without ordering or deformation. The fractional packing density for the random dense case is 0.637. Gamba [76] points out that this density equals $2/\pi$, which can be found frequently in theorems of integral geometry. Early experimental measurements of the packing density did not always correct for wall effects or other sources of error such as surface roughness on the spheres. Accordingly, reports of fractional density vary from 0.610 to 0.665, but largely cluster around 0.637; two-thirds of the reported values are between 0.636 and 0.644 [23,71-75,78,79,82-84,89-96]. The value of 0.637 is now widely accepted based on careful measurements by Scott [75] and Scott and Kilgour [71]. They performed density measurements in various container sizes and extrapolated to an infinite container to eliminate all wall effects. Subsequent analysis suggests that localized ordering occurs in a monosized sphere packing with a density above this value.

Early computer simulations of random dense packing in a gravitational field produced fractional densities in the range 0.57 to 0.61 [77]. Moreover there are systematic errors in the results.

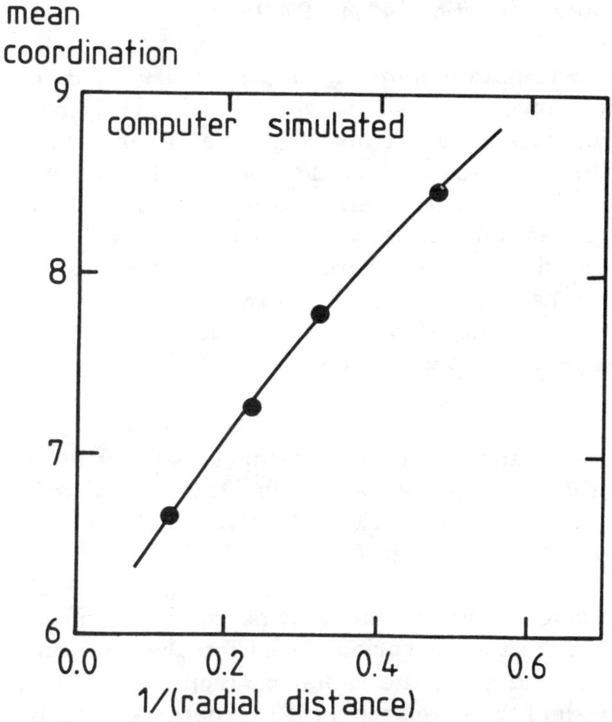

Figure 4.14 Computer simulated variation in the mean sphere coordination versus the inverse radial distance (normalized to the sphere radius) for a computer simulated random dense packing [95].

Figure 4.14 shows the predicted change in mean coordination number versus the size of the computational cluster as simulated by Bennett [95]. Here r is the distance from the center of the cluster. In these calculations the distribution is again nearly binomial, similar to that shown in Figure 4.13 for random loose packing. Recent computer simulations have produced packing densities near 0.637 for monosized spheres [47,89,95,97]. To obtain these higher computer simulated packing densities, some novel relaxation procedures have been required. Most notably, overlap of the particles initially has to be allowed while later the spheres are moved to point contact positions. Even so, the computer models are criticized for failing to replicate the details of the experimental radial distribution function [47,48,97,98].

The coordination number for the dense random packing is estimated to range from six to nine [15,36,37,53,57,84,99,100]. The most accurate measurements give a mean and mode value near six with a standard deviation near 1.1 contacts. As noted earlier, there is an experimental problem with determining the existence of an actual contact versus a near contact. The computer simulated coordination numbers range between six and seven contacts per sphere, and tend to be lower than some of the experimental determinations. However, the recent computer simulations which converge to a fractional density of 0.637 give a coordination number near six.

When a random dense packing is subjected to shear, the density increases slightly although the coordination number appears to remain constant. The fractional density at the onset of shear deformation is between 0.66 and 0.67 [73].

The pore structure for a random dense packing is often modeled as a collection of various sized tetrahedral units [15,60,62, 101]. This is valid since the radial position of neighboring spheres proves to be highly correlated [102]. There is a high probability that contacting spheres will have separation angles of π, $2\pi/3$, or $\pi/3$. The geometric model based on dilated tetrahedral units has proven successful in reducing the dense random packing into an easily analyzed structure. Figure 4.15 is a plot of the frequency of occurrence for various densities of tetrahedral unit cells [15]. This structure shows that packing densities range from 0.3 to 0.8 within the dense packing. Similar conclusions are apparent from the coordination number distributions; low coordination number regions correspond to low packing densities. Likewise, spheres with high coordination numbers will be in high density regions. In this sense, Figures 4.7, 4.10, 4.11, and 4.15 show that the dense random packing is composed of a mixture of packing structures.

A random dense packing occurs when the particles attain a close proximity to one another through agitation. Accordingly, the tap density is a good first approximation of the maximum loading attainable without application of an external stress. The tap or vibrated density depends on the material, vibration amplitude, vibration direction, frequency, shear, and test apparatus [10,103-

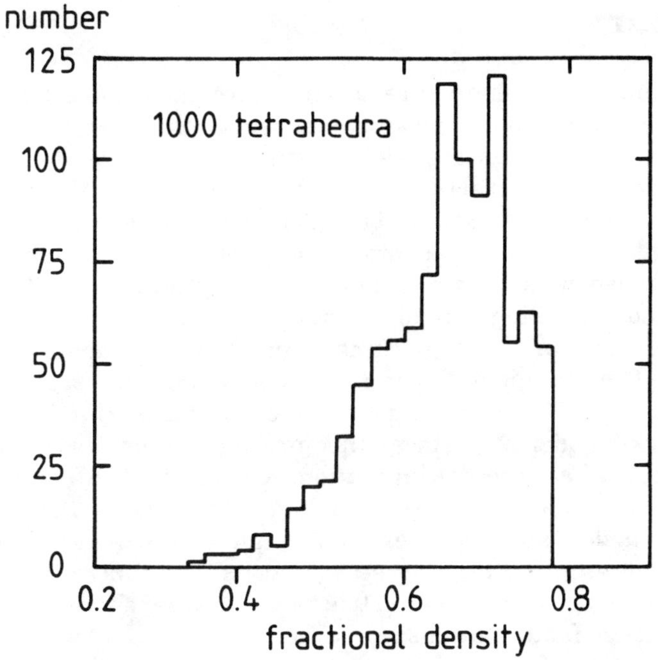

Figure 4.15 The variation in fractional densities for tetrahedral units within a dense random packing based on the model of Mason [15].

106]. During vibration the density varies with the number of vibrations. The random dense packing and random loose packing are fairly similar for large spherical particles. Consequently, spherical particles of sizes greater than approximately 100 μm undergo little densification during vibration. In contrast, smaller spheres and nonspherical particles exhibit a greater difference between the apparent and tap densities. These particles undergo a large packing increase with vibration. Indeed, the more irregular the particle shape, the greater the increase in packing density with vibration, but a longer vibration period is needed to reach the tap density.

Summary

Monosized spheres are the most frequent subject of packing studies. The fractional packing density is the first parameter of concern. Depending on the packing type (ordered, random loose, or random dense) the fractional packing density can range up to a maximum value of 0.7405. This density limit is rarely observed with powders. Rather, random structures are more common with fractional densities between approximately 0.60 and 0.64. The random loose packing roughly corresponds to the apparent density while the random dense packing roughly corresponds to the tap density. The accepted fractional density for the random dense packing is 0.637. There is more error in measuring the random loose packing density since the procedure for assembling the packing can affect the determination. Generally, the accepted fractional packing density for the random loose case is near 0.60. In contrast, under suitable vibration conditions, the random dense packing represents a limiting state and its experimental determination is more accurate. Both structures are sensitive to container size and other factors discussed earlier.

The radial distribution functions for random packings show small differences between the dense and loose random packings. Alternatively, the coordination number distribution is sensitive to the packing density. The coordination number increases with the fractional density. The upper limit, corresponding to the ordered close packed structure, is 12 contacts per sphere at a fractional density of 0.7405. Several mathematical models have been suggested for the relation between packing density and mean coordination number; however, most are rough approximations to actual behavior. In a random packing a binomial distribution function is a good approximation to the observed coordination number distribution. For randomly packed monosized spheres, approximately six contacts per sphere represents the stable assembly. The minimum number of contacts is three. Various computer models and experimental determinations have been applied to determining the actual coordinations. The difficulty with experimentation lies in the differentiating close from actual contacts. Alternatively, computer experiments often suffer fundamental problems, such as being unable to match the experimental fractional density of a random packing. Recent advances in computation techniques for assembling random packing have resorted to dilated structures to match the experimental behavior.

Associated with investigations of random packing is a desire to model the structure with simple unit cells. Early efforts assumed that the structure of a random packing could be subdivided between a few simple, ordered packings. With time it has been appreciated that the number of simple units needed to model observed behavior is quite large. Recent efforts have included over twenty different structural units [14]. As an alternative, Mason [15] and others [16,50,60-62,101] have suggested that dilated tetrahedral units can be used to model random packings of spheres. This approach has merit in that it shows the distributions of coordination number, local density, and pore size.

Observations of random packings show that there is a distribution to all of the characteristics, for example, local porosity, coordination number, and particle distances. Consequently, statistical assessments are most productive. Over the years there has been a better appreciation of the complexity of random packings. Consequently, the models have expanded to include more unit cells and more involved methods for computer simulating a random packing. Although simple models are useful, it is clear that the random packing of monosized spheres is actually complex.

References

1. D. R. Hudson, "Coordination Number in Solid Solutions - A Mathematical Paradox," *Journal of Applied Physics*, 1946, vol. 17, p. 1126.
2. S. Melmore, "The Densest and the Least Dense Packings of Equal Spheres," *Mineralogical Magazine*, 1949, vol. 28, pp. 479-485.
3. R. Ben Aim and P. Le Goff, "La Coordinance des Empilements Desordonnes de Spheres. Application aux Melanges Binaires de Spheres," *Powder Technology*, 1968, vol. 2, pp. 1-12.
4. M. Suzuki and T. Oshima, "Coordination Number of a Multiple Component Randomly Packed Bed of Spheres with Size Distribution," *Powder Technology*, 1985, vol. 44, pp. 213-218.
5. L. Oger, J. P. Troadec, D. Bideau, J. A. Dodds and M. J.

Powell, "Properties of Disordered Sphere Packings I. Geometric Structure: Statistical Model, Numerical Simulations and Experimental Results," *Powder Technology*, 1986, vol. 46, pp. 121-131.
6. E. Arzt, "The Influence of an Increasing Particle Coordination on the Densification of Spherical Powders," *Acta Metallurgica*, 1982, vol. 30, pp. 1883-1890.
7. D. P. Haughey and G. S. G. Beveridge, "Structural Properties of Packed Beds - A Review," *The Canadian Journal of Chemical Engineering*, 1969, vol. 47, pp. 130-140.
8. F. A. L. Dullien, "Structural Properties of Packings of Particles," *Handbook of Powder Science and Technology*, M. E. Fayed and L. Otten (eds.), Van Nostrand Reinhold, New York, NY, 1984, pp. 99-128.
9. R. L. Brown and J. C. Richards, *Principles of Powder Mechanics*, Pergamon Press, Oxford, UK, 1970.
10. G. D. Scott, A. M. Charlesworth and M. K. Mak, "On the Random Packing of Spheres," *Journal of Chemical Physics*, 1964, vol. 40, pp. 611-612.
11. T. M. Hare, "Statistics of Early Sintering and Rearrangement by Computer Simulation," *Sintering Processes*, G. C. Kuczynski (ed.), Plenum Press, New York, NY, 1980, pp. 77-93.
12. L. C. Graton and H. J. Fraser, "Systematic Packing of Spheres With Particular Relation to Porosity and Permeability," *Journal of Geology*, 1935, vol. 43, pp. 785-909.
13. I. J. Smalley, "Variations on the Particle Packing Theme of Graton and Fraser," *Powder Technology*, 1970, vol. 4, pp. 97-101.
14. H. J. Frost, "Cavities in Dense Random Packings," *Acta Metallurgica*, 1982, vol. 30, pp. 889-904.
15. G. Mason, "A Model of the Pore Space in a Random Packing of Equal Spheres," *Journal of Colloid and Interface Science*, 1971, vol. 35, pp. 279-287.
16. G. Mason, "General Discussion," *Discussions of the Faraday Society*, 1967, vol. 43, pp. 75-76.
17. D. P. Haughey and G. S. G. Beveridge, "Structural Properties of Packed Beds - A Review," *The Canadian Journal of Chemical Engineering*, 1969, vol. 47, pp. 130-140.
18. H. Scher and R. Zallen, "Critical Density in Percolation Processes," *Journal of Chemical Physics*, 1970, vol. 53, pp. 3759-3761.
19. R. M. German and Z. A. Munir, "Morphology Relations

During Surface-Transport Controlled Sintering," *Metallurgical Transactions B*, 1975, vol. 6B, pp. 289-294.
20. K. Ridgway and K. J. Tarbuck, "The Random Packing of Spheres," *British Chemical Engineering*, 1967, vol. 12, pp. 384-388.
21. A. N. Nikolenko and M. S. Kovalchenko, "Analysis of the Random Packing of Identical Particles III. Tapping of Spherical Powders," *Soviet Powder Metallurgy and Metal Ceramics*, 1986, vol. 25, pp. 24-26.
22. W. A. Gray, *The Packing of Solid Particles*, Chapman and Hall, London, UK, 1968.
23. P. C. Carman, "Fundamental Principles of Industrial Filtration," *Transactions of the Institution of Chemical Engineers*, 1938, vol. 16, pp. 168-188.
24. D. J. Cumberland and R. J. Crawford, *The Packing of Particles*, Elsevier Science, Amsterdam, The Netherlands, 1987.
25. E. Manegold, R. Hofmann and K. Solf, "Ueber Kapillarsystem XII. I. Die Matematische Behandlund Idealer Kugelpackungen and das Hohlramvolumen Realer Geruststrukturen," *Kolloid Zeitschrift*, 1931, vol. 55, pp. 142-159.
26. E. Klar and W. M. Shafer, "On Green Strength and Compressibility in Metal Powder Compaction," *Modern Developments in Powder Metallurgy*, vol. 9, H. H. Hausner and P. W. Taubenblat (eds.), Metal Powder Industries Federation, Princeton, NJ, 1977, pp. 91-113.
27. C. A. Rogers, "The Packing of Equal Spheres," *Proceedings of the London Mathematical Society*, 1958, vol. 8, ser. 3, pp. 609-620.
28. A. H. Boerdijk, "Some Remarks Concerning Close-Packing of Equal Spheres," *Philips Research Reports*, 1952, vol. 7, pp. 303-313.
29. N. J. A. Sloane, "The Packing of Spheres," *Scientific American*, 1984, vol. 250, no. 1, pp. 116-125.
30. F. A. Rocke, "The Cylindrically Ordered Packing of Equal Spheres," *Powder Technology*, 1970, vol. 4, pp. 180-186.
31. B. G. Bagley, "A Dense Packing of Hard Spheres with Five-Fold Symmetry," *Nature*, 1965, vol. 208, pp. 674-675.
32. J. A. R. Clarke, "A Dense Packing of Hard Spheres with Five-Fold Symmetry," *Nature*, 1966, vol. 211, p. 280.
33. S. Melmore, "Open Packing of Spheres," *Nature*, 1942, vol. 149, p. 412.
34. S. Melmore, "Open Packing of Spheres," *Nature*, 1942, vol.

149, p. 669.
35. M. Gardner, "Reflections on the Packing of Spheres," *Scientific American*, 1960, vol. 202, no. 5, pp. 174-184.
36. M. Suzuki, K. Makino, M. Yamada and K. Iinoya, "A Study on the Coordination Number in a System of Randomly Packed, Uniform-Sized Spherical Particles," *International Chemical Engineering*, 1981, vol. 21, pp. 482-488.
37. K. Gotoh, "Liquid Structure and the Coordination of Equal Spheres in Random Assemblage," *Nature Physical Science*, 1971, vol. 231, pp. 108-110.
38. K. Shinohara and T. Tanaka, "Ultimate Tensile Strength of Cohesive Powders Based on Non-Uniform Packing Model," *Journal of Chemical Engineering of Japan*, 1975, vol. 8, pp. 50-56.
39. J. P. Jernot, M. Coster and J. L. Chermant, "Model of Variation of the Specific Surface Area During Sintering," *Powder Technology*, 1981, vol. 30, pp. 21-29.
40. K. Shinohara, "Rheological Property of Particulate Solids," *Handbook of Powder Science and Technology*, M. E. Fayed and L. Otten (eds.), Van Nostrand Reinhold, New York, NY, 1984, pp. 129-169.
41. J. W. Ross, W. A. Miller and G. C. Weatherly, "Computer Simulation of Sintering in Powder Compacts," *Acta Metallurgica*, 1982, vol. 30, pp. 203-212.
42. M. Cross, "Prediction of the Voidage of Packed Beds," *Proceedings Fourth International Symposium on Agglomeration*, C. E. Capes (ed.), Iron and Steel Society, Warrendale, PA, 1985, pp. 713-717.
43. J. L. Finney, "Modeling the Structures of Amorphous Metals and Alloys," *Nature*, 1977, vol. 266, pp. 309-314.
44. J. C. Macrae and W. A. Gray, "Significance of the Properties of Materials in the Packing of Real Spherical Particles," *British Journal of Applied Physics*, 1961, vol. 12, pp. 164-172.
45. J. Eastwood, E. J. P. Matzen, M. J. Young and N. Epstein, "Random Loose Porosity of Packed Beds," *British Chemical Engineering*, 1969, vol. 14, pp. 1542-1545.
46. J. D. Bernal, I. A. Cherry, J. L. Finney and K. R. Knight, "An Optical Machine for Measuring Sphere Coordinates in Random Packing," *Journal of Physics E: Scientific Instruments*, 1970, vol. 3, pp. 388-390.
47. A. J. Matheson, "Computation of a Random Packing of Hard Spheres," *Journal of Physics C: Solid State Physics*, 1974, vol.

7, pp. 2569-2576.
48. T. Ichikawa, "The Assembly of Hard Spheres as a Structure Model of Amorphous Iron," *Physica Status Solidi*, 1975, vol. 29, pp. 293-302.
49. G. D. Scott and D. L. Mader, "Angular Distribution of Random Close-Packed Equal Spheres," *Nature*, 1964, vol. 201, pp. 382-383.
50. J. D. Bernal, S. V. King and J. L. Finney, "Random Close-Packed Hard-Sphere Model," *Discussions of the Faraday Society*, 1967, vol. 43, pp. 60-69.
51. J. D. Bernal, J. Mason and K. R. Knight, "Radial Distribution of the Random Close Packing of Equal Spheres," *Nature*, 1962, vol. 194, pp. 957-958.
52. G. D. Scott, "Radial Distribution of the Random Close Packing of Equal Spheres," *Nature*, 1962, vol. 194, pp. 956-957.
53. G. Mason, "Radial Distribution Functions from Small Packings of Spheres," *Nature*, 1968, vol. 217, pp. 733-735.
54. K. Gotoh, W. S. Jodrey and E. M. Tory, "Average Nearest Neighbor Spacing in a Random Dispersion of Equal Spheres," *Powder Technology*, 1978, vol. 21, pp. 285-287.
55. E. M. Tory, B. H. Church, M. K. Tam and M. Ratner, "Simulated Random Packing of Equal Spheres," *The Canadian Journal of Chemical Engineering*, 1973, vol. 51, pp. 484-493.
56. G. Mason and W. Clark, "Distribution of Near Neighbours in a Random Packing of Spheres," *Nature*, 1965, vol. 207, p. 512.
57. J. D. Bernal and J. Mason, "Coordination of Randomly Packed Spheres," *Nature*, 1960, vol. 188, pp. 910-911.
58. W. O. Smith, P. D. Foote and P. F. Busang, "Packing of Homogeneous Spheres," *Physical Review*, 1929, vol. 34, pp. 1271-1274.
59. S. Ergun, "Fluid Flow Through Packed Columns," *Chemical Engineering Progress*, 1952, vol. 48, pp. 89-94.
60. M. E. Wise, "Dense Random Packing of Unequal Spheres," *Philips Research Reports*, 1952, vol. 7, pp. 321-343.
61. H. Iwata and T. Homma, "Distribution of Coordination Numbers in Random Packing of Homogeneous Spheres," *Powder Technology*, 1974, vol. 10, pp. 79-83.
62. J. D. Bernal, "Geometry of the Structure of Monatomic Liquids," *Nature*, 1960, vol. 185, pp. 68-70.
63. N. Ouchiyama and T. Tanaka, "Porosity of a Mass of Solid Particles Having a Range of Sizes," *Industrial and Engineering*

Chemistry Fundamentals, 1981, vol. 20, pp. 66-71.
64. T. Tanaka and N. Ouchiyama, "Coordination Numbers and Porosity of Packing Calculated from Particle Size Distribution," *Proceedings of the Fourth International Symposium on Agglomeration*, C. E. Capes (ed.), Iron and Steel Society, Warrendale, PA, 1985, pp. 9-16.
65. M. J. Powell, "Site Percolation in Randomly Packed Spheres," *Physical Review B*, 1979, vol. 20, pp. 4194-4198.
66. M. C. Thadani and F. N. Peebles, "Variation of Local Void Fraction in Randomly Packed Beds of Equal Spheres," *Industrial and Engineering Chemistry Process Design and Development*, 1966, vol. 5, pp. 265-268.
67. R. H. Beresford, "Statistical Geometry of Random Heaps of Equal Hard Spheres," *Nature*, 1969, vol. 224, pp. 550-553.
68. B. Widom, "Random Sequential Addition of Hard Spheres to a Volume," *The Journal of Chemical Physics*, 1966, vol. 44, pp. 3888-3894.
69. W. M. Visscher and M. Bolsterli, "Random Packing of Equal and Unequal Spheres in Two and Three Dimensions," *Nature*, 1972, vol. 239, pp. 504-507.
70. W. S. Jodrey and E. M. Tory, "Simulation of Random Packing of Spheres," *Simulation*, 1979, vol. 32, pp. 1-12.
71. G. D. Scott and D. M. Kilgour, "The Density of Random Close Packing of Spheres," *British Journal of Applied Physics (Journal of Physics D)*, 1969, vol. 2, ser. 2, pp. 863-866.
72. M. Dixmier, "Une Nouvelle Description des Empilements Aleatoires et des Fluides Denses," *Journal de Physique*, 1978, vol. 39, pp. 873-895.
73. R. Rutgers, "Packing of Spheres," *Nature*, 1962, vol. 193, pp. 465-466.
74. D. I. Lee, "Packing of Spheres and Its Effect on the Viscosity of Suspensions," *Journal of Paint Technology*, 1970, vol. 42, pp. 579-587.
75. G. D. Scott, "Packing of Equal Spheres," *Nature*, 1960, vol. 188, pp. 908-909.
76. A. Gamba, "Random Packing of Equal Spheres," *Nature*, 1975, vol. 256, pp. 521-522.
77. E. M. Tory and W. S. Jodrey, "Comments on Some Types of Random Packing," *Advances in the Mechanics and the Flow of Granular Materials*, vol. 1, M. Shahipoor (ed.), Trans Tech Publ., Clausthal, Federal Republic of Germany, 1983, pp. 75-106.

78. C. Lemaignan, "Hard Spheres Simulation of the Size Effect in Liquid and Amorphous Metallic Alloys," *Acta Metallurgica*, 1980, vol. 28, pp. 1657-1661.
79. R. K. Bordia, "A Theoretical Analysis of Random Packing Densities of Mono-Sized Spheres in Two and Three Dimensions," *Scripta Metallurgica*, 1984, vol. 18, pp. 725-730.
80. M. Shahinpoor, "Statistical Mechanical Considerations on the Random Packing of Granular Materials," *Powder Technology*, 1980, vol. 25, pp. 163-176.
81. A. R. Dexter and D. W. Tanner, "Packing Density of Ternary Mixtures of Spheres," *Nature Physical Science*, 1971, vol. 230, pp. 177-179.
82. J. G. Berryman, "Random Close Packing of Hard Spheres and Disks," *Physical Review A*, 1983, vol. 27, pp. 1053-1061.
83. R. F. Fedors, "A Relationship Between Maximum Packing of Particles and Particle Size," *Powder Technology*, 1979, vol. 22, pp. 71-76.
84. K. Gotoh and J. L. Finney, "Statistical Geometrical Approach to Random Packing Density of Equal Spheres," *Nature*, 1974, vol. 252, pp. 202-205.
85. M. M. Levine and J. Chernick, "A Numerical Model of Random Packing of Spheres," *Nature*, 1965, vol. 208, pp. 68-69.
86. A. Bonissent and B. Mutaftschiev, "A Computer Built Random Model for Simulation of the Crystal-Melt Interface," *Philosophical Magazine*, 1977, vol. 35, pp. 65-73.
87. J. S. Goodling and M. S. Khader, "Coordination Number Distribution of Spherical Particles in a Packed Cylindrical Bed," *Powder Technology*, 1985, vol. 44, pp. 53-55.
88. E. M. Tory, N. A. Cochrane and S. R. Waddell, "Anisotropy in Simulated Random Packing of Equal Spheres," *Nature*, 1969, vol. 220, pp. 1023-1024.
89. J. L. Finney, "Structure and Properties of Granular Materials: Guidelines from Modeling Studies of Liquids and Amorphous Solids," *Advances in the Mechanics and the Flow of Granular Materials*, vol. 1, M. Shahipoor (ed.), Trans Tech Publ., Clausthal, Federal Republic of Germany, 1983, pp. 19-39.
90. D. J. Adams and A. J. Matheson, "Computation of Dense Random Packings of Hard Spheres," *Journal of Chemical Physics*, 1972, vol. 56, pp. 1989-1994.
91. J. L. Finney, "Random Packings and the Structure of Simple Liquids I. The Geometry of Random Close Packing,"

Proceedings of the Royal Society of London A, 1970, vol. 319, pp. 479-493.

92. J. G. Berryman, "Definition of Dense Random Packing," *Advances in the Mechanics and the Flow of Granular Materials*, vol. 1, M. Shahipoor (ed.), Trans Tech Publ., Clausthal, Federal Republic of Germany, 1983, pp. 1-18.
93. J. R. Parrish, "Packing of Spheres," *Nature*, 1961, vol. 190, p. 800.
94. S. Yerazunis, J. W. Bartlett and A. H. Nissan, "Packing of Binary Mixtures of Spheres and Irregular Particles," *Nature*, 1962, vol. 195, pp. 33-35.
95. C. H. Bennett, "Serially Deposited Amorphous Aggregates of Hard Spheres," *Journal of Applied Physics*, 1972, vol. 43, pp. 2727-2734.
96. L. V. Woodcock and C. A. Angell, "Diffusivity of the Hard Sphere Model in the Region of Fluid Metastability," *Physical Review Letters*, 1981, vol. 47, pp. 1129-1132.
97. W. S. Jodrey and E. M. Tory, "Computer Simulation of Isotropic, Homogeneous, Dense Random Packing of Equal Spheres," *Powder Technology*, 1981, vol. 30, pp. 111-118.
98. D. S. Boudreaux and J. M. Gregor, "Structure Simulation of Transition-Metal-Metalloid Glasses," *Journal of Applied Physics*, 1977, vol. 48, pp. 152-158.
99. E. H. Aigeltinger and R. T. DeHoff, "Quantitative Determination of Topological and Metric Properties During Sintering of Copper," *Metallurgical Transactions A*, 1973, vol. 6A, pp. 1853-1862.
100. H. F. Fischmeister, E. Arzt and L. R. Olsson, "Particle Deformation and Sliding During Compaction of Spherical Powders: A Study by Quantitative Metallography," *Powder Metallurgy*, 1978, vol. 21, pp. 179-187.
101. J. D. Bernal, "The Structure of Liquids," *Scientific American*, 1960, vol. 203, no. 2, pp. 124-134.
102. R. L. Eadie, W. A. Miller and G. C. Weatherly, "Random Close Packing and Sintering: Some Implications for Initial-Stage Shrinkage," *Scripta Metallurgica*, 1974, vol. 8, pp. 755-762.
103. R. K. McGeary, "Mechanical Packing of Spherical Particles," *Journal of the American Ceramic Society*, 1961, vol. 44, pp. 513-522.
104. T. G. Owe Berg, R. L. McDonald and R. J. Trainor, "The Packing of Spheres," *Powder Technology*, 1969, vol. 3, pp.

183-188.
105. J. C. Williams and G. Shields, "The Segregation of Granules in a Vibrated Bed," *Powder Technology*, 1967, vol. 1, pp. 134-142.
106. M. Takahashi and S. Suzuki, "Numerical Analysis of Tapping Behavior of Ceramic Powders," *Ceramic Bulletin*, 1986, vol. 65, pp. 1587-1590.

Chapter Five

Packing of Monosized Nonspherical Particles

Chapter Outline

Introduction

Particles with a rough surface texture or shape are more susceptible to adsorption of surface films and to agglomeration [1]. Along with an increase in surface texture goes an increase in the interparticle friction. Accordingly, the packing density and coordination number decrease. The surface friction between particles can be characterized by an angle of friction, one form of which is the angle of repose. This angle measures the natural stacking of a powder, as illustrated in Figure 5.1, and is the angle from the horizontal formed by a cone shaped pile of powder as it is discharged into a loose packing. A high angle of repose indicates considerable interparticle friction, low packing density, and low packing coordination [2,3].

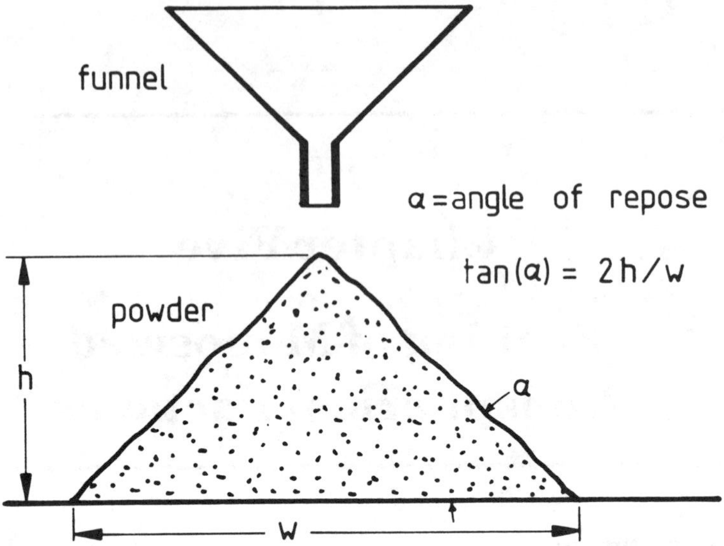

Figure 5.1 The angle of repose α for a powder is a rough measure of packing density changes due to interparticle friction and particle shape. It is defined by the loose packing angle of a powder.

Particle Shape Effects

Because of the higher interparticle friction, nonspherical particles give a lower random packing density and coordination number than spherical particles [4-8]. Figure 5.2 shows the fractional packing density for various monosized irregular particle shapes [9]. As the particle shape becomes more rounded (spherical) the observed packing density increases. Consequently, powders with a highly irregular particle shape do not match the packing density for spheres. Indeed, the range between random and ordered packing densities increases as the particle shape becomes nonspherical [10]. Spherical particles are more desirable in applications that require a high packing density or low interparticle friction.

If the particle has a regular geometric shape, then a decrement in packing density is not always observed even though the

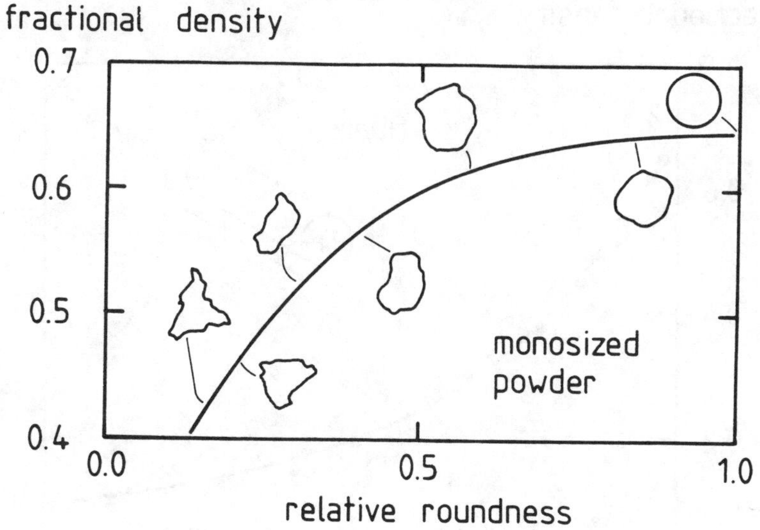

Figure 5.2 A plot of the fractional density versus the relative roundness for randomly packed monosized particles [6].

shape is not spherical. Nonisotropic particles can be packed to high densities if they are ordered. For example, cubic or rectangular particles can be packed to a fractional density of 1.00 when placed in perfect packing. Other polygonal shapes also exhibit high packing density structures [11]. Allen [12] shows that prolate spheroids of various axial ratios can be randomly packed to fractional densities as high as 0.74, depending on the axial ratio and degree of ordering attained in vibrating the particles.

The random packing of fibers provides an illustration of a decreasing packing density as the particles have a larger length-to-diameter ratio L/D. Figures 5.3 and 5.4 plot the fractional packing density versus the length-to-diameter ratio for fibers [13-15]. The first figure is scaled linearly to better emphasize the behavior of short fibers, while the second figure extends to large values using a logarithmic scaling. These figures show that fractional densities as low as 0.10 result from the random packing of fibers with length-to-diameter ratios over approximately 50. The highest packing

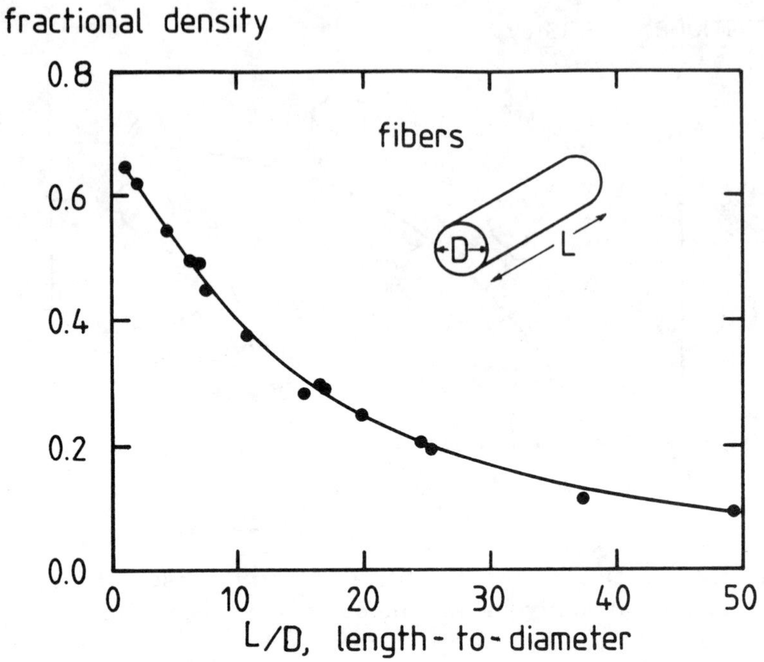

Figure 5.3 The fractional packing density of fibers decreases as the length-to-diameter ratio increases [13,14].

densities and the most isotropic structures are observed with equiaxed particles. In the extreme example of very long fibers with a length-to-diameter ratios over 100, the packing density decreases to 0.005 or less [15].

Generally, the wall effect on packing density is less pronounced with nonspherical particles. Figure 5.5 shows the variation in density with distance from the container wall for irregular berl saddles [8]. A comparison with the wall effect for spherical particles shows less of an effect. The packing density f depends on the relative container size D_c as follows [16-18]:

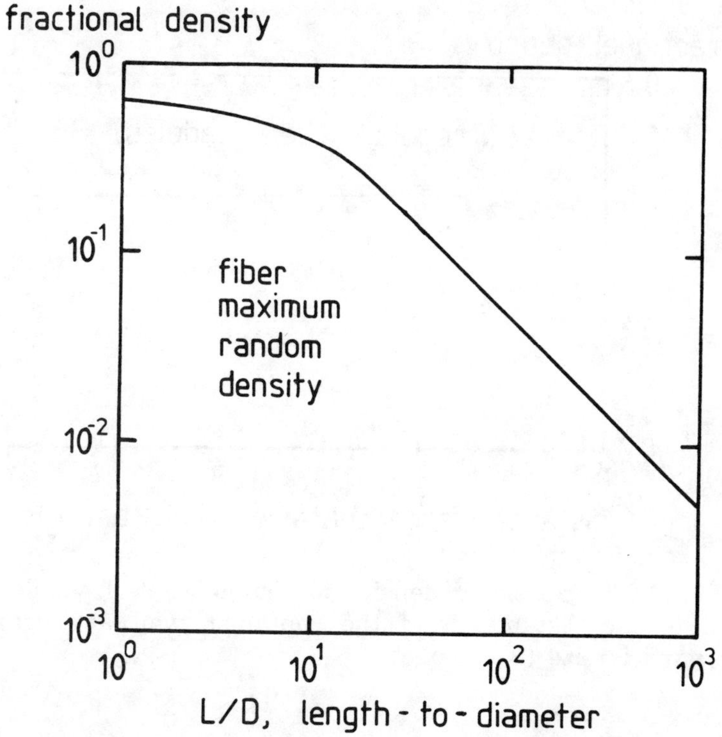

Figure 5.4 A logarithmic plot of the fractional density versus the length-to-diameter ratio for fibers in the random dense packing condition [15].

$$f = A - (K/D) - B\,exp(-C\,D_c/D) \qquad (5.1)$$

where D is the particle size and the constants A, B, C and K depend on the particle shape. This equation involves both the particle size and container size effects.

Obviously, there is an influence of particle size on packing density for angular particles. For angular particles in an infinite container, Equation 5.1 can be modified to give

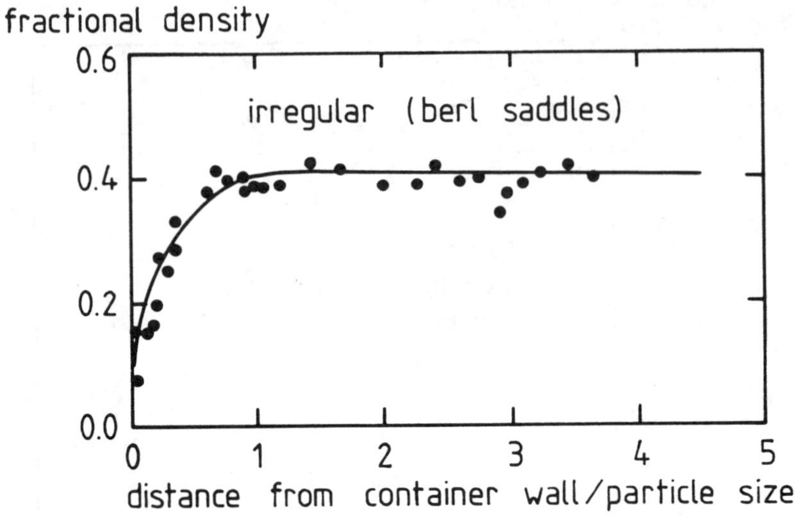

Figure 5.5 The fractional density for irregular berl saddles as a function of the distance from the container wall as normalized by the particle size [8].

$$f = A - (K/D) \tag{5.2}$$

with A approximately equal to 0.58. As illustrated in Figure 5.6 for feldspar and quartz particles (equiaxed and angular), Equation 5.2 provides a reasonable model for the behavior. In this case the particle size is expressed in millimeters. The parameter K varies with particle shape. For angular, equiaxed particles it is approximately equal to 7 μm, and for tetragonal particles it is near 32 μm. Equation 5.2 clearly demonstrates the particle size effect on packing density. Larger particles and spherical particle shapes give higher packing densities. For fibers, the density decreases as the length-to-diameter ratio (L/D) increases [19-21]. Figure 5.3 shows the systematic decrease in packing density with fiber length-to-diameter ratio. Such behavior has been approximated by the model [21]

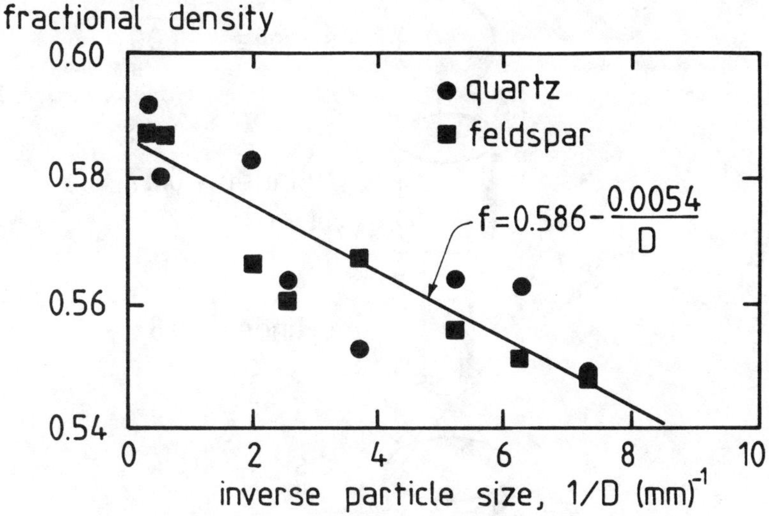

Figure 5.6 Particle size effect on the fractional packing density for irregular particles of feldspar and quartz. The apparent functional link between density and particle size is indicated by the solid line.

$$f = 1/[1.98 + 0.038\,(L/D)^2]^{1/2}\ . \tag{5.3}$$

Clearly, changes in particle size and particle shape both affect packing, but particle shape is usually most important.

For the typical case of a nonisotropic irregular solid, the nonspherical particle shape gives a lower packing density compared to a spherical shape. Wadell [22,23] defined the sphericity index Ψ for a particle as the ratio of surface areas;

$$\Psi = S_s/S \tag{5.4}$$

where S_s is the surface area of a sphere with the same volume as the test particle, and S is the actual surface area of the particle. Figure 5.7 sketches six simple shapes and tabulates the corresponding sphericity index. The effect of a nonspherical shape on packing

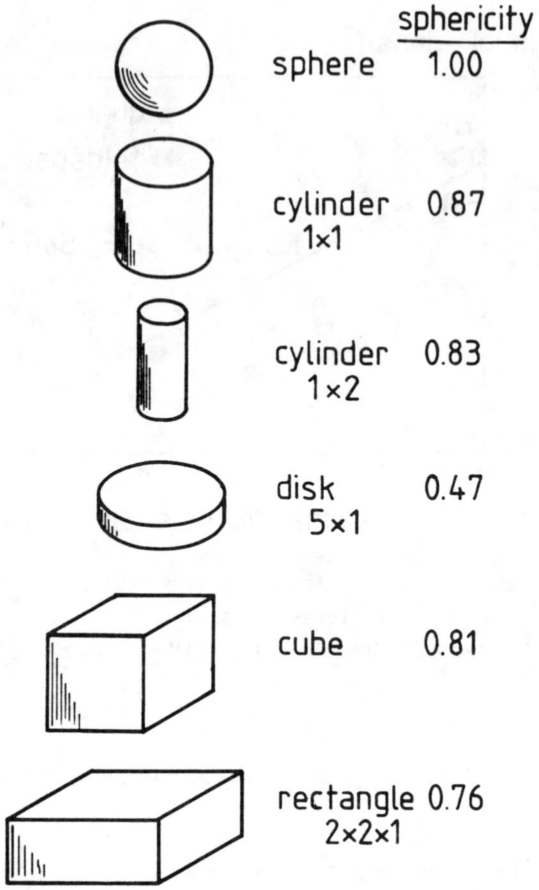

Figure 5.7 Sketches of six simple particle shapes with the sphericity index indicated for each.

density is illustrated in Figure 3.4. Warren and German [24] have linked the sphericity index to the fractional tap density f_t for a powder as follows:

$$\Psi = 0.079 + 0.831\, f_t + 1.53\, f_t^3 . \qquad (5.5)$$

Table 5.1

Particle Shape and Random Dense Packing [3,8,14-19,25-31]

particle shape	fractional density
spheres	0.64
cubes	0.72 to 0.79
rectangles (2x5x10)	0.51
plates (4x4x1)	0.67
plates (8x8x1)	0.59
tetrahedron	0.41 to 0.64
cylinders (L/D=167)	0.03
cylinders (L/D=60)	0.09
cylinders (L/D=25)	0.21
cylinders (L/D=15)	0.28
cylinders (L/D=10)	0.48
cylinders (L/D=5)	0.52
cylinders (L/D=1)	0.60 to 0.67
disks (L/D=0.5)	0.62 to 0.64
dagger shaped fibers	0.33
irregular filaments	0.06 to 0.35
irregular, isotropic	0.55 to 0.63
irregular, nonisotropic	0.30 to 0.50

Thus, the packing density follows a pattern similar to that observed with the interparticle friction; a lower density correlates with a lower coordination number and a higher angle of repose [25]. The random dense packing of spheres is contrasted with various other geometries in Table 5.1 [3,8,14-19,25-31]. The simple and equiaxed geometric shapes, such as cylinders and cubes, have random dense packings that rival those of spheres. Alternatively, the more irregular particle shapes have much lower packing densities. As with all packing structures, there are numerous factors which influence the actual density; however, the role of particle shape is dominant, as is evident from Table 5.1.

Summary

The packing of monosized nonspherical particles is dependent upon the details of the particle shape. Those particles which have simple equiaxed geometric shapes can be expected to pack to densities nearly equivalent to those of monosized spheres. The ordered packing of cubes, rectangles, and short cylinders can be better than that of spheres. However, as the length-to-diameter ratio increases the packing density decreases. A larger problem arises with the irregular or sponge particles such as those in Figure 5.8. These particles are difficult to describe quantitatively in terms of both size and shape. As a consequence, generalization about the packing behavior is even more difficult. These nonisotropic irregular shapes always exhibit inferior packing in comparison to spheres. The interparticle friction is a good gauge of the packing density, and correlations exist between the angle of repose and the packing density for several

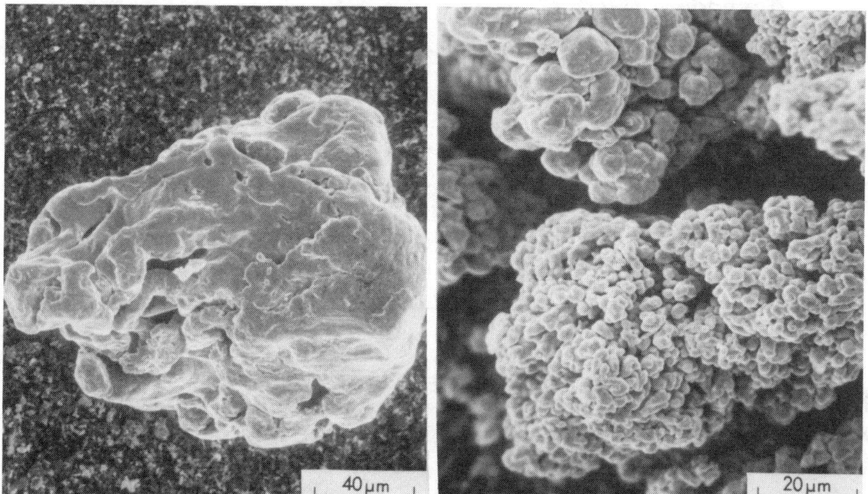

Figure 5.8 Scanning electron micrographs of irregular, sponge particles that prove difficult to describe in terms of simple size and shape parameters.

powder shapes. It is desirable to have a low interparticle friction and a low angle of repose to maximize the packing density of a powder. Unfortunately, this prevents the use of irregular particles in situations requiring a high packing density. Further, an irregular powder has more surface concavities for vapor adsorption, and because of this agglomeration and water adsorption tend to be larger problems with the irregular particles.

References

1. E. A. Boucher, "Porous Materials: Structure, Properties and Capillary Phenomena," *Journal of Materials Science*, 1976, vol. 11, pp. 1734-1750.
2. M. Suzuki and T. Oshima, "Estimation of the Coordination Number in a Two Component Mixture of Cohesive Spheres," *Powder Technology*, 1983, vol. 36, pp. 181-188.
3. R. Aoki and M. Suzuki, "Effect of Particle Shape on the Flow and Packing Properties of Non-Cohesive Granular Materials," *Powder Technology*, 1970, vol. 4, pp. 102-104.
4. K. Ridgway and K. J. Tarbuck, "Particulate Mixture Bulk Densities," *Chemical and Process Engineering*, 1968, vol. 49, no. 2, pp. 103-105.
5. J. P. Jernot, M. Coster and J. L. Chermant, "Model of Variation of the Specific Surface Area During Sintering," *Powder Technology*, 1981, vol. 30, pp. 21-29.
6. G. G. Brown, *Unit Operations*, John Wiley & Sons, New York, NY, 1950.
7. F. A. Shergold, "The Percentage Voids in Compacted Gravel as a Measure of Its Angularity," *Magazine of Concrete Research*, 1953, vol. 13, pp. 3-10.
8. D. J. Cumberland and R. J. Crawford, *The Packing of Particles*, Elsevier Science, Amsterdam, The Netherlands, 1987.
9. N. Peronius and T. J. Sweeting, "On the Correlation of Minimum Porosity with Particle Size Distribution," *Powder Technology*, 1985, vol. 42, pp. 113-121.
10. Y. Tsutsumi, "Arrangements of Uniformly Sized Particles in a Powder Bed," *Powder Technology*, 1973, vol. 7, pp. 181-188.
11. R. M. German, "The Two-Dimensional Connectivity of Liquid Phase Sintered Microstructures," *Metallurgical Transactions A*, 1987, vol. 18A, pp. 909-914.
12. J. R. L. Allen, "Notes Towards a Theory of Concentration of

Solids in Natural Sands," *Geological Magazine*, 1969, vol. 106, pp. 309-321.
13. J. V. Milewski, "The Combined Packing of Rods and Spheres in Reinforcing Plastics," *Industrial Engineering Chemistry Product Research and Development*, 1978, vol. 17, pp. 363-366.
14. J. V. Milewski, "Packing Concepts in the Utilization of Filler and Reinforcement Combinations," *Handbook of Fillers and Reinforcements for Plastics*, H. S. Katz and J. V. Milewski (eds.), Van Nostrand Reinhold, New York, NY, 1978, pp. 66-78.
15. D. M. Bigg, "Electrical Properties of Metal-Filled Polymer Composites," *Metal-Filled Polymers*, S. K. Bhattacharya (ed.), Marcel Dekker, New York, NY, 1986, pp. 165-226.
16. R. M. Koerner, "Limiting Density Behavior of Quartz Powders," *Powder Technology*, 1969, vol. 3, pp. 208-212.
17. A. N. Patankar and G. Mandal, "The Packing of Some Non-Spherical Solid Particles," *Transactions Journal of the British Ceramic Society*, 1980, vol. 79, pp. 59-66.
18. J. E. Ayer and F. E. Soppet, "Vibratory Compaction: II, Compaction of Angular Shapes," *Journal of the American Ceramic Society*, 1966, vol. 46, pp. 207-210.
19. M. Nardin, E. Papirer and J. Schultz, "Contribution a l'Etude des Empilements au Hasard de Fibres et ou de Particules Spheriques," *Powder Technology*, 1985, vol. 44, pp. 131-140.
20. J. V. Milewski, "Packing Concepts in the Use of Filler and Reinforcement Combinations," *Handbook of Reinforcements for Plastics*, J. V. Milewski and H. S. Katz (eds.), Van Nostrand Reinhold, New York, NY, 1987, pp. 14-33.
21. J. V. Milewski and D. E. Davenport, "Short Metal Fibers and Flakes," *Handbook of Reinforcements for Plastics*, J. V. Milewski and H. S. Katz (eds.), Van Nostrand Reinhold, New York, NY, 1987, pp. 103-114.
22. H. Wadell, "Volume, Shape, and Roundness of Quartz Particles," *Journal of Geology*, 1932, vol. 40, pp. 250-280.
23. H. Wadell, "Sphericity and Roundness of Rock Particles," *Journal of Geology*, 1933, vol. 41, pp. 310-331.
24. J. Warren and R. M. German, "The Effect of Powder Characteristics on Binder Incorporation for Injection Molding Feedstock," *Modern Developments in Powder Metallurgy*, vol. 18, P. U. Gummeson and D. A. Gustafson (eds.), Metal Powder Industries Federation, Princeton, NJ, 1988, pp. 391-402.
25. R. L. Brown and P. G. W. Hawksley, "Packing of Regular (Spherical) and Irregular Particles," *Nature*, 1945, vol. 156, pp.

421-422.
26. H. Y. Sohn and C. Moreland, "The Effect of Particle Size Distribution on Packing Density," *Canadian Journal of Chemical Engineering*, 1968, vol. 46, pp. 162-167.
27. K. E. Meiners, *Process Handbook, Complex Metal Powder Shapes by Injection Molding*, Battelle Columbus Laboratories, Columbus, OH, 1985.
28. J. R. Parrish, "Packing of Spheres," *Nature*, 1961, vol. 190, p. 800.
29. W. A. Gray, *The Packing of Solid Particles*, Chapman and Hall, London, UK, 1968.
30. J. M. Coulson, "The Flow of Fluids Through Granular Beds: Effect of Particle Shape and Voids in Streamline Flow," *Transactions of the Institution of Chemical Engineers*, 1949, vol. 27, pp. 237-157.
31. J. M. Campbell and R. L. Huntington, "Heat Transfer and Pressure Drop in Fixed Beds of Spherical and Cylindrical Solids," *Petroleum Refiner*, 1951, vol. 30, no. 12, pp. 127-133.

Chapter Six

Bimodal Mixtures of Spherical Particles

Chapter Outline

Introduction

Figure 6.1 shows a simple experiment involving two different sized particles demonstrating the improved packing that can be attained using mixed sizes. Two graduated cylinders are shown in Figure 6.1a. One cylinder is filled with large dark spherical particles, while the second cylinder is half full of smaller white particles. By pouring the smaller particles into the bed of large particles, the small particles are able to cascade through the large pores and fill previously empty space. Figure 6.1b is a photograph taken during filling and Figure 6.1c is a photograph of

Figure 6.1 These three photographs show the increased packing density possible by mixing two spherical particles of differing sizes: a) prior to mixing, b) during mixing, and c) after mixing.

the graduated cylinder after filling. The original space occupied by the two powders has been reduced by placing the small particles into the interstitial voids between the large particles.

Such simple experiments illustrate that bimodal particle combinations can pack to higher densities than can monosized particles. The key to improved packing is the particle size ratio. Small particles are selected such that they fit into the interstices between larger particles without forcing the large particles apart. In turn, even smaller particles can be selected to fit into the next level of pores, giving a corresponding improvement in packing density. Figure 6.2 further illustrates the basic concept using an ordered packing of two-dimensional disks to represent three-dimensional spherical packing. In the first case, shown in Figure 6.2a, monosized disks are packed into an orderly two-dimensional array with pores between them. In the binary packing case of Figure 6.2b, the secondary disks are of a smaller size selected to fill the pores between the larger disks without distorting the original packing. Accordingly, a ternary packing, such as shown in Figure

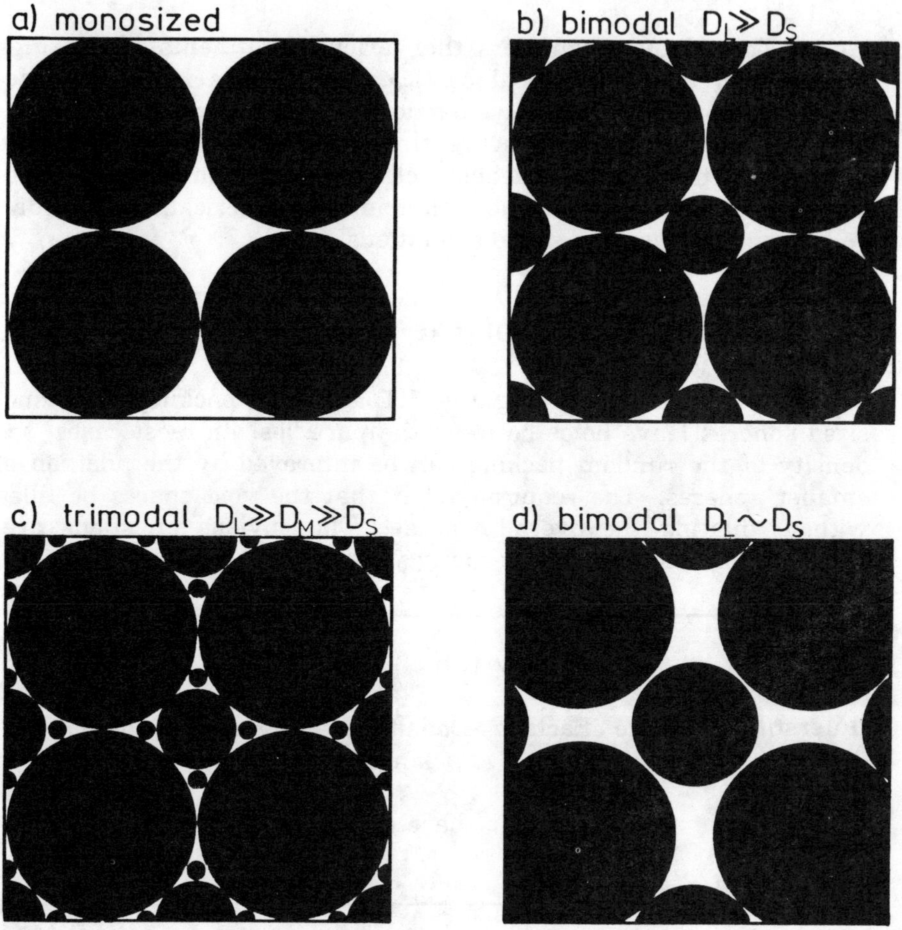

Figure 6.2 Two-dimensional representations of the effects from combinations of differing sizes; a) monosized, b) bimodal with a large difference, c) trimodal with a large difference, and d) bimodal with a small difference.

6.2c, can be created using even smaller disks that fill the pores in the binary case. However, as noted in the lower right drawing, Figure 6.2d, the selection of too large a disk will not improve the packing density since the added disks force the primary structure apart.

This chapter considers the various treatments of packings composed of mixed spherical particles. The focus is on the fractional packing density versus particle size ratio and the composition of the mixture. The first treatment will be for binary mixtures of spherical particles. Subsequent chapters build on these concepts and discuss bimodal nonspherical, trimodal spherical, multimodal, and continuous particle size distributions.

Qualitative Description

As illustrated in Figures 6.1 and 6.2, packings of monosized spheres leave holes between the particles. In most cases, the density of the random packing can be improved by the addition of smaller spheres. The requirement is that the void spaces be filled without dilating the overall volume. The obvious questions are, what size of sphere and how many yield optimal packing?

Table 6.1

Interstitial Sphere Packing Density Improvements for Ordered Packings

(diameter of base sphere = 1.0)

base structure	initial porosity	diameter added	volume added	mixture porosity
cubic				
$N_c = 6$	0.476	0.723	0.391	0.271
orthorhombic				
$N_c = 8$	0.395	0.528	0.147	0.307
rhombohedral				
$N_c = 12$	0.260	0.225	0.019	
		0.414	0.070	0.190

For an ordered packing, the diameter of the sphere that will fill the interstitial pore decreases as the packing coordination increases. Table 6.1 summarizes the effect of adding selective small spheres to three ordered packings. In the case of a coordination number of twelve, two different interstitials exist (octahedral and tetrahedral); thus, two sphere sizes and contents are needed for minimal porosity. For optimal packing the number of small particles increases as their relative size decreases. The rules for constructing ordered packings from various size ratios of particles follow exactly from those deduced for ionic crystals [1,2]. In both cases, the porosity is minimized by adding a specific amount and size of interstitial spheres. However, if the packing is not ordered, then the actual packing density tends to be less than that predicted by Table 6.1. Indeed, this provides a basis for constructing models of ionic glasses using randomly oriented unit cells containing the appropriate stoichiometry [3,4].

The basic behavior for random dense packing is sketched in Figure 6.3. The packing volume, termed the specific volume (volume to mass ratio), is shown as a function of the composition for a mixture composed of large and small spheres. There is a composition of maximum packing density [5-9], the mathematical concept of which is referred to as the Furnas relation. According to Furnas [6], at the maximum packing composition there is a greater volume of large particles than small particles. The relative improvement in packing density depends on the particle size ratio of the large and small particles. Within a limited range, the greater the size ratio of large to small particles, the higher the maximum packing density. This is true up to a limiting size ratio of approximately 20:1, but requires at least a 20% difference in particle sizes to occur [10-12]. Using Figure 6.3, the specific volume versus composition can be analyzed to determine the optimal composition. Assuming dense particles, the specific volume of a powder V is defined as the inverse of the apparent density

$$V = 1/(\rho f) \tag{6.1}$$

where ρ is the theoretical density for the powder and f is the fractional density for the powder.

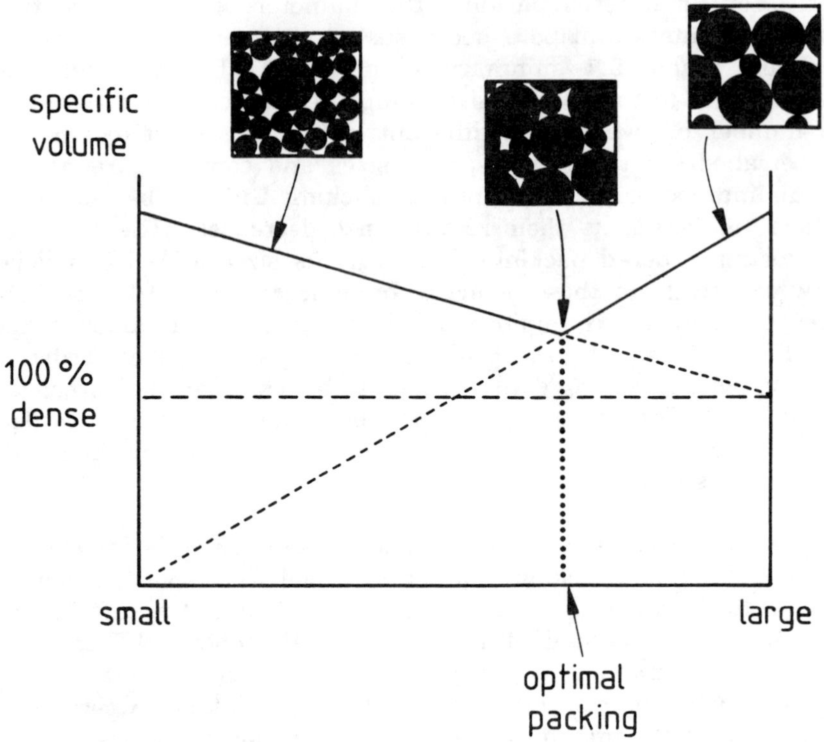

Figure 6.3 The reduction in specific volume for mixed large and small spheres, showing the condition of optimal packing where the small spheres fill all voids in the large sphere packing.

Beginning with the large particles, the specific volume initially decreases as small particles are added because the smaller particles fill the unoccupied voids between the large particles. This behavior can be seen in Figure 6.1b. A plot of the specific volume versus composition for this process is given in the right-hand portion of Figure 6.3. Eventually, beyond the optimal packing, the quantity of small particles becomes too great since all of the voids are filled. Accordingly, further additions force the large particles apart and no longer improve the packing density.

In contrast, starting with a container filled with the small particles, clusters of small particles and their associated voids can be removed and replaced by large particles. Since the large particles are fully dense, this process leads to an increase in density because of the simultaneous elimination of voids between small particles. Hence, a porous region is replaced with a fully dense region everywhere a large particle is added. The packing benefit from replacing small particles with large particles continues until a concentration where the large particles contact one another. Figure 6.3 shows this process as the left-hand plot. The point of minimum specific volume for the mixture corresponds to the intersection of these two curves.

Quantitative Model for Optimal Packing

The maximum packing density for a bimodal mixture of spheres corresponds to the minimum specific volume. This is termed the saturation point. At saturation the large particles are in point contact with one another and all of the interstitial voids are filled with small particles. A mathematical description of bimodal packing starts by designating the large particles with the subscript "L" and the small particles with the subscript "S". Calculation of the saturation composition in terms of the weight fraction of large particles X_L^* is an obvious goal of the mathematical treatments. The weight fraction of large particles at any composition depends on the following equation:

$$X_L = W_L/(W_L + W_S) \tag{6.2}$$

with W indicating the weight. The weight of large particles is calculated from the theoretical density ρ_L and fractional packing density f_L as

$$W_L = f_L\ \rho_L\,. \tag{6.3}$$

For maximum packing density the desire is to add sufficient small particles to just fill the void space between the large particles without forcing the large particles apart. The amount of void space equals $1 - f_L$ and the fractional packing density for the small particles times the small particle theoretical density gives its weight fraction as

$$W_S = (1 - f_L) f_S \, \rho_S \, . \tag{6.4}$$

Thus, for saturation X_L^* is given as follows:

$$X_L^* = f_L \, \rho_L \, /[f_L \, \rho_L + (1 - f_L) f_S \, \rho_S] \, . \tag{6.5}$$

In the typical case where the two particles composing the bimodal mixture are of the same composition, the density of the small and large particles will be the same, giving

$$\rho_L = \rho_S \, . \tag{6.6}$$

For this case, the calculation of X_L^* at the saturation point reduces to a dependence on the packing fractions for the two component powders:

$$X_L^* = f_L \, / \, (f_L + (1 - f_L) f_S) \, . \tag{6.7}$$

This equation can be further simplified for the case where the fractional packing densities for the two powders are the same (further assuming the theoretical densities are the same). In this case saturation is given by the simple relation:

$$X_L^* = 1/(2 - f_L) \, . \tag{6.8}$$

Equations 6.5, 6.7, and 6.8 give the compositions in terms of the proportional weight of large particles for the cases of 1) differing fractional packing densities and differing theoretical densities, 2) equal theoretical densities and differing fractional packing densities, and 3) equal theoretical densities and equal fractional packing densities, respectively.

As illustrated in Figure 6.4, the fractional density is dependent on the composition and particle size ratio [13]. The larger the particle size ratio, the higher the packing density at all compositions. For compositions away from the maximum packing density composition, Mangelsdorf and Washington [8] use the composition and

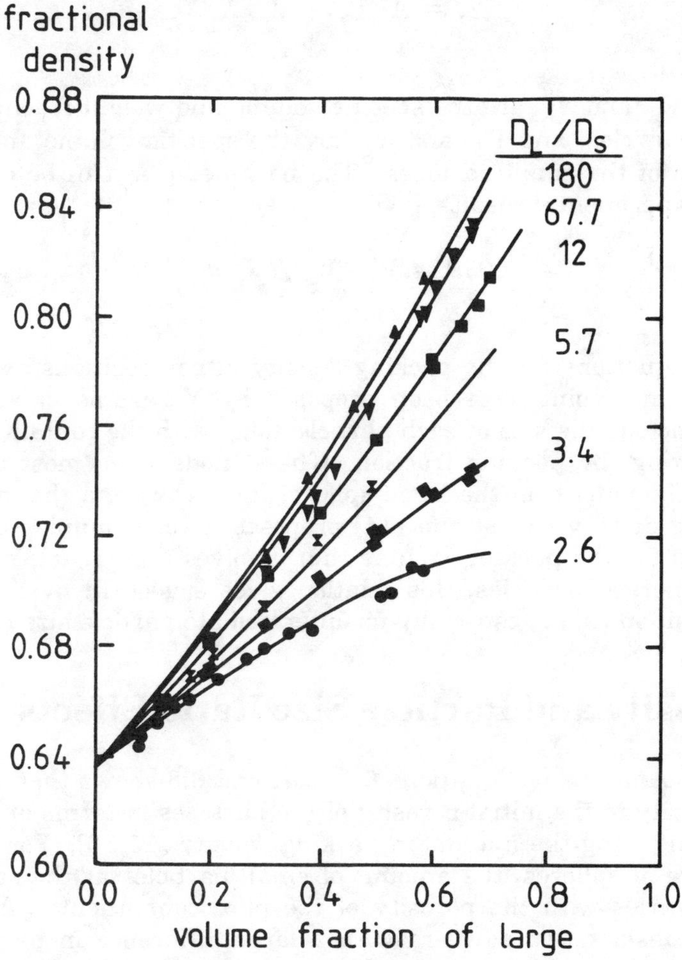

Figure 6.4 A plot of the fractional density versus composition for six particle size ratios [13].

volumes of the two powders to calculate the change in packing volume upon mixing. The decrease in packing volume Δf depends on the amount of each constituent as follows:

$$\Delta f = A\,(V_L + V_S)\,X_S X_L \tag{6.9}$$

where V_L and X_L are the specific volume and weight fraction of the large particles, and V_S and X_S are the specific volume and weight fraction of the small particles. The parameter A can be estimated from experimental results, giving

$$A = 0.039\,[4(V_L - V_S)/(V_L + V_S)]^2\,. \tag{6.10}$$

Other equations for the packing density at compositions away from the optimal point have been proposed by Yerazunis et al. [9,13]. They include the size of each particle along with the container size in calculating the packing fraction. These models are most useful at compositions rich in the large sphere, far away from the maximum packing density. Westman [14] proposes a very complex empirical relation for the packing volume that involves the particle size ratio. For spherical particles, this relation gives a good fit over the total composition range, especially for intermediate particle size ratios.

Porosity and Particle Size Ratio Effects

Analysis of Equations 6.5, 6.7, and 6.8 shows that there is a sensitivity to the initial porosity of the particles in terms of the composition giving the maximum packing density [15,16]. For a binary mixture of spheres, the amount of small particles at the saturation point varies with the porosity of the pure constituents. As a first case, consider two powders with a large difference in particle size that exhibit random dense packing with the usual fractional density of 0.637. The corresponding weight fraction of large particles for maximum packing is 0.734, or 26.6 wt.% of the smaller particles.

For the more general cases, Figure 6.5 plots the porosity of mixed powders versus the percentage of the large powder (assuming the powders have the same inherent fractional packing density of 0.50). This figure shows the simultaneous effects of composition and particle size ratio (large particle size divided by small particle size) on the porosity of the packing [6]. A minimum porosity occurs for each mixture at a composition rich in the large particles. The minimum porosity is larger and shifts toward the small particle axis as

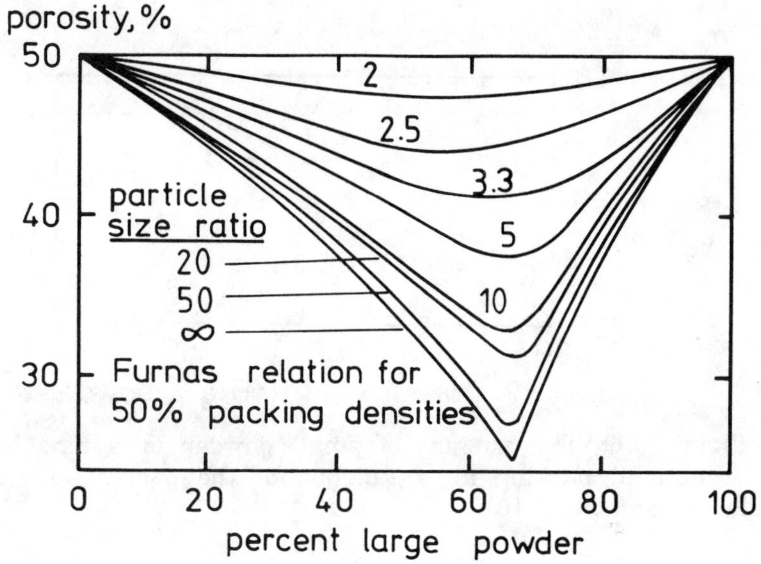

Figure 6.5 The porosity versus composition for bimodal mixtures of varying particle size ratios, assuming an inherent fractional packing density of 0.50 [6].

the particle size ratio decreases. The fractional packing density variation with the particle size ratio and composition has been curve fit as follows for the compositions rich in larger particles [13]:

$$f = 0.64/[1 - (0.362 - 0.315 (D_S/D_L)^{0.7}) X_L + 0.955 (D_S/D_L)^4 (X_L^2/(1 - X_L))] . \qquad (6.11)$$

In this equation, D is the particle diameter with the subscript denoting either the large or small sphere, and X_L is the fraction of large spheres. For the case of an infinite particle size ratio, then the relation becomes much simplified:

$$f = 0.64/(1 - 0.362 X_L) . \qquad (6.12)$$

This assumes that the particles are spherical with a monosize

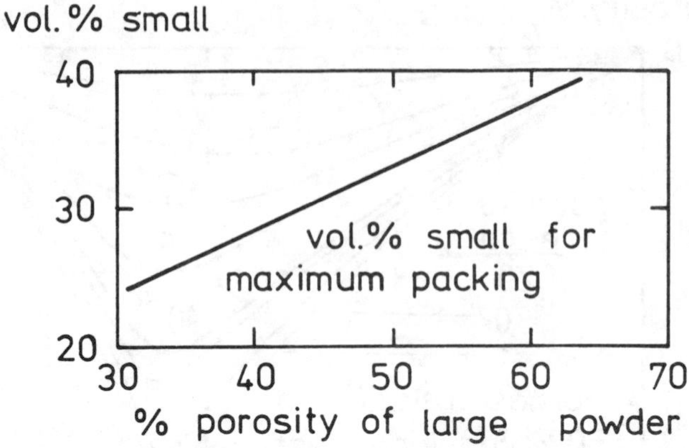

Figure 6.6 The amount of small powder in a bimodal mixture at optimal packing as a function of the packing density of the large component [6].

random dense packing fraction of 0.64.

Figure 6.6 gives the relation between the inherent packing density of the large particles and the amount of small particles needed to attain maximum packing density as calculated by Furnas [6]. As expected, the amount of small powder in the optimal mixture increases linearly with the porosity of the large particles. Essentially, this assumes that the large particles act as a filter bed where the small particles fill the voids.

There is no gain in packing density if the particles have the same size. Alternatively, the packing benefit is maximized by a large difference in particle sizes. Figure 6.7 shows experimental results using mixtures of spherical lead particles of varying size ratios [7]. The fractional packing density is plotted against the inverse particle size ratio (D_L/D_S). In many instances the inverse ratio is used in analyzing packing effects since this compresses the size scale at large size differences. Since the highest binary packing density occurs when the larger particles provide a fixed bed for the smaller particles to fill, the smaller the relative size of the small particles the better the packing.

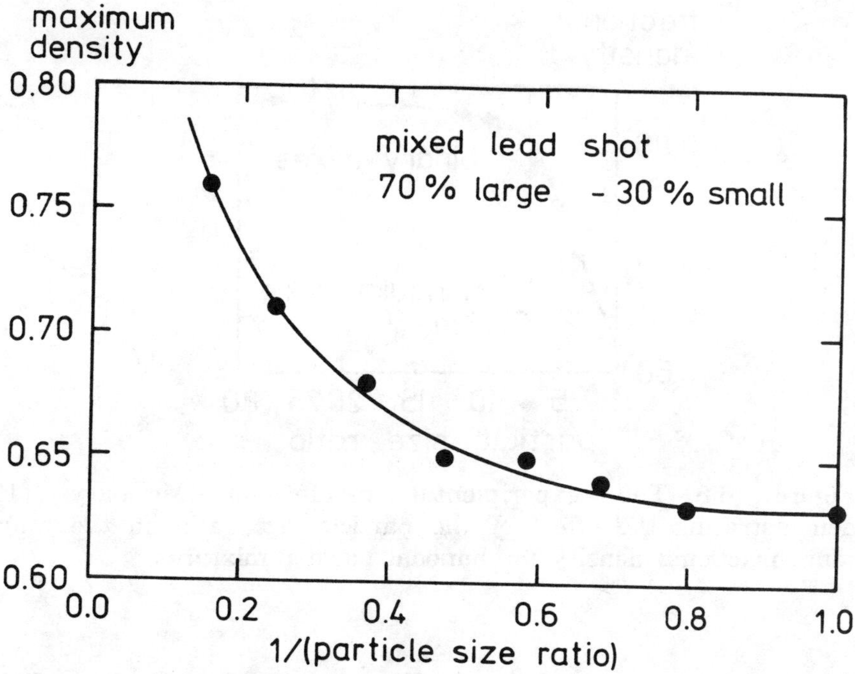

Figure 6.7 Experimental results for spherical lead shot at a composition of 70% large and 30% small, showing the change in fractional packing density with the inverse of the particle size ratio [7].

Although the absolute particle sizes are relatively unimportant, the particle size ratio is very important. Figure 6.8 shows the measurements by McGeary [17] for the random dense packing of mixtures of various model packings. The fractional packing density increases as the particle size ratio increases up to a ratio of 15:1 or so. Note the dramatic change in behavior at the particle size ratio corresponding to one particle filling the triangular pores between the coarse particles; this occurs at roughly a 7:1 size ratio. Experiments by Ayer and Soppet [18] show that the packing density is unchanged by particle size ratios larger than 20:1. Alternatively, there is some evidence that at small particle size ratios there is a density loss. For example, a mixture of sizes with a ratio of 1.24 reportedly gave a lower density than the constituent powders for all compositions [11].

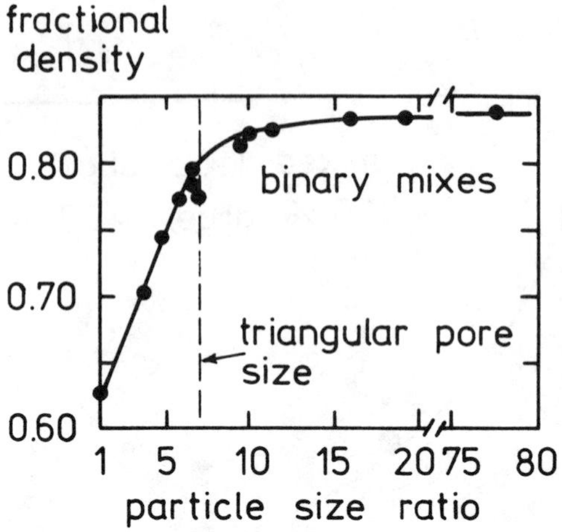

Figure 6.8 The experimental results of McGeary [17] demonstrating the effect of the particle size ratio on the maximum fractional density for bimodal powder mixtures.

Figure 6.9 graphs the packing density gain as a function of the inverse particle size ratio D_S / D_L at the maximum packing density [10,19-21]. There is no packing gain predicted when the particle sizes are within approximately 20% of each other.

In randomly mixed particles there is a complication in predicting the mixture density because of localized distortions [16,22-24]. The small spheres preferentially collect on the surface of the large spheres. As a consequence, the small sphere packing structure is distorted at the interface in the otherwise random packing. The large particle surface acts as a curved boundary that induces nonrandom structure in the small spheres near the interface. Associated with the nonrandom packing structure at the interface is a decrease in the local density similar to that encountered at a container wall. Since the small spheres do not efficiently cover the large sphere surface, the localized fractional density is approximately 67% of the inherent density for the small particles. The actual distortion in the small sphere packing depends on the particle size ratio. Because of this localized surface effect, the composition for maximum packing density in a random mixture tends to be slightly shifted to a lower content of large particles.

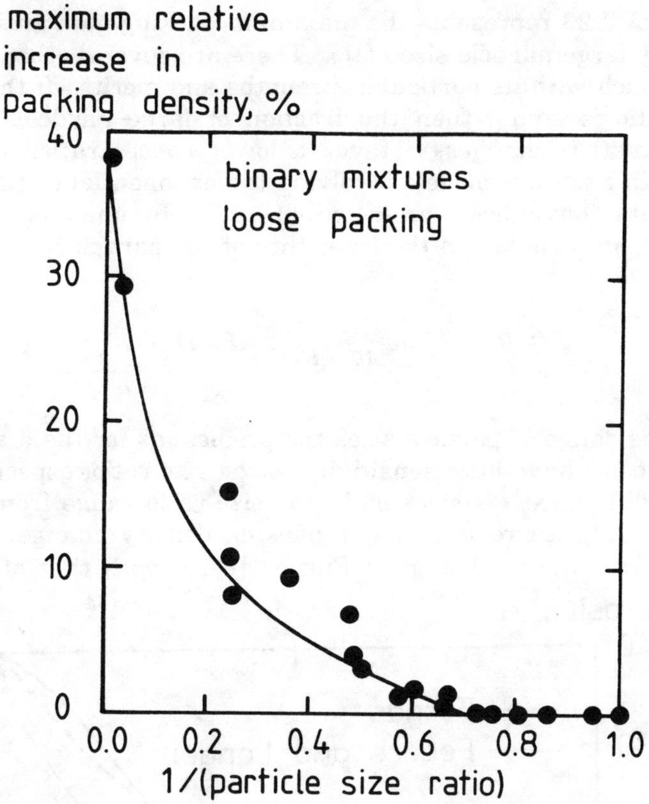

Figure 6.9 The relative packing density increase versus the inverse particle size ratio for binary mixtures of spherical powders [10,19-21].

Various models have been proposed to account for the particle size ratio effect on the maximum packing density [10-21,25,26]. At large particle size differences the maximum relative gain in packing density increases and approaches a fractional packing density of 0.87 with an infinite size ratio. The contraction at the maximum packing density can be estimated from the following empirical quadratic equation [26]:

$$\Delta f = 0.23\,[1 - 2.35\,(D_S/D_L) + 1.35\,(D_S/D_L)^2]\,. \tag{6.13}$$

The value of 0.23 represents the maximum packing gain anticipated from a very large particle size ratio. There are several other models available, each with its particular strengths and merits. If the particle size ratio is small, then the fraction of large particles at the optimal condition changes to favor a lower concentration of large particles. In contrast, models involving the exponential of the particle size ratio have been predicted [18,27]. In one version, the packing density is linked to the logarithm of the particle size ratio as [11,25]

$$\Delta f = 0.1737 \log_{10}[D_L/(1.239\ D_S)]\,. \tag{6.14}$$

Over a wide range of particle sizes the predictions for the maximum packing density have little sensitivity to the size ratio, especially for large size differences. However, in the size ratio range from 1:1 to 20:1 for D_L/D_S there is a major packing density change. Figure 6.10 compares the prediction of Furnas [6,19] with that of Fedors

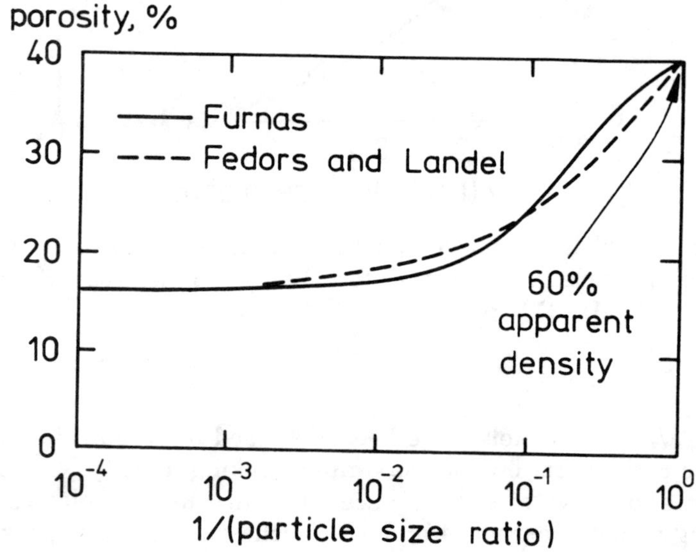

Figure 6.10 The predictions of Furnas [6,19], and Fedors and Landel [28,29] for the porosity versus the inverse particle size ratio of bimodal sphere mixtures with inherent fractional densities of 0.6.

and Landel [28,29] for particles with fractional densities of 0.6. Shown is the porosity of the mixture versus the inverse particle size ratio. The two relations are similar, but the Furnas predictions are generally better known. Both models show that major packing benefits are associated with large differences in sphere sizes, but very large differences have no further benefit.

Coordination Numbers

Under optimal conditions, the maximum fractional packing density for a binary mixture of spheres has been estimated to be between 0.84 and 0.88 [9,10,18,27]. The coordination of the two constituents depends on the amount and size ratio of the particles. As the particle size ratio of large to small particles D_L/D_S goes up, the packing coordination of the large particles increases [30-34]. For example, with a particle size ratio of 1000, the coordination number of a large particle surrounded by small particles exceeds 3 million contacts. At a particle size ratio of 100, the large particle will experience over 31,000 contacts with small particles. Based on computer simulations, the coordination number N_c of the large sphere at optimal packing can be estimated as follows [34,35]:

$$N_c = 3.0 + 3.16\,(D_L/D_S)^2 \,. \qquad (6.15)$$

This equation gives a coordination number of 6.16 when the particles are of equal size. Figure 6.11 plots this coordination number function for various particle size ratios. One interesting result of the computer simulations is that the overall average coordination remains nearly constant at 6.0 for all particle size ratios [36]. Figure 6.12 illustrates this behavior for a particle size ratio of 3.0 by showing the packing coordinations for both the large and small particles versus composition. The behavior has been fit with the following equation for the number of contacts on the large sphere:

$$N_c = 3.5 + 1.25\,[D_L/(n_S D_L + n_L D_L)]\,[1 + D_L/\,(n_S D_L + n_L D_L)] \qquad (6.16)$$

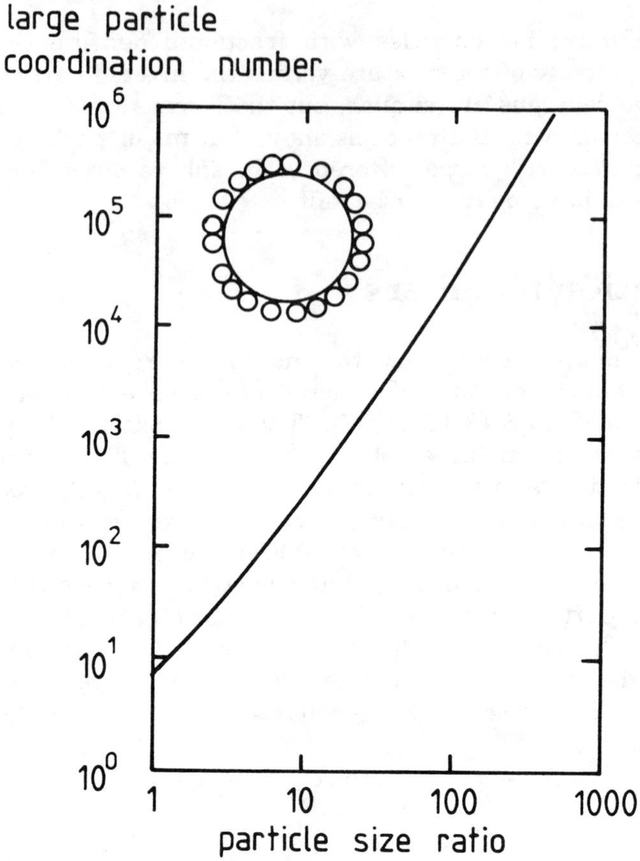

Figure 6.11 The computer calculated coordination number of the large sphere in a bimodal mixture versus the particle size ratio [34].

where n_S is the number fraction of small particles and n_L is the number fraction of large particles. Across the binary mixture the coordination number for the small particles increases from 3 to 6, and the coordination number for the large particles increases from 6 to 25.

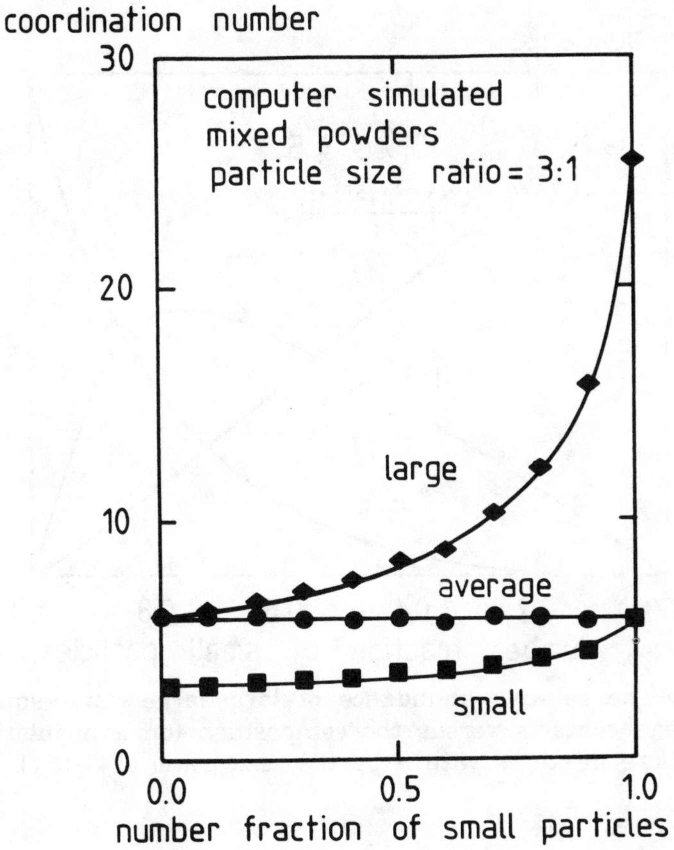

Figure 6.12 The coordination numbers for bimodal mixtures of large and small spheres in a size ratio of 3:1 versus the mixture composition [36].

Both experimental and theoretical treatments can predict the coordination number distributions, including the number of small-small, small-large, and large-large contacts versus size ratio and composition [31-33]. Figure 6.13 shows the variation in the relative abundance of large-large, large-small, and small-small contacts for a homogeneous mixture of spheres with a particle size ratio of five [37]. The abundance of each contact type is plotted against the number fraction. Because of the size difference, the population and weight distributions are very different. As an example, for a

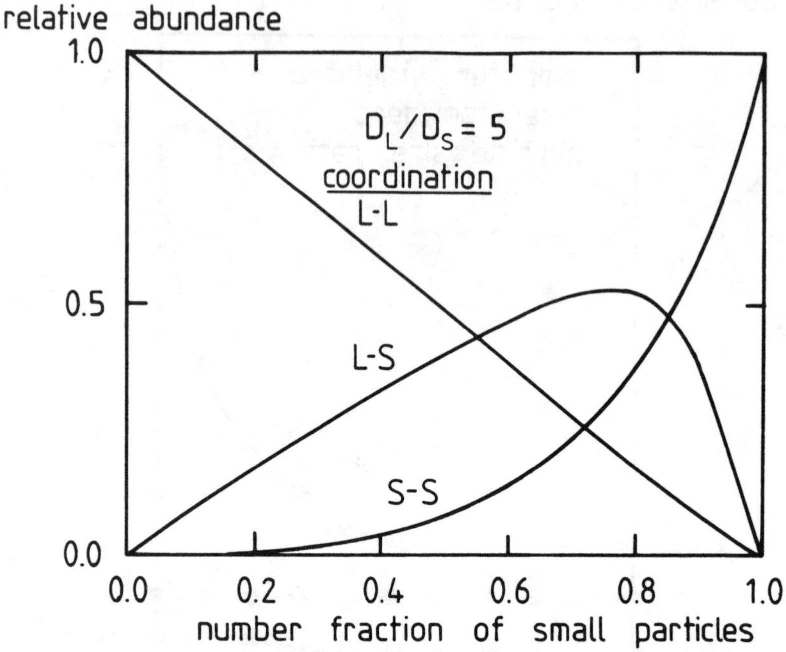

Figure 6.13 The relative abundance of large-large, large-small, and small-small contacts versus the composition (on a population basis) for sphere mixtures with a particle size ratio of 5 [37].

particle size ratio of 5, a population of 93% small particles corresponds to approximately 10% by volume. In real bimodal powder mixtures, the coordination number for a sphere of a given size will actually be variable within the packing. Although Figures 6.12 and 6.13 represent the mean behavior, packing inhomogeneities will contribute to a distribution in the number of contacts per particle throughout the compact. Often mixed particles will exhibit clustering of similar sized neighbors. One model [33] includes a factor for the decreased packing density at the interface between large and small particles.

Homogeneity Effects

The homogeneity of a mixture has a major influence on the packing density [38-44]. It is clearly established that the density varies with the composition for mixtures of differing particle sizes. Consider that in an inhomogeneously mixed packing some regions will have more large spheres than others. Consequently, the fractional density will vary from point to point in proportion to the local composition. Most of the models cited in this chapter have inherently assumed orderly positioning of the particles. In reality this is a difficult state to attain. The typical structure of mixed particles will have random packing and the resulting segregation or fluctuations in local composition will result in fluctuations in density. Because of this influence, models of packing density for bimodal mixtures overestimate that attainable in random mixtures.

There is a particular problem in mixtures formed without vibration since the small spheres may actually cause a dilation of the mixture, as sketched in Figure 6.14. The bimodal mixtures introduced for high packing densities assume the small spheres are sited in the interstitial voids. This becomes a special problem for those mixtures in which the particles are similar in size since the small particles can not fall through the constricted openings between

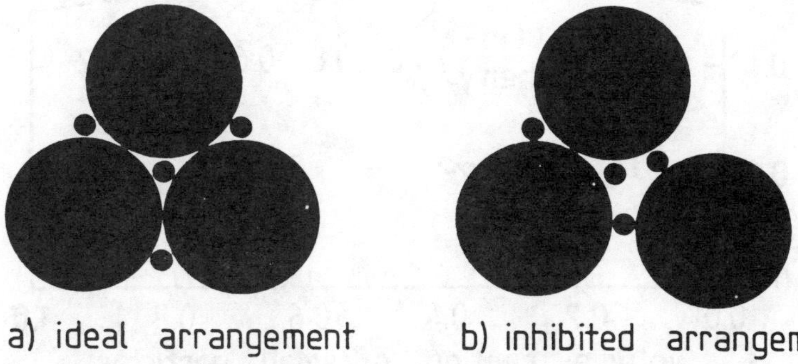

Figure 6.14 A sketch showing how small particles can inhibit optimal packing due to improper vibration into interstitial sites.

the large particles. As the fractional packing density of the large particles increases, the size of the opening decreases; for a simple cubic lattice the small sphere size must be less than 1/2.4 of the large sphere size to pass through the interparticle throat constriction. If the large sphere packing is close-packed cubic, then the small sphere must be less than 1/6.5 of the large sphere to pass through the interparticle throat constriction. In both cases, the sphere size that can fit into the pores is larger than the size of the sphere that can pass through the constrictions between the large spheres.

The packing density will increase with the homogeneity of the mixture [42,43,45]. Figures 6.15 and 6.16 show how composition and the degree of mixing influence the packing density [43]. Figure 6.15 is a plot of the fractional packing density versus composition for several levels of mixture homogeneity. This plot assumes an infinite particle size ratio. Note the similarity between the homogeneity effect in Figure 6.15 and the particle size ratio

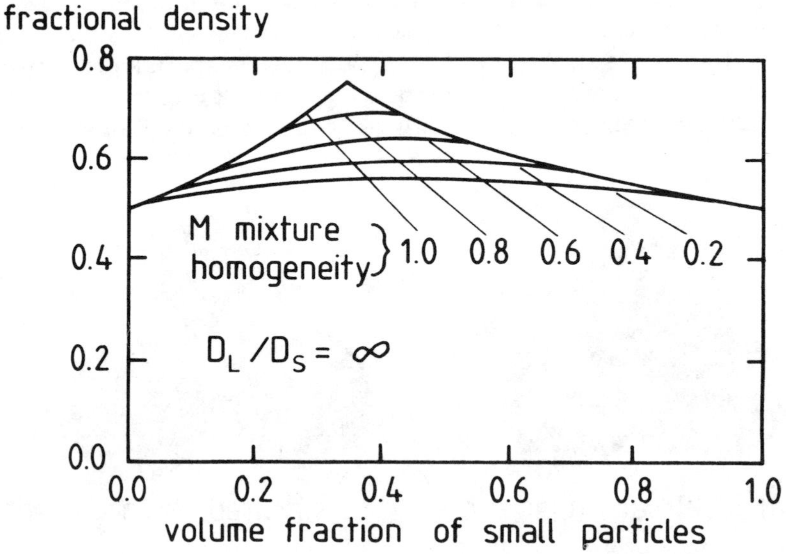

Figure 6.15 A plot of the fractional packing density versus composition for binary mixtures with an infinite particle size ratio at various levels of mixture homogeneity [43].

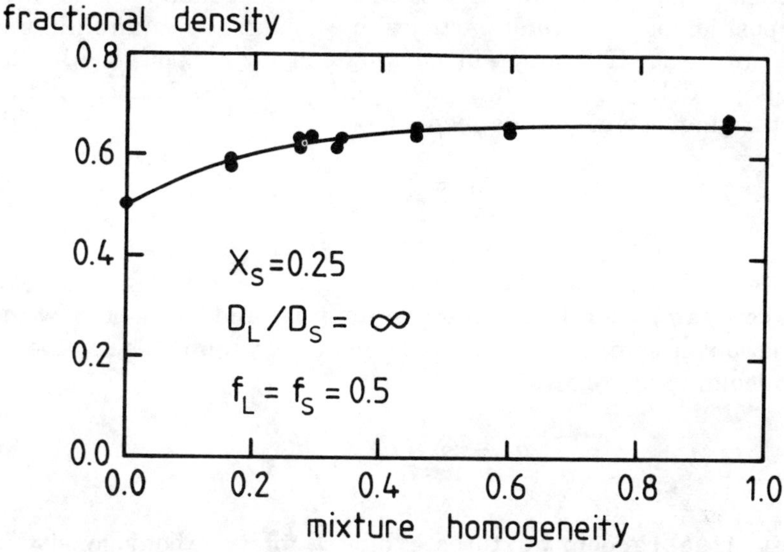

Figure 6.16 A plot of fractional packing density versus mixture homogeneity for a mixture consisting of 25% small particles and an infinite particle size ratio [43].

effect in Figure 6.5. The density improves with the degree of mixing. The upper limit is based on perfect mixing, while the lower limit assumes a totally segregated structure. In the unmixed or segregated condition, the fractional packing density is linearly dependent on the composition and weight fractions of the two powders

$$f_u^{-1} = X_L/f_L + X_S/f_S \tag{6.17}$$

where f_u is the fractional density for the unmixed system, composed of X_L weight fraction of large particles with a fractional density of f_L and X_S as the volume fraction of small particles with f_S as the fractional density. If the fractional packing densities for the large and small particles are the same, then in an unmixed system the packing density will be unaffected by composition. Randomly mixed systems will range between the unmixed and fully mixed limits, depending on the mixture homogeneity. In Figure 6.16 the fractional density is shown versus the homogeneity. This is for a composition of 25% small particles in a mixture with an infinite particle size ratio. The inherent packing density for both the large and small particles is assumed to be 0.5. In these experiments the mixture homogeneity is defined as

$$M = 1 - (\sigma/\sigma_o) \tag{6.18}$$

where σ is the standard deviation in the composition between random samples taken from the mixture and σ_0 is the standard deviation in composition for the unmixed system. The value of σ_0 is calculated as follows:

$$\sigma_o = (X_L(1 - X_L))^{1/2} \,. \tag{6.19}$$

Most actual random mixtures exhibit a mixture homogeneity index near 0.7.

Summary

The optimal packing is generally the main concern in dealing with a bimodal mixture, although for those cases away from optimal packing the above treatment outlines the procedure for estimating the packing density from the composition and particle sizes. Optimal packing requires a large particle size ratio, with values of D_L/D_S over 20 being most useful. The particle size ratio dominates the optimal packing density as demonstrated in Figure 6.9. There is no density gain for mixtures involving particles with a particle size ratio near unity.

A high packing density for the individual component particles is helpful in attaining a high packing density for the mixture. In the case of randomly packed monosized spheres with a fractional density of 0.637, a mixture of 73% large and 27% small will give a fractional packing density for the mixture approaching 0.86 [28]. The greater the size difference, the closer the packing density of the mixture will approach this limiting value. For powder mixtures with a large size ratio (D_L/D_S approaching infinity), the maximum packing density can be estimated as follows:

$$f_{max} = f_L + (1 - f_L) f_S \, . \tag{6.20}$$

The composition for the optimal packing density occurs when

$$X_L^* = f_L / f_{max} \, . \tag{6.21}$$

For a finite particle size ratio, the maximum packing density is given approximately as

$$f_{max} = f_L + (1 - (D_L/D_S)^{1/2})(1 - f_L) f_S \, . \tag{6.22}$$

For the simple case of bimodal mixtures of spheres of the same composition and inherent packing density, the maximum packing occurs at

$$X_L^* = 1/(2 - f_L) \, . \tag{6.23}$$

In the optimal case of homogeneously mixed spherical particles the fractional packing density will approach approximately 0.86 in the case of a large particle size difference. These relations assume optimal positioning of the small and large spheres. In reality, inhomogeneities will degrade the packing density of the mixture. As the homogeneity of a random mixture improves, the packing density will increase.

References

1. M. T. Kamel, E. M. Tory and W. S. Jodrey, "The Distribution of kth Nearest Neighbors and its Application to Cluster Settling in Dispersions of Equal Spheres," *Powder Technology*, 1979, vol. 24, pp. 19-34.
2. D. R. Hudson, "Density and Packing in an Aggregate of Mixed Spheres," *Journal of Applied Physics*, 1949, vol. 20, pp. 154-162.
3. R. J. Bell and P. Dean, "Properties of Vitreous Silica: Analysis of Random Network Models," *Nature*, 1966, vol. 212, pp. 1354-1356.

4. D. L. Evans and S. V. King, "Random Network Model of Vitreous Silica," *Nature*, 1966, vol. 212, pp. 1353-1354.
5. T. L. Starr, "Packing Density of Fiber/Powder Blends," *Ceramic Bulletin*, 1986, vol. 65, pp. 1293-1296.
6. C. C. Furnas, "The Relations Between Specific Volume, Voids and Size Composition in Systems of Broken Solids of Mixed Sizes," Serial 2894, Report of Investigations, U.S. Bureau of Mines, North Central Experiment Station, Minneapolis, MN, October 1928.
7. A. E. R. Westman and H. R. Hugill, "The Packing of Particles," *Journal of the American Ceramic Society*, 1930, vol. 13, pp. 767-779.
8. P. C. Mangeldorf and E. L. Washington, "Packing of Mixtures of Hard Spheres," *Nature*, 1960, vol. 187, p. 930.
9. S. Yerazunis, J. W. Bartlett and A. H. Nissan, "Packing of Binary Mixtures of Spheres and Irregular Particles," *Nature*, 1962, vol. 195, pp. 33-35.
10. K. Ridgway and K. J. Tarbuck, "Particulate Mixture Bulk Densities," *Chemical and Process Engineering*, 1968, vol. 49, no. 2, pp. 103-105.
11. N. Epstein and M. J. Young, "Random Loose Packing of Binary Mixtures of Spheres," *Nature*, 1962, vol. 196, pp. 885-886.
12. G. D. Scott and G. J. Kovacs, "The Radial Distributions of a Binary Mixture of Spheres," *Journal of Physics D: Applied Physics*, 1973, vol. 6, pp. 1007-1010.
13. S. Yerazunis, S. W. Cornell and B. Wintner, "Dense Random Packing of Binary Mixtures of Spheres," *Nature*, 1965, vol. 207, pp. 835-837.
14. A. E. R. Westman, "The Packing of Particles: Empirical Equations for Intermediate Diameter Ratios," *Journal of the American Ceramic Society*, 1936, vol. 19, pp. 127-129.
15. J. A. Dodds, "The Porosity and Contact Points in Multicomponent Random Packings Calculated by a Simple Statistical Geometric Model," *Journal of Colloid and Interface Science*, 1980, vol. 77, pp. 317-327.
16. D. J. Cumberland and R. J. Crawford, *The Packing of Particles*, Elsevier Science, Amsterdam, The Netherlands, 1987.
17. R. K. McGeary, "Mechanical Packing of Spherical Particles," *Journal of the American Ceramic Society*, 1961, vol. 44, pp. 513-522.
18. J. Ayer and F. E. Soppet, "Vibratory Compaction: I,

Compaction of Spherical Shapes," *Journal of the American Ceramic Society*, 1965, vol. 48, pp. 180-183.
19. C. C. Furnas, "Grading Aggregates I - Mathematical Relations for Beds of Broken Solids of Maximum Density," *Industrial and Engineering Chemistry*, 1931, vol. 23, pp. 1052-1058.
20. C. Lemaignan, "Hard Spheres Simulation of the Size Effect in Liquid and Amorphous Metallic Alloys," *Acta Metallurgica*, 1980, vol. 28, pp. 1657-1661.
21. D. I. Lee, "Packing of Spheres and Its Effect on the Viscosity of Suspensions," *Journal of Paint Technology*, 1970, vol. 42, pp. 579-587.
22. H. J. Fraser, "Experimental Study of the Porosity and Permeability of Clastic Sediments," *Journal of Geology*, 1935, vol. 43, pp. 910-1010.
23. J. V. Milewski, "Packing Concepts in the Use of Filler and Reinforcement Combinations," *Handbook of Reinforcements for Plastics*, J. V. Milewski and H. S. Katz (eds.), Van Nostrand Reinhold, New York, NY, 1987, pp. 14-33.
24. Y. V. Maksimov, "Packing Density of Highly Dispersed Particles on the Surface of Coarsely Dispersed Inclusions," *Colloid Journal of the USSR*, 1986, vol. 48, pp. 693-694.
25. J. Eastwood, E. J. P. Matzen, M. J. Young and N. Epstein, "Random Loose Porosity of Packed Beds," *British Chemical Engineering*, 1969, vol. 14, pp. 1542-1545.
26. A. B. Yu and N. Standish, "Porosity Calculations of Multi-Component Mixtures of Spherical Particles," *Powder Technology*, 1987, vol. 52, pp. 233-241.
27. W. A. Gray, *The Packing of Solid Particles*, Chapman and Hall, London, UK, 1968.
28. R. F. Fedors and R. F. Landel, "An Empirical Method of Estimating the Void Fraction in Mixtures of Uniform Particles of Different Size," *Powder Technology*, 1979, vol. 23, pp. 225-231.
29. R. F. Fedors and R. F. Landel, "Effect of Surface Adsorption and Agglomeration on the Packing of Particles," *Powder Technology*, 1979, vol. 23, pp. 219-223.
30. M. Dixmier, "Une Nouvelle Description des Empilements Aleatoires et des Fluides Denses," *Journal de Physique*, 1978, vol. 39, pp. 873-895.
31. M. Suzuki and T. Oshima, "Estimation of the Coordination Number in a Two Component Mixture of Cohesive Spheres," *Powder Technology*, 1983, vol. 36, pp. 181-188.

32. S. V. Nair, B. C. Hendrix and J. K. Tien, "Obtaining the Radial Distribution of Random Dense Packed Hard Spheres," *Acta Metallurgica*, 1986, vol. 34, pp. 1599-1605.
33. N. Ouchlyama and T. Tanaka, "Porosity Estimation for Random Packing of Spherical Particles," *Industrial and Engineering Chemistry Fundamentals*, 1984, vol. 23, pp. 490-493.
34. R. Ben Aim and P. Le Goff, "La Coordinance des Empilements Desordonnes de Spheres. Application aux Melanges Binaires de Spheres," *Powder Technology*, 1968, vol. 2, pp. 1-12.
35. P. Le Goff, D. Leclerc and J. Dodds, "The Structure of Packed Beds: Continuity of Research in Nancy and Some New Results," *Powder Technology*, 1985, vol. 42, pp. 47-53.
36. M. Suzuki and T. Oshima, "Estimation of the Coordination Number in a Multi-Component Mixture of Spheres," *Powder Technology*, 1983, vol. 35, pp. 159-166.
37. J. Rodriguez, C. H. Allibert and J. M. Chaix, "A Computer Method for Random Packing of Spheres of Unequal Size," *Powder Technology*, 1986, vol. 47, pp. 25-33.
38. H. E. Rose and D. J. Robinson, "The Density of Packing of Two Component Powder Mixtures," *Powder Metallurgy*, 1965, vol. 8, pp. 20-37.
39. H. Deresiewicz, "Mechanics of Granular Matter," *Advances in Applied Mechanics*, vol. 5, H. L. Dryden and T. von Karman (eds.), Academic Press, New York, NY, 1958, pp. 233-306.
40. G. L. Messing and G. Y. Onoda, "Inhomogeneity-Packing Density Relations in Binary Powders," *Journal of the American Ceramic Society*, 1978, vol. 61, pp. 1-5.
41. G. L. Messing and G. Y. Onoda, "Inhomogeneity-Packing Density Relations in Binary Powders - Experimental Studies," *Journal of the American Ceramic Society*, 1978, vol. 61, pp. 363-366.
42. D. W. Fuerstenau and J. Fouladi, "Degree of Mixedness and Bulk Density of Packed Particles," *Ceramic Bulletin*, 1967, vol. 46, pp. 821-823.
43. G. Y. Onoda and G. L. Messing, "Packing and Sintering Relations for Binary Powders," *Processing of Crystalline Ceramics*, H. Palmour, R. F. Davis and T. M. Hare (eds.), Plenum Press, New York, NY, 1978, pp. 99-112.
44. J. T. Cartensen, F. Puisieux, A. Mehta and M. A. Zoglio, "Apparent Density Versus Composition Relations in Cascaded Binary Powder Beds," *Powder Technology*, 1978, vol. 20, pp. 249-255.

45. R. K. Bordia, "A Theoretical Analysis of Random Packing Densities of Mono-Sized Spheres in Two and Three Dimensions," *Scripta Metallurgica*, 1984, vol. 18, pp. 725-730.

Chapter Seven

Bimodal Mixtures Involving Nonspherical Particles

Chapter Outline

Introduction

The packing of mixtures involving nonspherical particles builds on the presentation in Chapter Six for binary mixtures of spheres. Analogous to the behavior of spheres, there is an increase in density associated with mixing different particle sizes of similar shapes. However, a major difference between spheres and nonspherical particle shapes is that the initial monosize packing is generally higher for the spheres. The greater the surface roughness, shape irregularity, or particle aspect ratio, then the lower the inherent packing density. Thus, although the relative gain in density is similar for spheres and nonspherical particles, the starting density for nonspherical particles is lower. Accordingly, at all compositions the nonspherical mixture will be lower in density. The relative gain in density for a binary mixture (of either spheres or nonspheres) increases as the difference in particles sizes between the two powders increases. A 20% difference in size is needed to attain a measurable benefit from such mixtures. In most cases, the composition with the highest packing density consists of at least 50% of the larger particles. Like the behavior for spheres, the relative gain in packing density is dependent on composition and particle size ratio [1]. However, with a mixture of spherical and nonspherical particles, the behavior has a sensitivity to shape that influences the size ratio for optimal packing. This behavior is especially important for the fabrication of composite materials. Hence, the first case considered in this chapter is the packing of spheres and fibers.

Mixtures of Spheres and Fibers

Mixtures of fibers and spherical particles provide the basis for fabricating many high performance composite materials, especially those with ceramic or metallic matrices. In those cases involving aligned fibers, the packing behavior can be treated as a two-dimensional problem [2]. It is beneficial to the consolidation process to start with a high packing density. The maximum packing for sphere-fiber mixtures requires attention to the particle size, fiber length, fiber diameter, and amount of fiber. Of particular importance is the fiber aspect ratio, defined as the length divided by the diameter. The peak in the packing density generally occurs at small concentrations of fiber, near 25 volume percent for fibers with aspect ratios over 10. Generally, the packing of high concentrations of long fibers requires a particle size much smaller than the fiber diameter. Furthermore, the packing density is increased as the fiber aspect ratio is reduced; this is often contrary to the requirements for optimal strength. If we let R be the ratio of sphere diameter D to fiber diameter D_f

$$R = D/D_f \tag{7.1}$$

then two combinations exist for high packing densities. The first case corresponds to spheres filling the voids between large fibers for a size ratio R less than unity. The second case is for the fibers filling the voids between large spheres for R greater than unity. The packing density for a sphere-fiber combination increases with very large or very small R values [3]. The cases of R near unity give the lowest packing densities. Figure 7.1 shows an example of the packing density variation with composition and size ratio R for a constant fiber length-to-diameter ratio of ten ($L/D_f = 10$) [3,4]. As the size ratio R increases, the packing density likewise increases at all compositions. There is a composition of maximum packing at high volume fractions of spheres, typically near 75% spheres.

In contrast, Figure 7.2 plots the fractional packing density for a fixed size ratio R of 10 and a varying fiber length-to-diameter ratio [3]. As noted in Chapter Five, the inherent packing density for the fibers (without added particles) decreases as the aspect ratio increases. Long fibers in a random orientation give a low packing density. For fibers with a high length-to-diameter ratio, all concentrations of fiber degrade the packing of the spheres. Alternatively, for relatively short fibers, there are some possible high packing compositions. Again the high packing density compositions occur at low

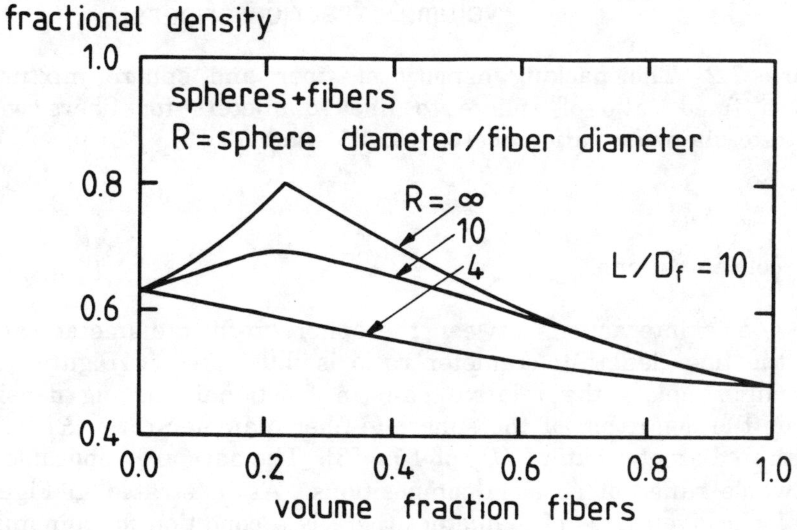

Figure 7.1 The packing density variation with composition for fiber-sphere mixtures with a fiber length-to-diameter ratio of 10 and various ratios R of sphere diameter to fiber diameter [4].

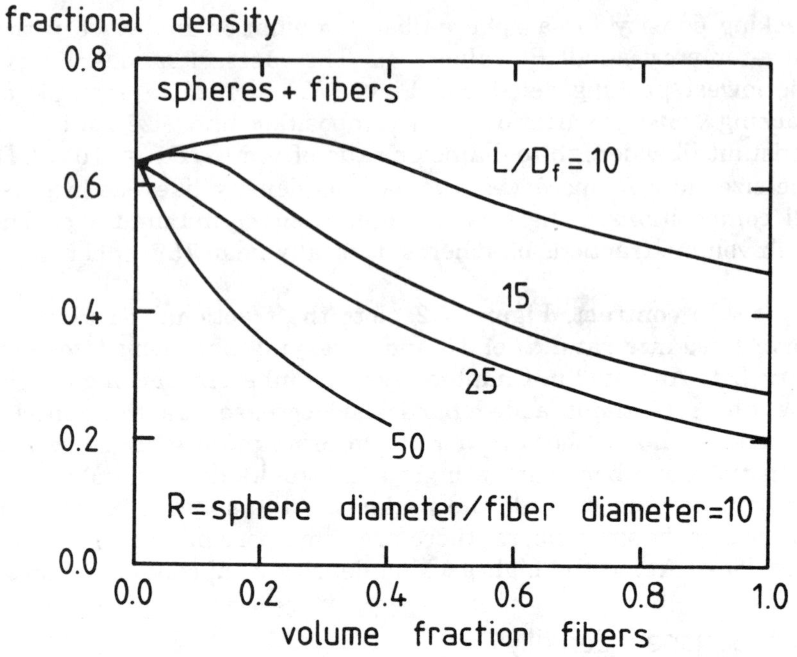

Figure 7.2 The packing density of fiber and sphere mixtures with a fixed ratio of sphere to fiber diameters for fibers with length-to-diameter ratios of 10, 15, 25, and 50 [3].

fiber concentrations.

The interaction between the sphere-to-fiber diameter ratio, and the fiber length-to-diameter ratio is illustrated in Figure 7.3. This figure plots the relative gain in fractional packing density versus the logarithm of the sphere-to-fiber diameter ratio (R) for a length to diameter ratio L/D_f of 15.5 [3]. The data are applicable to the whole range of possible compositions. As illustrated in Figure 7.3, for a given ratio of diameters there is a condition for minimum packing density. The combination of R and L/D_f for minimum packing is plotted in Figure 7.4. These results show that for long and thin fibers, the lowest packing density occurs with large particles. Alternatively, for short and squat fibers a low packing density is associated with particles nearly the same diameter as the fiber.

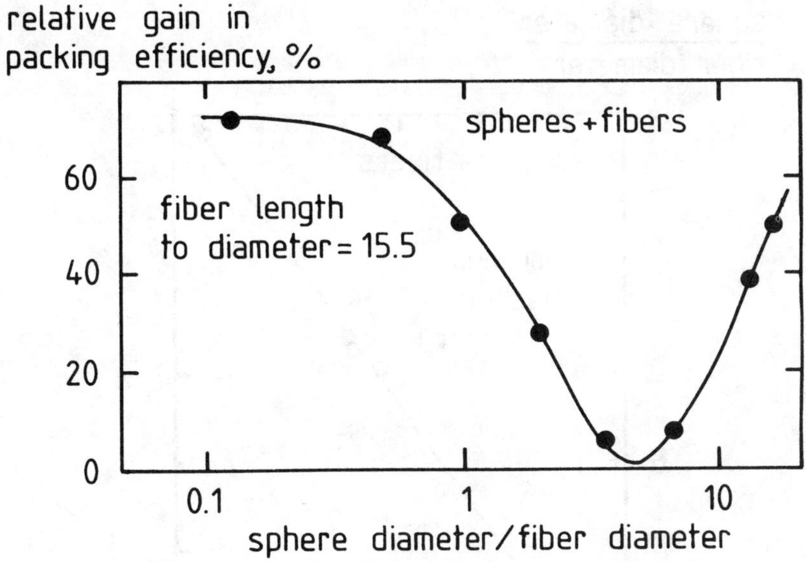

Figure 7.3 An example of the interaction between the fiber and particle characteristics with respect to the packing density for a range of sphere-to-fiber diameter ratios [3,4].

There is a relation between the fiber loading, the diameter ratio, and the fiber length-to-diameter ratio. Figure 7.5 shows the key interaction, where the choice of small particles improves the packing density with high fiber loadings and long fibers. Alternatively, large spheres are favored with short fibers, especially at low fiber concentrations. An approximate formula for the optimal fiber content X_f^* based on the fractional packing density for the spheres f_s and fractional packing density for fibers f_f is given as follows [5]:

$$X_f^* = f_f(2 - f_s)/[2 f_f + f_s - f_s f_f] . \qquad (7.2)$$

This model generally predicts optimal packing at relatively low fiber contents. For example, with a fiber length-to-diameter ratio of 25, the fiber fractional packing density f_f is approximately 0.2. Monosized spheres will pack at a fractional density f_s of

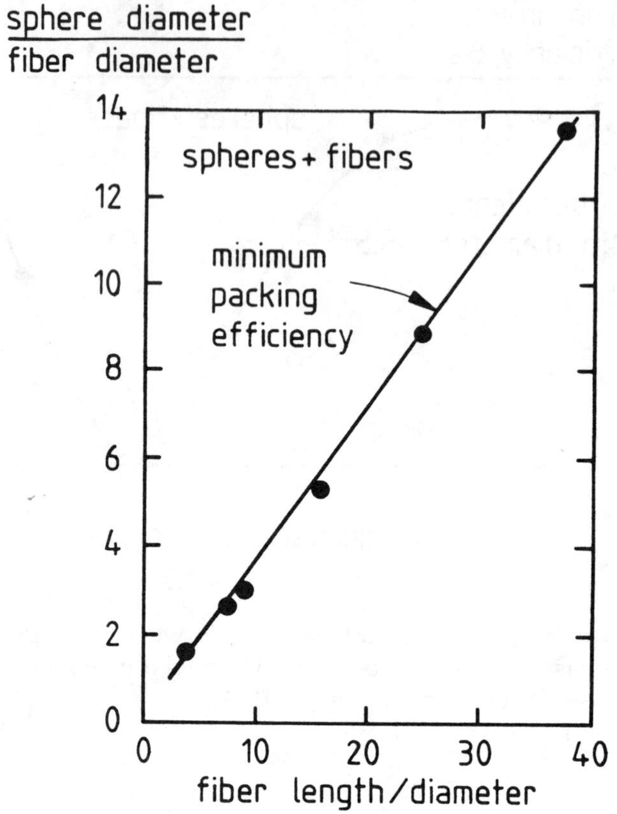

Figure 7.4 The interrelation between the sphere-to-fiber diameter ratio and the fiber aspect ratio for lowest packing densities; high densities are associated with combinations away from this line [3].

approximately 0.64. Accordingly, the optimal fiber concentration X_f^* from Equation 7.2 is 0.38, with the spheres being smaller in diameter than the fibers.

Table 7.1 gives a further illustration of these factors. Here the fractional packing density is shown as a function of the diameter ratio R, length-to-diameter ratio L/D_f, and fiber content [3]. Another form of this information is given in Table 7.2 for mixtures of 25% fiber and 75% spheres. For improved strength in a

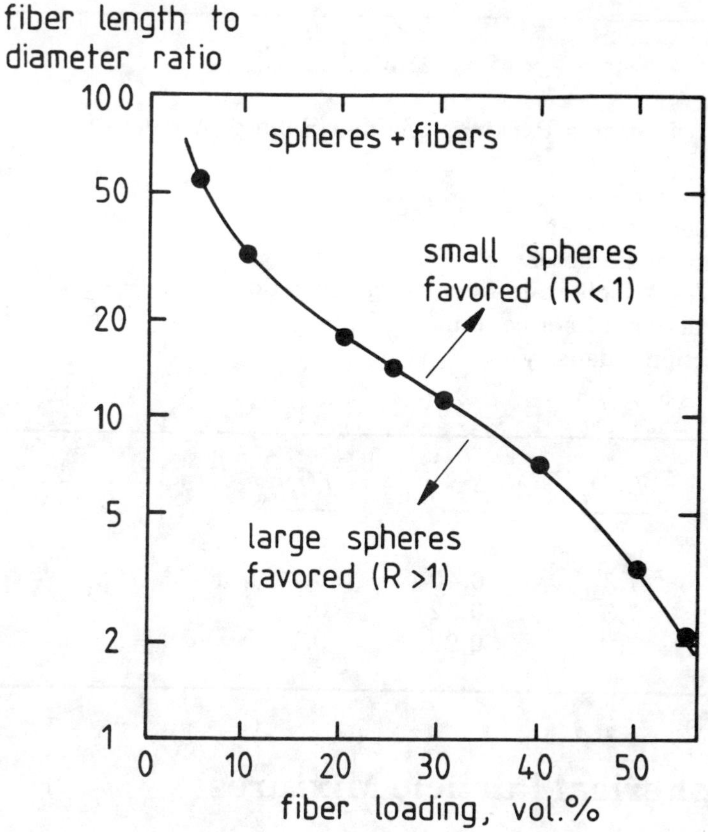

Figure 7.5 The effect of the fiber content and fiber length-to-diameter ratio on the choice of either relatively large or small spheres for optimal packing [3].

composite, a long fiber is generally beneficial; as a rule of thumb, a fiber length-to-diameter ratio in the range from 80 to 200 is desired for significant strengthening [6]. The best packing for long fibers occurs with relatively small particles.

Table 7.1

Example Packings of Fiber-Sphere Mixtures [3]

L = fiber length
D_f = fiber diameter
D = sphere diameter
$R = D/D_f$ = sphere-to-fiber diameter ratio
X_f = fractional fiber content
f = fractional density

R	L/D_f	X_f	f
0.11	7.3	0.40	0.73
7.0	7.3	0.22	0.65
17.4	7.3	0.26	0.75

Nonspherical Particle Mixtures

Irregular, small particles can be mixed with spheres at low concentrations without significantly degrading the packing density. Mixtures of spheres with low concentrations (less than 10%) of smaller irregular particles do not exhibit significant packing density degradation [7], though at high concentrations of irregular particles the packing structure is degraded by the irregular particles. This is especially true if the irregular particles approach the size of the spheres. Consequently, the density decreases in proportion with the concentration of irregular particles when the particles are of similar sizes. Figure 7.6 illustrates this behavior by plotting the porosity versus composition for mixtures of angular and round particles [8]. In this instance the fractional density is a linear function of the composition,

Table 7.2

Fractional Solids Contents for Fiber-Sphere Mixtures at a Composition of 25% Fibers

L = fiber length
D_f = fiber diameter
D = sphere diameter
$R = D/D_f$ = sphere-to-fiber diameter ratio

L/D_f	R values 0.11	0.94	3.7	7.0	17.4
3.9	0.685	0.617	0.645	0.700	0.764
15.5	0.667	0.599	0.503	0.505	0.575
37.1	0.480	0.420	0.377	0.338	0.392

$$f = A - B X_a \qquad (7.3)$$

where f is the fractional density of the mixture, X_a is the fractional amount of angular particles in the mixture, and A and B are system dependent constants. For the mixture of 10 mm angular and smooth particles in Figure 7.6, the constant A is 0.663 and B is 0.067.

The packing of monosized nonspherical particles typically is inferior to that for spheres. Accordingly, there is interest in increasing the packing density of nonspherical particles by mixing differing particle sizes. Unfortunately, volume expansion is sometimes observed in mixtures of irregular particles [9,10]. Thus, a significant problem arises in determining the particle shape effect on packing. In general, the packing density increases with large differences in particle sizes. At small particle size ratios there is no benefit from mixtures of the two component powders [1,11,12]. Under

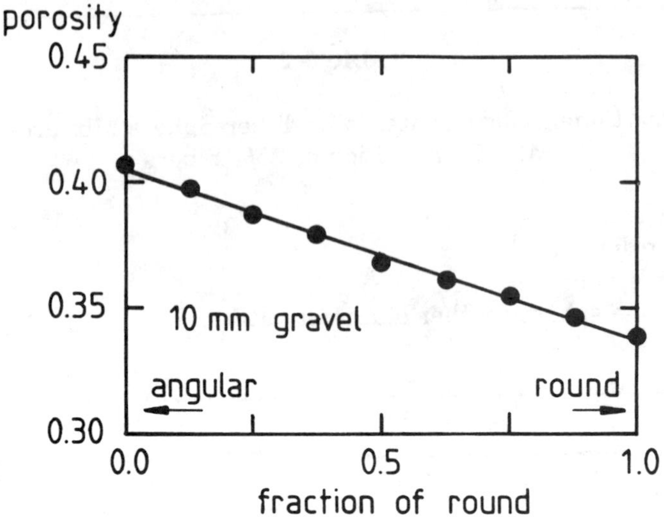

Figure 7.6 Mixture porosity shown as a function of composition for angular and round 10 mm gravel particles [8].

optimal conditions, the improvement in packing density Δf depends on the size ratio as follows [12-14]:

$$\Delta f = A - B\,exp(-C\,D_L/D_S) \qquad (7.4)$$

where A, B, and C are constants determined by the particle shapes and inherent packing characteristics. The particle size ratio D_L/D_S is the size of the large particle divided by the size of the small particle. Figure 7.7 plots the observed fractional density versus particle size ratio for irregular feldspar and quartz particles [12]. This plot is at a composition of 77% large particles, and it shows that packing density improves as the particle size ratio becomes larger. However, increasing the particle size ratio over approximately 8 does not further improve the packing density. Optimal packing conditions occur over a narrow composition range near 25 to 30% of the small particles. Other models for the bimodal nonspherical case include a particle size effect in the A and B constants [9].

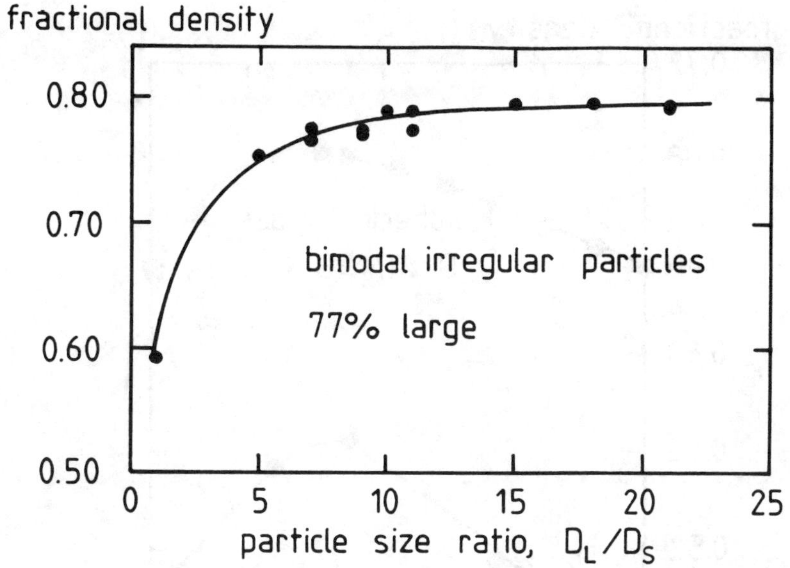

Figure 7.7 The maximum packing density (at the 77% large composition) versus the particle size ratio for bimodal mixtures of irregular mineral particles [12].

With respect to the composition influence, the behavior is similar to that noted in Chapter Six for bimodal sphere mixtures. Figure 7.8 shows the packing behavior of mixed crushed glass particles versus the composition [15]. The large particles were 3.4 times the size of the small particles. For contrast, the packing of mixed spheres is also shown in this figure. The particle size ratio was similar for both particle mixtures. These results show better packing of spheres in comparison with irregular particles. The peak in packing density occurs at roughly the same composition of large and small particles, irrespective of the particle shape. However, the relative change in packing density is larger with the irregular particles, although the spheres have a higher packing density at all compositions.

In other examinations of nonspherical particle packing, the optimal packing condition has been as high as 50% small particles. This is because the optimal packing composition depends on the particle size ratio as illustrated in Figure 7.9 [11]. This figure shows

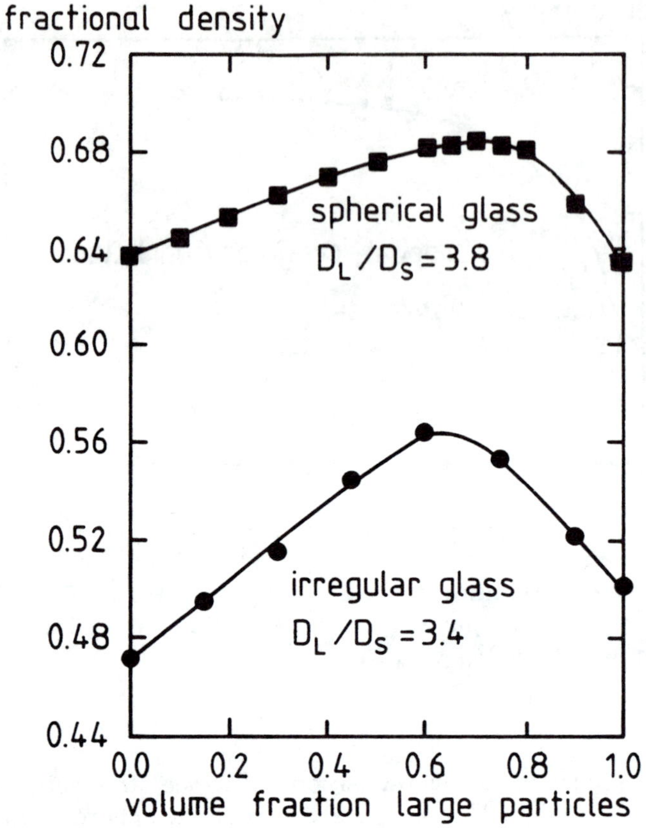

Figure 7.8 The dependence of the fractional packing density on composition for mixtures of irregular (crushed) and spherical glass with particle size ratios of 3.4 and 3.8 [15].

the optimal composition in terms of the amount of small particles versus the particle size ratio for angular limestone particles. The initial, unmixed packing density was 0.46, but this improved to a density of 0.63 in a mixture of approximately one-third small particles at a size ratio of 20. The optimal packing composition shifts toward a lower content of large particles as the particle size ratio increases. Most likely, there is also a mixture homogeneity effect on the optimal packing composition analogous to that presented in Chapter Six for bimodal spheres.

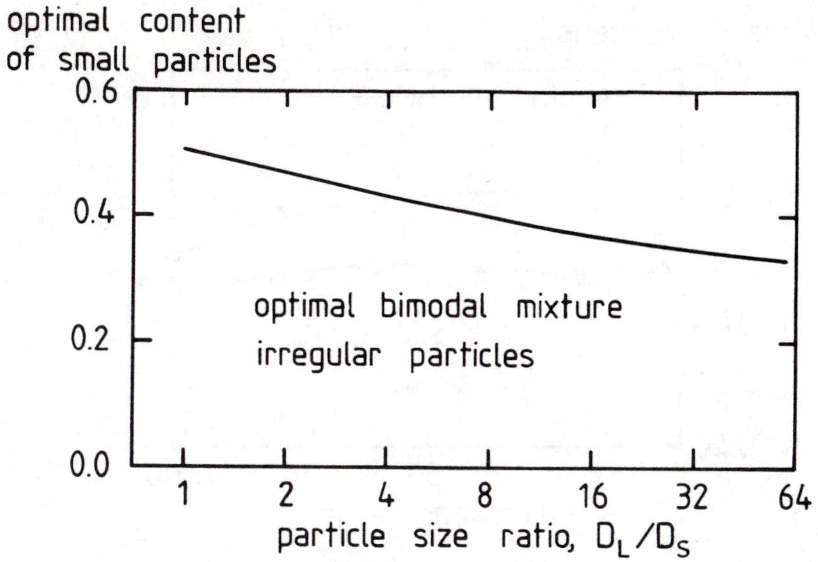

Figure 7.9 The variation in composition for maximum packing density as a function of the particle size ratio for irregular particles with an inherent packing density of 0.5 [11].

Measurements on coke particles further demonstrate the packing behavior of irregular particles. Depending on the particle shape and size ratio, the peak in packing density can occur over a broad range of compositions. This range is broader than observed with spherical particles [16]. Such behavior is illustrated in Figure 7.10 for two particle size ratios. In contrast with Figure 7.8, the relative packing density improvement with a mixture of particle sizes is less. The most likely cause for the difference between Figures 7.8 and 7.10 is the particle size ratio. This behavior illustrates a basic problem in dealing with irregular particles. Each case must be treated on an individual basis.

Because of the wide range of possible shape combinations, it is difficult to predict the specific behavior for mixtures involving nonspherical particles. Typically, the principles associated with bimodal sphere mixtures are observed, but the combination of two different shapes is untreated by theory. The few experimental observations fail to demonstrate any universal character.

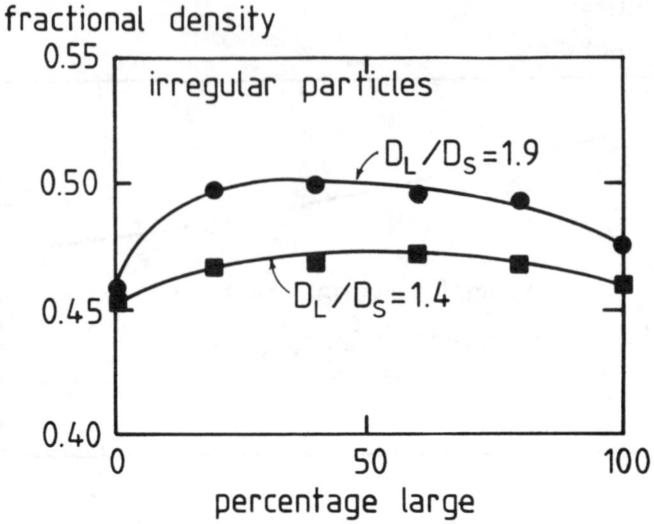

Figure 7.10 Fractional packing densities of mixtures of irregular coke particles with particle size ratios of 1.4 and 1.9 [16].

Summary

The packing behavior of nonspherical particles is an empirical area. The behavior patterns are similar to those found in the packing of spheres. With particles of greatly differing shapes and similar sizes, the density will decrease when the particles are mixed together. But in most cases involving similar particle shapes and differing particles sizes, improved packing occurs. The optimal composition for a high packing density has a high proportion of larger particles. The relative gain in density is proportional to the composition, like that observed for spherical particles. However, a major difference with nonspherical particle shapes is that the initial monosize packing is generally inferior to that of spheres. Thus, although the relative gain in density is similar for spherical and nonspherical particles, the starting density for nonspherical particles is lower.

The packing of powder mixtures involving nonspherical particles has three general aspects. The first aspect is the packing of spheres with nonspherical shapes, which is largely focused on mixtures using cylindrical fibers. The second aspect of packing nonspherical particles focuses on the packing of geometrically similar shapes of differing sizes, while the third aspect deals with dissimilar shapes and sizes.

In the case of fiber reinforced composites using spheres and cylinders, the fibers strengthen the consolidated structure. Here relatively long fibers are needed for optimal properties in the final composite. Such composites typically attain a high packing density with relatively small particles. There is an interaction between the particle diameter, fiber diameter, fiber content, and fiber length in determining the packing density. This chapter reviews the first considerations in treating these interactions.

For the packing of nonspherical particles of similar shapes but differing sizes, the behavior is analogous to that of mixed spheres. The greatest improvement in packing density occurs with large differences in particle size. As the particle size ratio increases, the density is more sensitive to the mixture composition. Generally, optimal packing occurs at compositions containing a majority of the large particles. The optimal composition shifts with the particle size ratio. Likewise, inhomogeneities in the bimodal mixture will decrease the packing density. The mixing of nonspherical particles of differing sizes and differing shapes is poorly understood.

References

1. K. Ridgway and K. J. Tarbuck, "Particulate Mixture Bulk Densities," *Chemical and Process Engineering*, 1968, vol. 49, no. 2, pp. 103-105.
2. A. N. Nikolenko and M. S. Kovalchenko, "Analysis of the Random Packing of Identical Particles. V. Morphological Properties of Transversely Isotropic Fiber Composites," *Soviet Powder Metallurgy and Metal Ceramics*, 1987, vol. 26, pp. 106-111.
3. J. V. Milewski, "Packing Concepts in the Utilization of Filler and Reinforcement Combinations," *Handbook of Fillers and Reinforcements for Plastics*, H. S. Katz and J. V. Milewski (eds.),

Van Nostrand Reinhold, New York, NY, 1978, pp. 66-78.
4. J. V. Milewski, "The Combined Packing of Rods and Spheres in Reinforcing Plastics," *Industrial Engineering Chemistry Product Research and Development*, 1978, vol. 17, pp. 363-366.
5. T. L. Starr, "Packing Density of Fiber/Powder Blends," *Ceramic Bulletin*, 1986, vol. 65, pp. 1293-1296.
6. K. K. Chawla, *Composite Materials Science and Engineering*, Springer-Verlag, New York, NY, 1987.
7. J. R. Parrish, "Packing of Spheres," *Nature*, 1961, vol. 190, p. 800.
8. F. A. Shergold, "The Percentage Voids in Compacted Gravel as a Measure of Its Angularity," *Magazine of Concrete Research*, 1953, vol. 13, pp. 3-10.
9. D. J. Cumberland and R. J. Crawford, *The Packing of Particles*, Elsevier Science, Amsterdam, The Netherlands, 1987.
10. K. Karlsson and L. Spring, "Packing of Irregular Particles," *Journal of Materials Science*, 1970, vol. 5, pp. 340-344.
11. P. M. Worthington, "An Experiment on the Theory of Voids in Granular Materials," *Magazine of Concrete Research*, 1953, vol. 4, no. 12, pp. 121-126.
12. A. N. Patankar and G. Mandal, "The Packing of Some Non-Spherical Solid Particles," *Transactions Journal of the British Ceramic Society*, 1980, vol. 79, pp. 59-66.
13. J. E. Ayer and F. E. Soppet, "Vibratory Compaction: II, Compaction of Angular Shapes," *Journal of the American Ceramic Society*, 1966, vol. 46, pp. 207-210.
14. W. A. Gray, *The Packing of Solid Particles*, Chapman and Hall, London, UK, 1968.
15. S. Yerazunis, J. W. Bartlett and A. H. Nissan, "Packing of Binary Mixtures of Spheres and Irregular Particles," *Nature*, 1962, vol. 195, pp. 33-35.
16. N. Standish and D. E. Borger, "The Porosity of Particulate Mixtures, *Powder Technology*, 1979, vol. 22, pp. 121-125.

Chapter Eight

Multiple Mode and Continuous Particle Distributions

Chapter Outline

Introduction

The ideas developed in the previous two chapters for bimodal spherical and nonspherical particle packings are extended in this chapter to multimodal systems. The first step is to consider trimodal mixtures, in which there are three particle sizes. Subsequently, the trimodal systems form the basis for considering multimodal mixtures (mixtures of several different monosized powders). Extension of the multimodal concepts provides a basis for analyzing continuous particle size distributions. The continuous particle size distribution is of great interest since this is a practical approach to creating high packing density systems directly in powder fabrication. As with bimodal packing, mixture heterogeneity will have an influence on packing density for multiple mode or continuous distributions. For the present treatment a homogeneous mixture is assumed.

Before detailing the packing characteristics of the various trimodal and multimodal packings, a few comments about the practical aspects of multimodal systems are in order. At least a 7:1 particle size ratio is needed for optimal packing. For a trimodal packing, this corresponds to a size ratio of 49:7:1. Assuming the smallest particles are 1 μm in diameter, then the largest particles will be 49 μm in diameter. Such a size is within a realistic working range for many particulate material systems. However, the classification of particles into sizes of 1, 7, and 49 μm is difficult, though not impossible. If the smallest particles are increased to 5 μm, then the largest particle size is 245 μm. The separation of particles into discrete size classes is aided by these larger particle sizes, but the largest particles may not suit the application. It is this rapid compounding of particle size that poses a major processing problem with trimodal mixtures. Likewise, in analyzing the packing options for submicron particles, it is difficult to attain monosized particles with large relative size differences. Obviously, as the size ratio becomes larger, the practical problems increase. Accordingly, the problems in controlling size segregation and powder classification become large with systems made from four or five components.

Whereas the maximum fractional packing density of monosized spheres is 0.64, that of bimodal mixtures is probably as large as 0.86. In the best case, this corresponds to a 34% improvement in packing. For a trimodal mixture, the maximum packing density is approximately 0.95. This is slightly more than a 10% improvement in packing over the bimodal situation. The addition of more components will yield a decreasing benefit in packing, though the practical problems in forming such mixtures may be insurmountable. Hence, a clear case can be made for bimodal mixtures because of the large packing gain and relative ease of finding adequately different sizes; whereas mixtures formed from more components may not be fruitful.

In spite of this shortfall in packing benefit, the study of trimodal and other multimodal packing systems is useful. The generalizations that arise from considering the multimodal systems provide a basis for understanding the character of continuous particle size distributions. It is in the interest of generalizing results and understanding continuous distributions that the current chapter is organized.

Trimodal Mixtures

Studies of ternary mixtures generally focus on spherical particles, although other particle shapes give similar results. A trimodal mixture of monosized spheres is very sensitive to particle size ratio, with wide size differences being most advantageous. In essence, an examination of a trimodal mixture begins by considering the three binary mixtures possible from the combinations of the constituents. For this discussion, the ternary sizes will be termed large, medium, and small. Figure 8.1 illustrates the combination of three powders to form a trimodal mixture. The height at any point corresponds to the fractional packing density. Ideally, there is a combination of the three component powders that results in a high packing situation. Figure 8.2 is a plot of the same type of behavior. This plot is analogous to a ternary phase diagram, where isodensity contours are projected onto the equilateral triangle. The

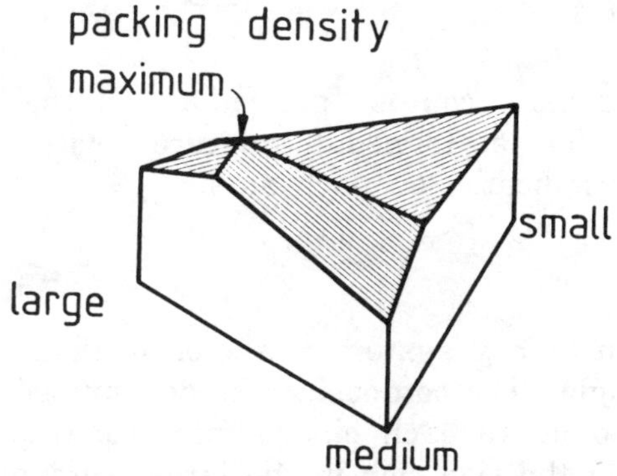

Figure 8.1 A sketch of an idealized trimodal powder mixing, where the height indicates the packing density, with an optimal intermediate composition.

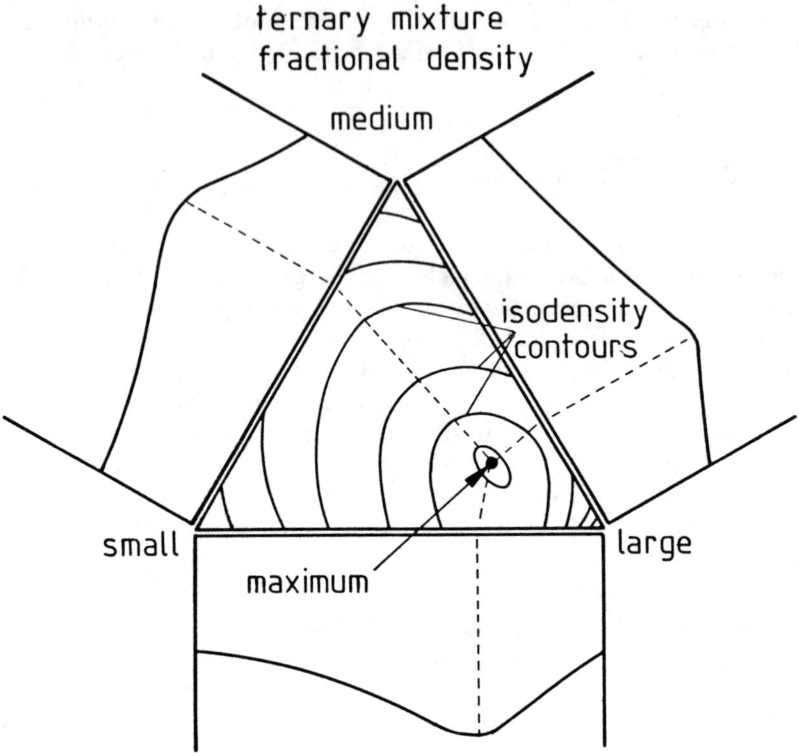

Figure 8.2 Isodensity contour projections on the ternary composition diagram for a trimodal mixture, with the component binary packings indicated on each axis.

behavior of each binary system is sketched along the sides of the equilateral triangle. The composition is determined by the distance from each endpoint. In each binary mixture, the highest packing density occurs on the side rich in the larger component. As expected, a high packing density occurs in the large-small binary. Depending on the ratio of sizes, an intermediate ternary mixture may have a higher packing density than the large-small binary. The composition of maximum packing density depends on the size ratio of the three component powders.

A key characteristic of all multimodal powder mixtures is that a large difference in particle sizes aids packing. Furnas [1,2] calculated whether optimal packing would occur in the binary, ternary, or higher order system based on the particle size ratio. His model results are shown in Figure 8.3. This plot indicates the number of components corresponding to optimal packing for a given particle size ratio and inherent packing fraction. With a small particle size ratio, optimal packing would occur with two components. Any particles which differ sufficiently in size will exhibit improved packing in the bimodal mixture of the largest and smallest particles. For a trimodal mixture, the intermediate particle size would fail to improve packing beyond that available with the large-small combination. Alternatively, with more substantial particle size differences, such as 1000:1, Figure 8.3 indicates that a ternary mixture would yield better packing.

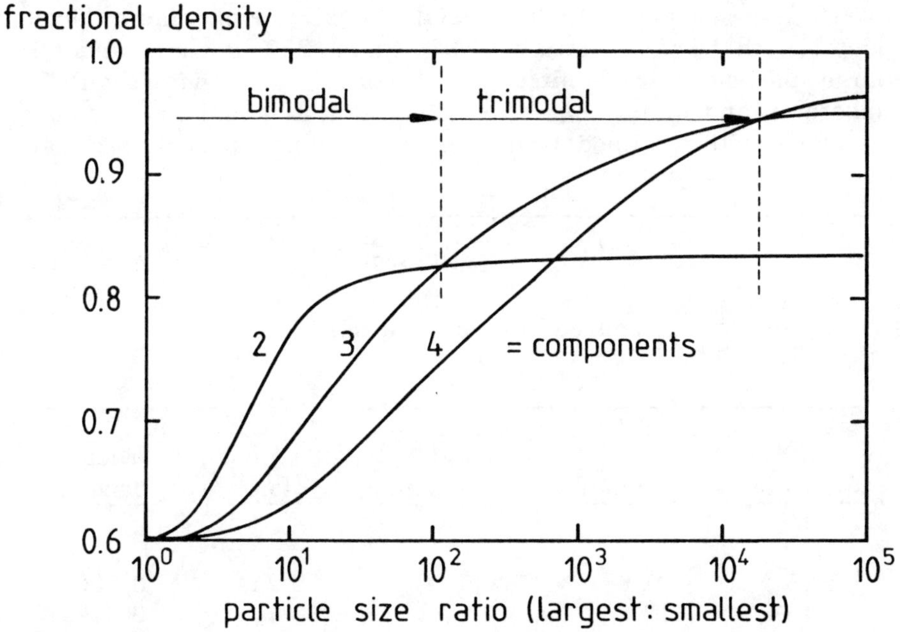

Figure 8.3 The effect of particle size ratio on the optimal packing density. Shown are the number of components for optimization assuming an inherent packing density of 0.6 [2].

It is also desirable to know the composition corresponding to the maximum packing density. As illustrated in Table 8.1, various optimal trimodal packings have been reported in the past [3-6]. It is worthwhile to examine the number of particles corresponding to these packings. For the size ratio of 1:7:77, the volume percentages are 10, 23 and 67, respectively. The approximate number of particles corresponds to 69000, 490, and 1 for the small, medium, and large size classes. This means that a few large particles go a long way toward filling space. The calculated packing density dependence on composition for trimodal sphere mixtures is illustrated in Figure 8.4 [4]. This plot corresponds to a size ratio of 25:5:1 giving optimal packing at a composition of approximately 69% large, 9% medium, and 22% small, with a maximum density of 0.85. In the limiting case of an infinite particle size ratio, Lee [4] projects the maximum packing density to be 0.926, with volume fractions of 8.7% small, 24.2% medium, and 67.1% large particles. In other analyses [2,7], a maximum density is projected from a mixture of 9% small, 25% medium and 66% large particles with infinite size ratios. For spheres with an inherent random packing density of 0.64, the maximum density available with a trimodal mixture at an infinite particle size ratio

Table 8.1

Some Optimized Trimodal Packings of Spheres [3-6]

size ratio	% small	% medium	% large	fractional density
1:5:25	21.6	9.2	69.2	0.850
1:7:49	13.2	20.7	66.1	0.878
1:7:49	11.0	14.0	75.0	0.950
1:7:77	10.0	23.0	67.0	0.900
1:10:100	11.2	22.5	66.3	0.892
1:100:10000	10.0	23.4	66.6	0.916

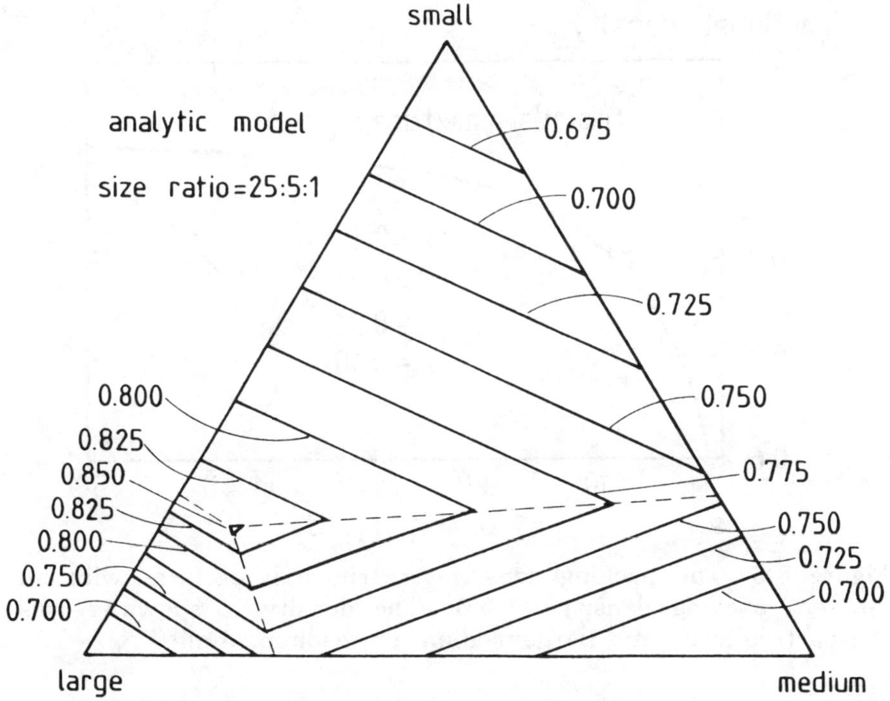

Figure 8.4 Calculated isodensity contours for a trimodal mixture of spheres with a size ratio of 25:5:1 [4]. The optimization point occurs at a composition rich in large particles.

is estimated to be 0.95 [8].

The level of density attainable and the optimal composition both depend on the homogeneity, initial packing density for the powders, and particle size ratio. A high inherent packing density powder requires fewer small particles to fill the interparticle void spaces. Figure 8.5 shows the fractional density versus the particle size ratio for a trimodal mixture of spheres with an initial packing density of 0.6 [2]. The scaling is in terms of the size ratio between the large and medium sizes, which is equivalent to the medium to small particle size ratio. The greatest improvement in packing density occurs with size increments between the large and medium or medium and small up to approximately 30; larger differences in

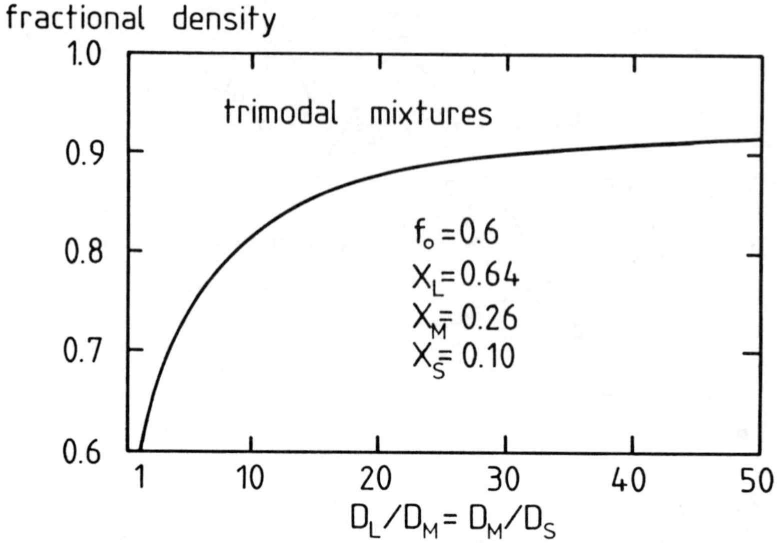

Figure 8.5 The packing density for trimodal mixtures with an inherent packing density of 0.6. The density is shown versus the particle size ratio (large:medium or medium:small) [2].

particle sizes do not greatly improve packing, while small particle size ratios have minimal benefit.

As noted in Figures 8.3 and 8.5, trimodal mixtures involving small particle size ratios are not productive. Indeed, binary mixtures of the largest and smallest particles often exceed the packing density of all possible ternary mixtures. Figure 8.6 illustrates this for mixtures composed of 7, 14, and 28 mm spheres (4:2:1 size ratio) [9]. This figure plots the isodensity contours versus composition for the ternary system. Optimal packing occurs in the large-small binary (28 and 7 mm sizes) at the compositions between approximately 60 and 70% large. All of the trimodal compositions fail to match or exceed the binary packing density, as predicted in Figure 8.3.

Alternatively, with sufficiently large particle size ratios the maximum packing density occurs at an intermediate composition involving all three powders, as predicted by Figure 8.3. Figure 8.7

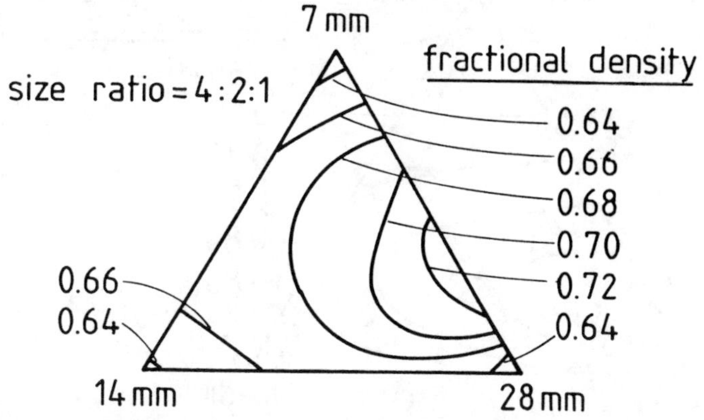

Figure 8.6 Fractional density contours on the ternary composition plot for mixed spheres with sizes in the ratio 4:2:1 [9]. Note that the highest packing density occurs in the large-small binary.

plots the results of a study involving particles with a size ratio of 51:8:1 [7]. In this instance, the maximum packing density occurs inside the ternary diagram. Various models have been suggested for predicting the packing density of trimodal mixtures of particles [6,10-14]. The models require information on the composition, particle sizes, and packing densities of the individual components. For example, at the optimal composition, the packing density is predicted to depend on the particle size ratios as follows [6]:

$$f = 0.951 - 0.098\, A_1 + 0.098\, A_2 - 0.198\, A_1 A_2 \qquad (8.1)$$

where

$$A_1 = \exp(-\,0.201\, D_L/D_M) \qquad (8.2)$$

and

$$A_2 = \exp(-\,0.201\, D_M\, /D_S)\,. \qquad (8.3)$$

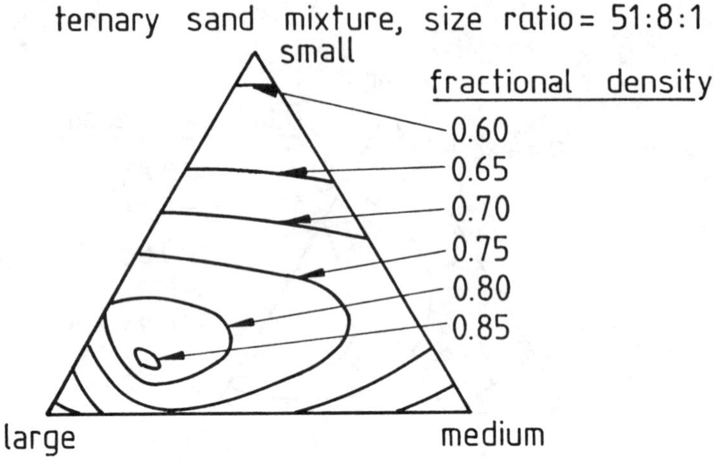

Figure 8.7 The isodensity contours for trimodal sand mixtures with a size ratio of 51:8:1 [7].

In these equations, D is the particle size and the subscripts denote large (L), medium (M), and small (S) particle sizes. Such models have merit for both spherical and nonspherical particle mixtures. One suggestion is that packings of particle mixtures is analogous to the thermodynamics of chemical mixtures, where each component has a partial specific volume [12,15]. Thus, the behavior of binary mixtures provides a basis for projecting the behavior of ternary mixtures.

The coordination number varies with the particle size in a trimodal mixture. The packing coordination has recently been examined using computer simulations. Figure 8.8 illustrates the behavior for a trimodal mixture with a size ratio of 3:2:1 [16]. The contours show the mean packing coordination number for each component versus the mixture composition for the large, medium, and small particles. The overall mean coordination number is approximately six at all compositions. Note the large particles generally have a high packing coordination while the small particles have a low packing coordination.

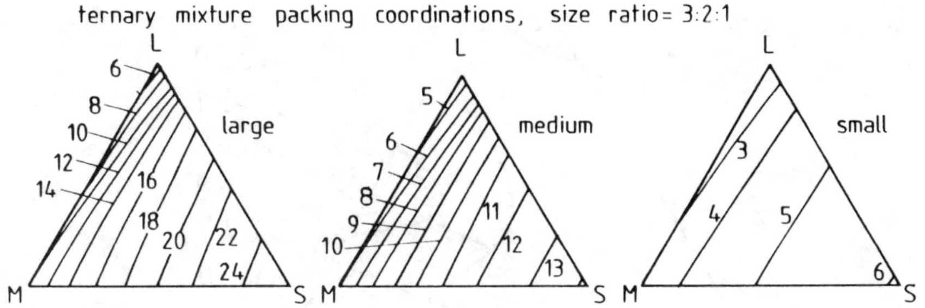

Figure 8.8 Packing coordination number contours shown on ternary composition plots for trimodal mixtures with a size ratio of 3:2:1 [16]. The mean coordination number is approximately six for all compositions.

Trimodal particle packings involving nonspherical particles show less of a benefit in density as well as a more subdued change in density with composition, indicating less sensitivity to composition [17-20]. Otherwise, there is similar behavior between the trimodal spherical and nonspherical cases. As an illustration of this point, Figure 8.9 plots the packing density contours for trimodal mixtures of irregular particles [17]. For these measurements the size ratio of particles was 5.1:3.7:2.7, which is very small (approximately 2:1.5:1). As expected, the minimum porosity occurs in the large-small binary system. Furthermore, the packing densities are low in comparison to spherical particles.

In addition to the effect of particle size, the friction between particles has an influence on the packing and coordination number of trimodal particle mixtures. For a high packing density, minimal interparticle friction is most desirable. Particles that exhibit cohesion tend to have lower packing densities and lower coordination numbers. A long time is needed to vibrate the particles into a configuration of maximum packing density because of the large number of small particles in an optimal trimodal mixture.

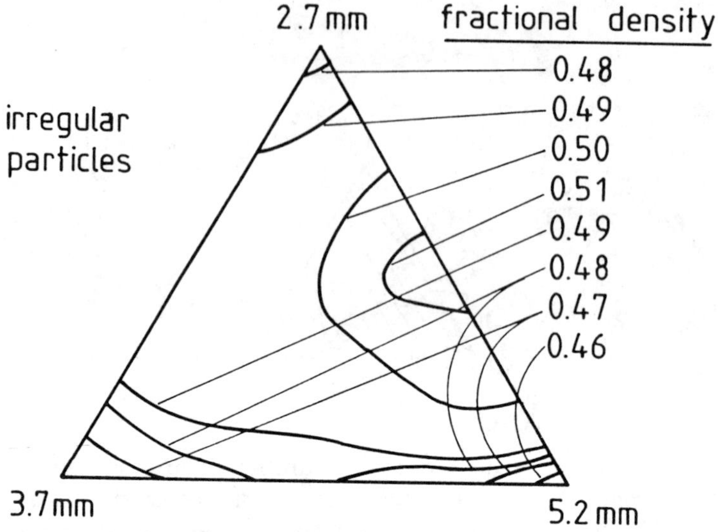

Figure 8.9 Isodensity contours for a trimodal mixture of irregular particles with a size ratio of 5.1:3.7:2.7 [17].

Multimodal Packings

The effect of mixed particle sizes on packing density is clearly demonstrated by bimodal and trimodal packings. Further consideration of mixed particle sizes leads to four, five, and more components. In this section general concepts dealing with multimodal packings are developed. The next section uses these ideas as a basis for study of continuous particle size distributions. As the number of components in a mixture increases, the experimentation necessary to find an optimal packing composition becomes much more difficult. Accordingly, computer models increase in usefulness as the complexity of the system increases.

Horsfield [21] performed one of the early treatments of multimodal packing. He demonstrated, for a mixture of four or five components, packing densities near 0.85. Table 8.2 presents the effects of the number of components on the packing density for an ordered packing with up to five components [21-23]. A six component mixture with a large percentage of small particles proposed by Horsfield has a limiting fractional density of 0.961. In experiments involving a 316:1 size range, McGeary [3] experimentally produced a packing density of 0.951. The maximum packing attainable from a random mixture of particle sizes is illustrated in Table 8.3 for size ratios of 7:1 at each step [5]. In this table, the weight percent of the largest particle size fraction is given first. Other treatments have given slightly different values for the optimal compositions, with correspondingly different packing fractions [2,3,7,21-24]. Obtaining particles with similar shapes over a large size range is difficult. The benefits from a high packing fraction are offset by the difficulties associated with fabrication, classification, and handling of large size differences.

Efforts to model the packing behavior have established the need for large particle size differences [6,23,24]. For a system consisting of n components with very large size differences, and fractional densities of f_1, f_2 ... f_n, the projected optimal fractional packing density of the mixture f_{mix} is given as follows [24]:

Table 8.2

Some Example Multimodal Ordered Packings [21-23]

number of components	size ratio	relative number	relative weight	fractional density
1	1.000	1	1.000	0.740
2	0.414	1	0.071	0.793
3	0.225	2	0.022	0.810
4	0.175	8	0.043	0.842
5	0.117	8	0.013	0.851

Table 8.3

Packing Densities for Spherical Particles in Size Ratios of 7:1 [5]

number of components	weight percents	packing fraction
1	100	0.64
2	84, 15	0.86
3	75, 14, 11	0.95
4	72, 14, 10, 3	0.98

$$f_{mix} = f_1 + (1 - f_1)f_2 + (1 - f_1)(1 - f_2)f_3 + \ldots + (1 - f_1)(1 - f_2)(1 - f_3) \ldots f_n \tag{8.4}$$

where the subscripts correspond to 1 = largest particle size, 2 = next largest, and so on to n = smallest. This assumes that the small particles just fill the voids between the large particles. The volumetric composition X_i of the mixture for the i component at maximum packing density is estimated as follows:

$$X_1 = f_1 / f_{mix}$$

$$X_2 = (1 - f_1)f_2 / f_{mix}$$

$$X_n = (1 - f_1)(1 - f_2) \ldots f_n / f_{mix} . \tag{8.5}$$

From such models, it is evident that the relative increase in density is small for n greater than three. Unfortunately, most of the models of an optimized multimodal packing are based on ordered structures, while little attention has been directed at the practical problems related to multimodal packing of random structures.

In these models, the structure takes its basic form from the packing of the largest particles. The remaining volume of voids is filled by the smaller particles. In turn, the remaining smaller voids are filled by subsequently smaller particles. This concept can be extended to include smaller particles with no volume dilation, as each component fills the voids left by the larger particles. On this basis, fractional densities of 0.99 have been estimated for a five component system [4,7,24]. The effect of a small particle size ratio is to decrease the maximum packing fraction for a given number of components. Expansion has been observed in some packings involving small particle size ratios [8]. As shown in Figure 8.3, the number of components needed to form an optimal packing depends on the particle size ratio. Figure 8.10 shows the relation between the optimal number of components, particle size ratio (large:small), and inherent fractional density. Note that small size ratios are not productive. Alternatively, a three component mixture increases density for particle size ratios of 100 or more.

Furnas [2] provides a formula for predicting the maximum packing density for a mixture composed of n sizes of spheres. The packing density is given by a model similar to Equation 8.4. Such an approach assumes that the particles are very different in size and that the size ratio is constant between each successive size level. For spheres with inherent fractional packing densities of 0.64, this formula predicts maximum fractional packing densities of 0.87, 0.95, 0.98, and 0.99 for systems with two, three, four, and five components (binary, ternary, quaternary, and quinary mixtures). As emphasized in Figure 8.3, the optimal number of components needed for maximum packing depends on the particle size ratio. In many experiments, the size ratio has been too small to truly optimize packing.

The main parameters for determining the packing density for a multimodal mixture are the particle size ratio, number of size classes, composition of the mixture, and inherent porosity of the component powders. Tables 8.3, 8.4, and 8.5 illustrate these factors. The final packing density and optimal composition are given in Table 8.4 for binary, ternary, and quaternary mixtures of spheres of infinite size ratios. In Table 8.5 the maximum densities are given for ternary, quaternary, and quinary mixtures of spheres with different size ratios [4]. Large size ratios are capable of yielding packing densities approaching 0.95 with a five component mixture.

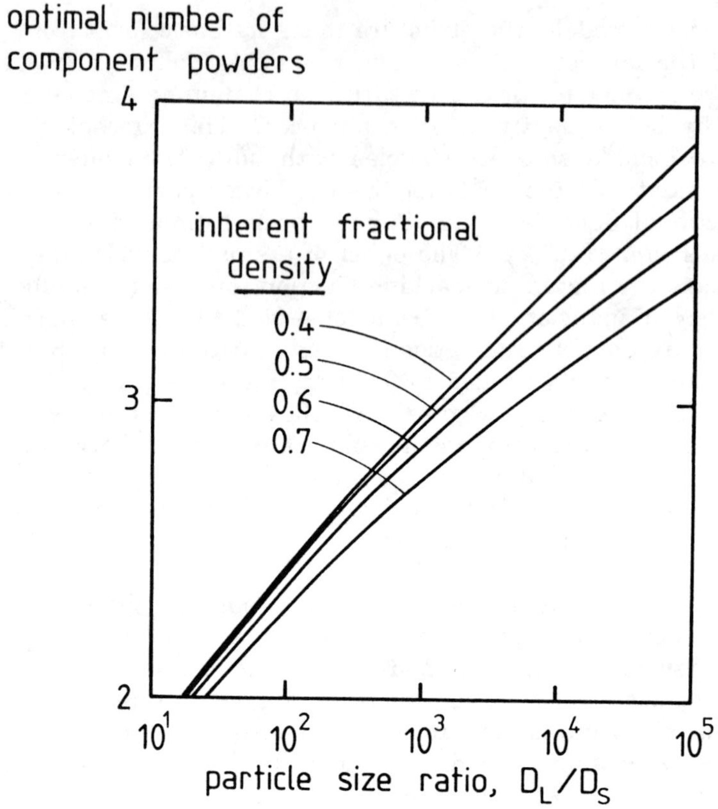

Figure 8.10 The optimal number of components for the maximum fractional density as a function of the particle size ratio and the inherent packing density [2].

As a practical example, a particle size ratio of 7:1 gives a packing density of 0.93. This mixture would be composed of spheres in the size classes having a ratio of 1:7:49:343:2401 with weight fractions of 4.3, 7.0, 9.7, 18.5 and 60.5%, respectively.

Continuous Particle Size Distributions

The packing density of a multiple mode mixture increases as the number of components increases, as long as the particles are

Table 8.4

The Characteristics of Multimodal Packings with Infinite Particle Size Ratios [2]

inherent density	number of components	volume fractions	mixture density
0.70	2	0.770, 0.230	0.905
0.70	3	0.720, 0.215, 0.065	0.966
0.70	4	0.707, 0.211, 0.063, 0.019	0.973
0.60	2	0.715, 0.285	0.835
0.60	3	0.642, 0.256, 0.102	0.949
0.60	4	0.617, 0.246, 0.098, 0.039	0.964

very different in size and an optimal composition is maintained. A wide particle size distribution gives a higher packing density compared to a narrow particle size distribution.

Early work concerned with developing a particle size distribution for optimal packing was done by Fuller and Thompson in 1907 [25]. They were concerned with developing a concrete, and reported that optimal strength and density occurred in mixtures containing high concentrations of large particles. They observed that the strength correlated with the packing density. From their research it is evident that a high packing density requires a distribution with a high proportion of large, round particles. They also found that the shape of the optimal particle size distribution was independent of the mean particle size.

Figure 8.11 shows the similar results of Sohn and Moreland [26] for various Gaussian particle size distributions and two levels of sphericity. This figure expresses the fractional density versus the

Table 8.5

Particle Size Ratio Effects on Maximum Densities for Multimodal Packings [4]

diameter ratio	ternary	quaternary	quinary
5	0.850	0.884	0.912
7	0.878	0.911	0.930
10	0.892	0.923	0.938
100	0.916	0.938	0.946
∞	0.926	0.943	0.949

coefficient of variation, which is defined as the standard deviation divided by the mean particle size. As shown, the random packing density improved with wide distributions, independent of the mean particle size. However, as indicated in Figure 8.11, there was a particle shape effect, with the less spherical particles giving a lower packing density.

Random packing of spheres with various distributions have been computer simulated as a basis for understanding and optimizing the use of continuous particle size distributions. Various sized fractions of spheres can be combined to improve packing [27]. Starting with ordered packings, fractional densities as high as 0.81 have been achieved using mixed particle sizes for filling pores. The more orderly the packing of the largest sphere, then the narrower the distribution of additive spheres needed to fill the interparticle voids.

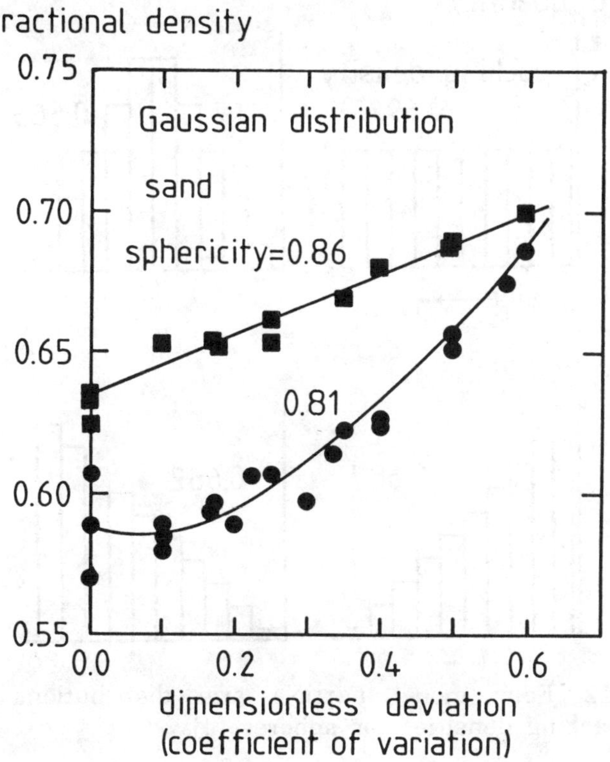

Figure 8.11 Fractional packing density shown as a function of the coefficient of variation for continuous Gaussian particle size distributions of two sands of differing sphericities [26].

The shape of the particle size distribution affects the packing density [4,25-33]. Figure 8.12 illustrates some of the distributions considered by Lee [4]. Shown with each histogram of the particle size distribution is the resulting fractional packing density. The highest density resulted from the flat distribution. Although the shape of the particle size distribution plays a role, the magnitude is not large.

Various investigations have concluded that a broad log-normal distribution gives a high packing density. This is attributed

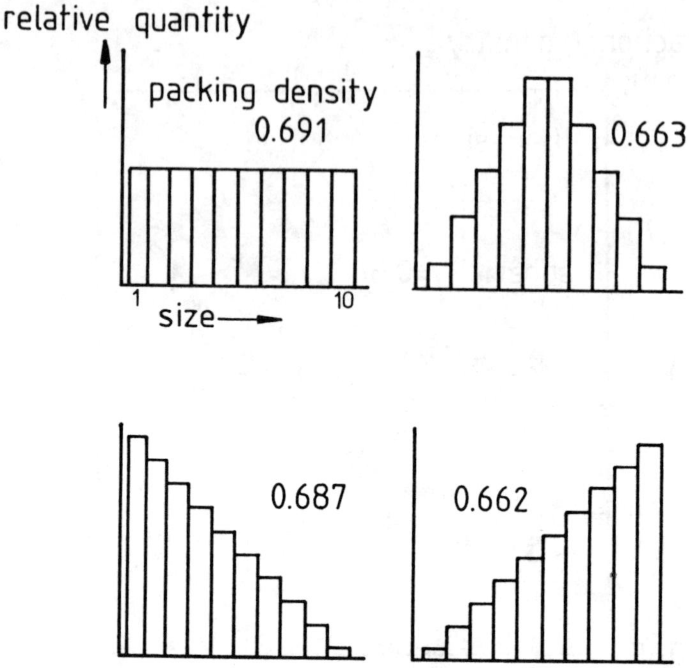

Figure 8.12 Four model particle size distributions and the resulting packing densities for spheres [4].

to the large population of small particles (although the weight fraction is small). Generally, the normal or Gaussian distribution is not a good description of a particle size distribution [34-36]. A normal distribution predicts a finite probability of particles with a negative size. Alternatively, many naturally occurring particles form a log-normal distribution. The log-normal distribution is described in terms of the mean particle size $\bar{D}$ and the geometric standard deviation as follows [35-39]:

$$P(x) = \frac{1}{(2 \pi)^{1/2} \sigma} \exp\left[- \frac{(x - \bar{x})^2}{2 \sigma^2} \right] \qquad (8.6)$$

where $P(x)$ is the probability of finding a particle with size $x = ln(D)$ where D is the particle size. The mean particle size gives $\bar{x} = ln(\bar{D})$. The distribution spread parameter σ is defined as

$$\sigma = ln(D_{84} / D_{50}) \tag{8.7}$$

where D is the particle size and the subscripts indicate the percentage points in the cumulative particle size distribution (on the cumulative particle size distribution, 84.13% corresponds to one standard deviation from the mean at 50%). Thus, σ is the measure of the distribution width expressed in terms of the logarithm of the size at one standard deviation divided by the size at the mean. This is the logarithm of the size ratio giving one geometric standard deviation. Only two parameters are needed to describe a log-normal distribution, the mean size and the deviation. Figure 8.13 shows four model particle size distributions both in terms of the cumulative and standard deviations versus the logarithm of the particle size. The latter plot is linear, emphasizing the simplicity of the log-normal distribution.

For spheres, the packing density varies with the spread σ of the log-normal particle size distribution [15,19,26,40-46]. With many distributions, the larger the spread, the higher the packing density. For relatively narrow distributions, the packing density varies linearly with the distribution spread according to the relation

$$f = a_1 + b_1 \sigma \tag{8.8}$$

where a_1 and b_1 are empirical constants that vary with the particle sphericity. Figure 8.14 illustrates such behavior for small spreads in the size distribution of random packings [26,40]. This trend cannot extend to very wide distributions, since the fractional density is limited by an upper bound of 1.00. Instead, with wide log-normal particle size distributions the packing density approaches a limiting value. Based on experimental observations, an alternative form of Equation 8.8 is used to represent the limiting density condition:

$$f = a_2 - b_2 / \sigma \tag{8.9}$$

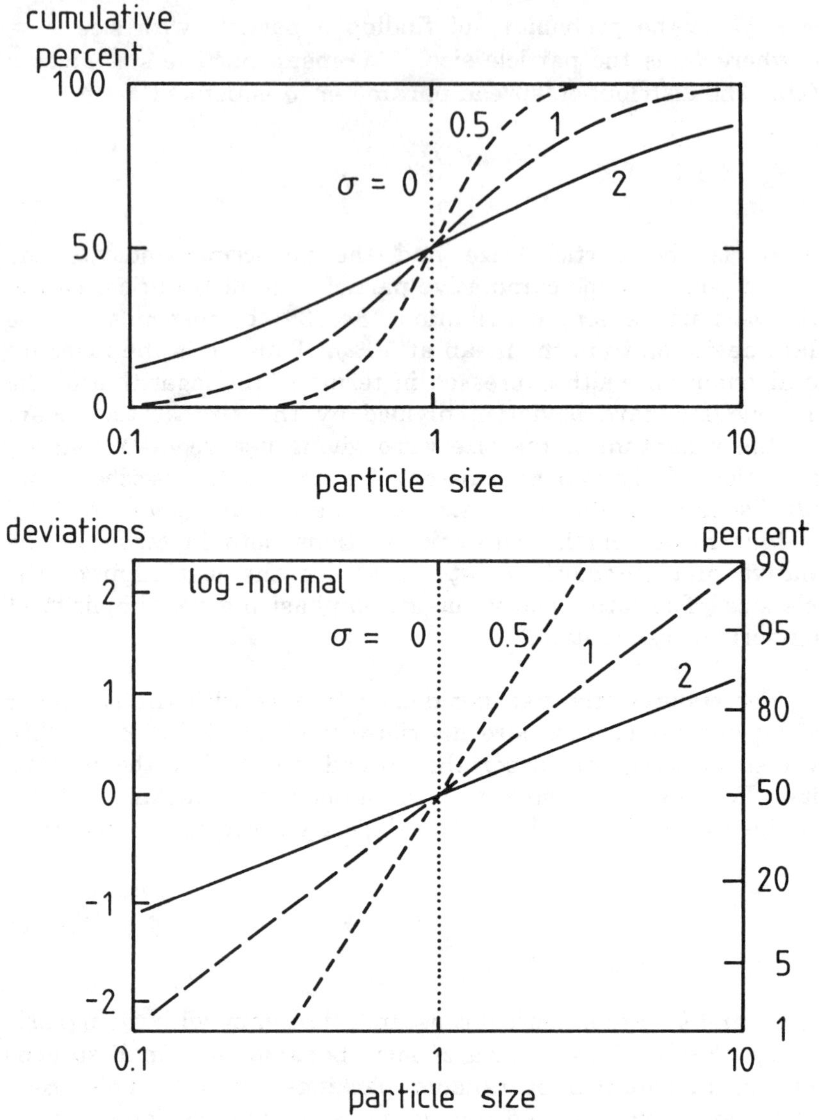

Figure 8.13 The basic log-normal particle size distribution. The cumulative size distribution is shown on the top and the corresponding plots of standard deviation versus size are shown below.

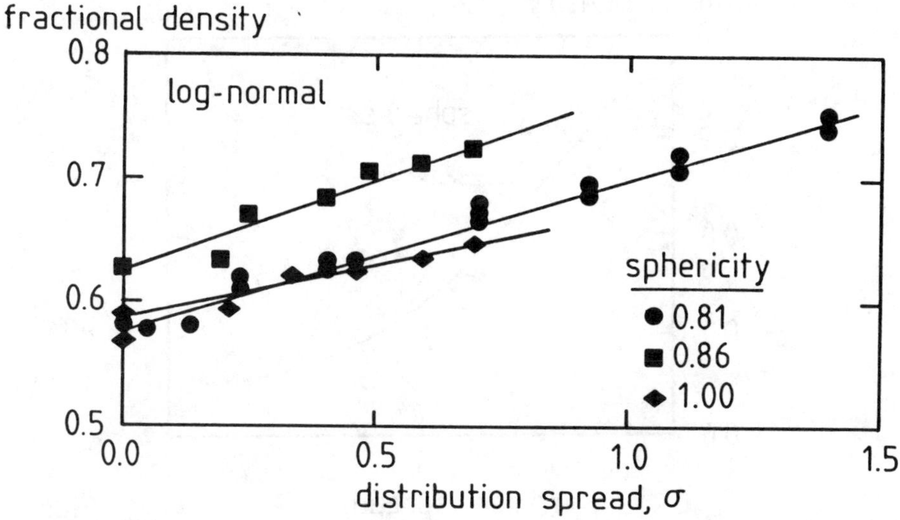

Figure 8.14 The variation in the fractional packing density for narrow log-normal distributions versus the distribution spread as measured by σ [26,40].

again, a_2 and b_2 are empirical constants.

For an infinitely wide particle size distribution, the projected fractional packing density has been estimated to range from 0.82 to 0.96. These values are comparable with the fractional densities attainable with bimodal and trimodal mixtures with large particle size ratios. There is a maximum benefit to having a wide particle size distribution. As an illustration of this fact, Figure 8.15 plots the random packing density versus the width of the distribution as measured by σ [32]. The packing density improves with wider particle size distributions up to a point, beyond which there is no further gain [40]. The results presented in Figure 8.15 illustrate that an upper density of 0.96 should be attainable from a wide particle size distribution; however, no experimental verification is available for this prediction. In contrast, Figure 8.16 shows an experimental optimal packing condition at an intermediate spread [31]. These distributions were approximately log-normal with varying degrees of spread in the particle size distributions. As discussed later,

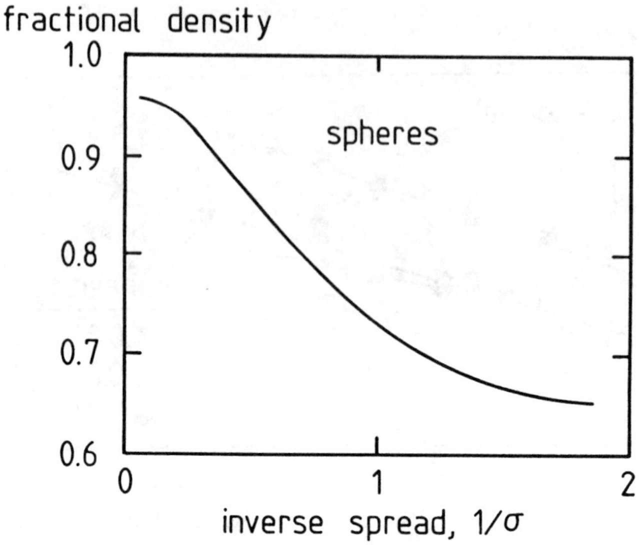

Figure 8.15 The fractional packing density as an inverse function of the distribution spread (σ) for broad log-normal distributions [32].

there is empirical evidence of similarities in the distribution shapes of high packing density continuous systems.

Furnas [2] considered the possibility of using screened particle fractions varying in size by a constant factor of 1.414. He predicted that optimal packing will occur when each screen fraction has 1.10 times the amount of powder on the lower sized screen fraction. This behavior is influenced by the inherent porosity of the powder and by the particle shape, and it provides a first estimate of the distribution shape that gives best packing. Subsequent measurements confirm the optimal behavior of this type of distribution [29]. Alternatively, it has been suggested that optimal packing occurs when the particle size distribution can be described by the Andreasen equation [30]

$$W = A\,D^{q} \tag{8.10}$$

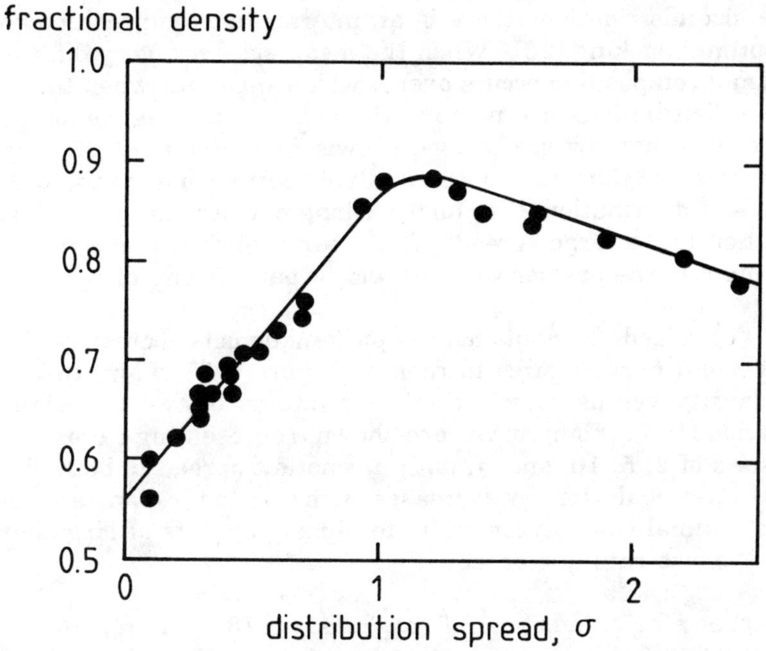

Figure 8.16 The relation between the packing density and spread of approximate log-normal particle size distributions, showing optimized packing at an intermediate spread [31].

where W is the weight fraction of particles less than size D, and A and q are empirical constants used to fit the particle size distribution. Generally, the porosity decreases with a wide range of particle sizes. The maximum density occurs for q values from 0.5 to 0.67 [30,31,47]. The high proportion of small particles helps to fill the voids between the nearly continuous matrix of large particles.

Continuous particle size distributions can be mixed to improve the packing density. A density benefit from such a mixture requires a large difference in mean sizes of the two distributions. Should one powder have a wide particle size distribution, then it is possible that the mixture will not improve packing [48]. Generally, the particle size distribution of the smaller powder should be wider than the distribution of the larger powder to assure a benefit from mixing the powders. For mixtures of either Gaussian or log-normal particle size distributions there is an intermediate composition that gives optimal packing [26]. When the mean sizes are very different, the optimal composition occurs over a wide range, but when the particle size distributions are narrow, the optimal packing composition occurs over a narrow size range. Lewis and Goldman [49] argue that the mixture that gives optimal mixing corresponds to the widest particle size distribution. For high packing density, the particle size distribution has a large spread. The optimal composition contains a high content of the powder with the widest particle size distribution.

For mixed distributions, the packing density increases as the ratio of mean particle sizes increases. Figure 8.17 shows the fractional density versus composition for mixtures of two particle size distributions [26]. Four curves are shown, representing ratios in the mean sizes of 2, 5, 10, and ∞, with a constant spread to both distributions. The peak density increases with the ratio of mean sizes, and the optimal composition shifts to higher contents of large particles as the size ratio increases.

Fractional densities of at least 0.78 are reported for mixtures of log-normal distributions with mean particle sizes in a ratio of 10:1. In this case, the maximum packing density occurs between 55 and 75% of the larger powder, and the density peak is broader than observed with mixed bimodal powders. Also, under optimal conditions there is less sensitivity to the exact composition. A similar result has been reported for multimodal packings when each size class was not truly monodisperse [15]. For mixed log-normal particle size distributions, various combinations of distribution width and composition can give the same packing density [50]. This indicates that packing combinations for mixtures of continuous particle size distributions are not unique. In general, for widely different mean particle size distributions, the maximum packing density occurs at approximately 70% of the large powder and 30% of the smaller powder. When the mean particle sizes approach one another, the maximal density decreases and the optimal composition shifts toward the 50% composition. The packing density is inherently high for broad particle size distributions, so mixing with another distribution does little good. Alternatively, for narrow initial particle size distributions, the packing density is more like the case of the monosized sphere. Here the mixing of distributions of different mean sizes aids in attaining a high packing density, especially if the sizes are very different. An upper limiting

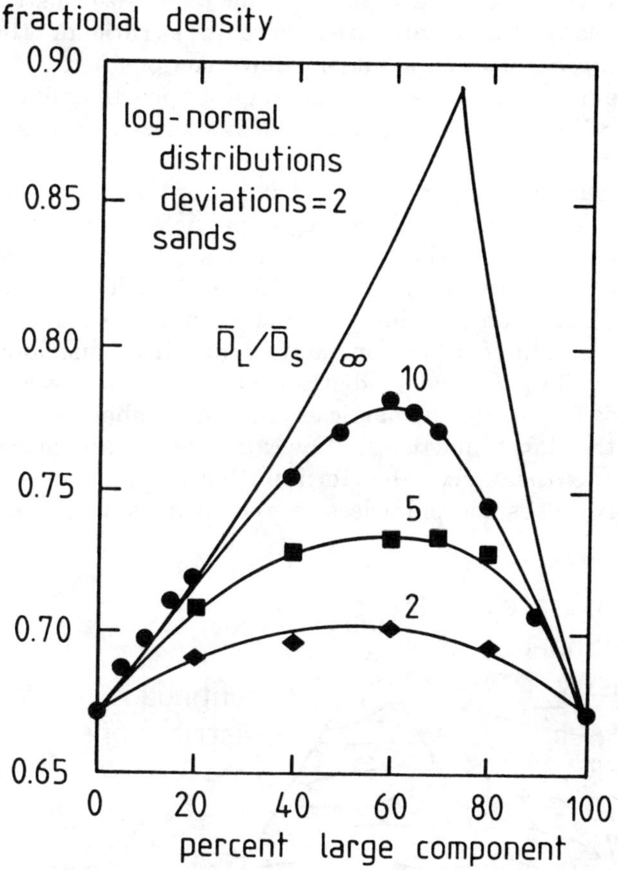

Figure 8.17 Fractional density versus composition for binary mixtures of log-normal distributions [26]. The spread in the distributions was constant while the ratio of mean sizes was varied.

fractional density of 0.90 is projected for mixtures of wide particle size distributions with a ratio of mean sizes of 10:1 [48].

Ternary mixtures of continuous particle size distributions have also been considered. Similar to the behavior of trimodal mixtures, ternary mixtures require a substantial mean particle size difference to improve packing. Figure 8.18 shows the results for a ternary mixture of relatively narrow particle size distributions [32]. In this case, the mean sizes were in a ratio of 100:10:1. Optimal packing occurs at an intermediate composition rich in the large component. Indeed, the general shape appears similar to Figures 8.6 and 8.7 for trimodal mixtures of monosized spheres.

The packing coordination number has been an object of concern in several studies. The large particles have a greater number of contacts than the small particles. As the particle size distribution width increases there are more contacts for the large particles and fewer contacts for the smaller particles. Figure 8.19 plots the variation in coordination number for three distributions of varying widths [33]. The wider distribution gives the lowest coordination number for the small particles and the highest coordination number for the large particles. Overall, the mean coordination number remained near six. One formula for the packing coordination number N_c uses the particle size and mean surface density of

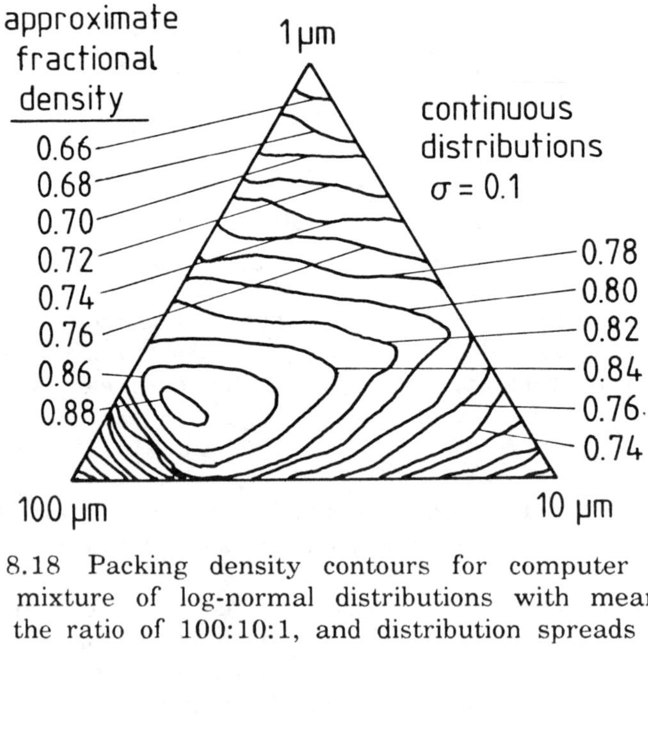

Figure 8.18 Packing density contours for computer simulated ternary mixture of log-normal distributions with mean particle sizes in the ratio of 100:10:1, and distribution spreads of $\sigma = 0.1$ [32].

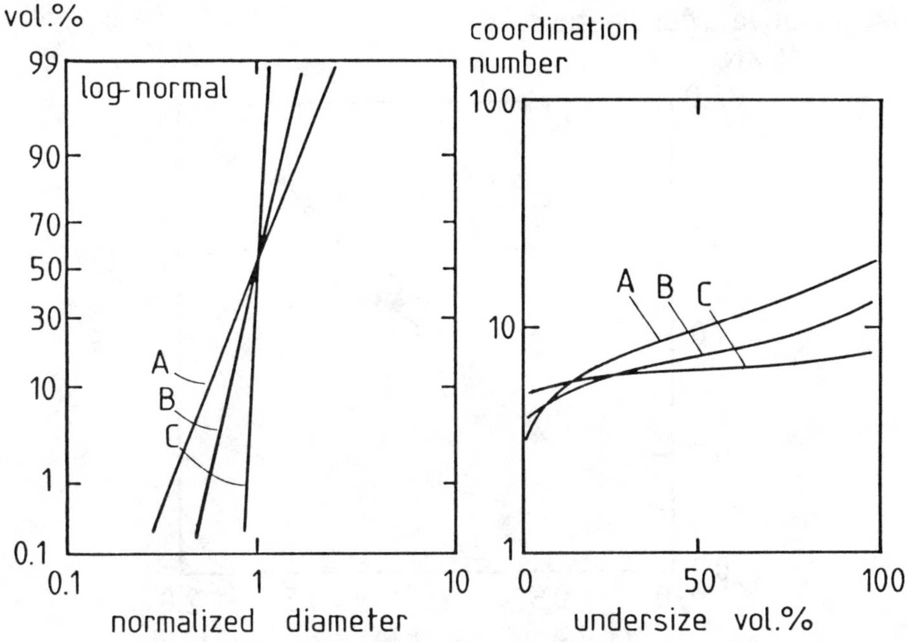

Figure 8.19 The coordination number distributions versus composition for three log-normal particle size distributions [33].

the packing f_s as follows [51,52]:

$$N_c = 16\, f_s\, [(D + \bar{D})/(2\,\bar{D})]^2 \tag{8.11}$$

where D is the particle diameter and $\bar{D}$ is the mean particle diameter. The mean surface density of the packing is defined as the relative density of spheres on a given sphere surface. This formula indicates that the coordination number increases with particle size. Particles larger than the mean size will have a higher coordination number, as indicated in Figure 8.20 [51], where $\bar{N}_c$ is the mean coordination number for the packing. The data in Figure 8.20 show the measured coordination number divided by the value for the mean particle size versus the relative particle size $D/\bar{D}$. Typically, the mean coordination number for the distribution is near six

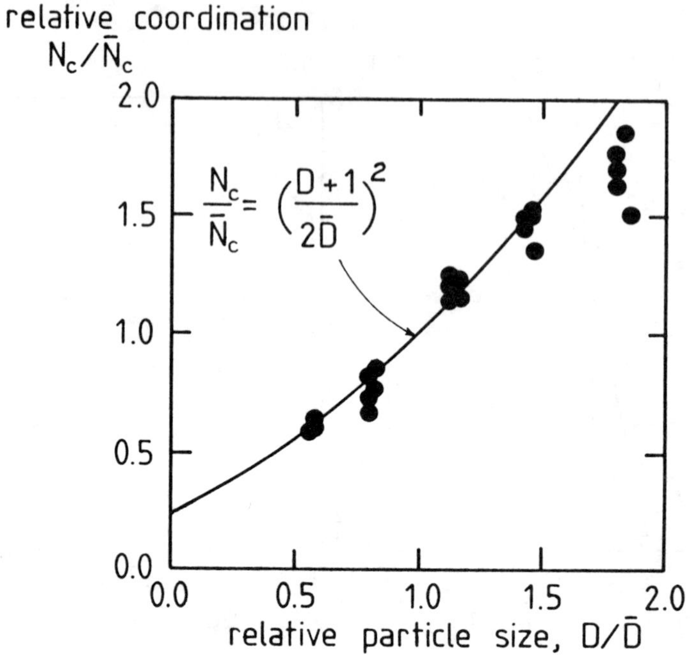

Figure 8.20 Coordination number expressed as a function of the relative particle size for continuous particle size distributions [51].

[28,53]. Consequently, the distribution in coordination numbers varies with the spread in the particle size distribution.

Summary

Mixing different particle sizes increases the packing density, especially when the particles are very different in size. For every ratio of particle sizes, there is a composition that will optimize the packing density. As the number of component powders is increased, the relative change in the packing density becomes progressively smaller. Many of the existing models of optimized packing with multimodal particle mixtures assume an ordered structure. For optimal packing, the largest particles form a fixed skeleton so that each successively smaller particle fills the remaining voids [54].

Each of the remaining (smaller) voids is then filled by an even smaller particle. This concept has merit for analyzing the packing structure of random particle arrays. Unfortunately, in random mixtures of particles there is no method for placing the particles in their optimal sites. Thus, actual packing densities fall short of the projected values due to lack of order and homogeneity. Small errors in placing the particles degrade the packing density, making it difficult to obtain the model packing densities in practice [55]. Additionally, there are practical difficulties in obtaining and handling particles with large particle size differences [56].

The packing density of a multimodal mixture may not be optimal if the particle size ratio is small. Furnas [2] analyzed the possible benefits from various multimodal mixtures. As shown in Figures 8.3 and 8.10, the optimal number of component particles depends on the inherent packing density of the powders and the particle size ratio of the mixture. Bimodal mixtures give higher packing densities than trimodal mixtures for particle size ratios less than approximately 100. In this chapter, data showing the required particle size ratios and compositions needed for optimal packings with up to five components have been presented. The optimal composition can be estimated from the inherent packing density of the powders. In all cases, the optimal packing condition has a high proportion (by volume or weight) of the largest particles. With large particle size ratios, the actual number of large particles may be small, but their volume will be sufficient to create a continuous skeleton.

Based on the concept of a multimodal packing, attention has been directed to the problem of a continuous particle size distribution. Here density depends on the shape of the particle size distribution. A higher packing density is attained with wide particle size distributions. Additionally, the more equiaxed and spherical the particles, the higher the packing density.

There is a maximum attainable density. Various theories predict maximum fractional densities as high as 0.96 for continuous particle size distributions. Near optimal distributions have been empirically discovered in several industries [2,25,39-31,42,57-60]. Figure 8.21 illustrates how several high packing density systems have similar particle size distributions. This plot shows the normalized optimal distributions for cases of spherical and irregular

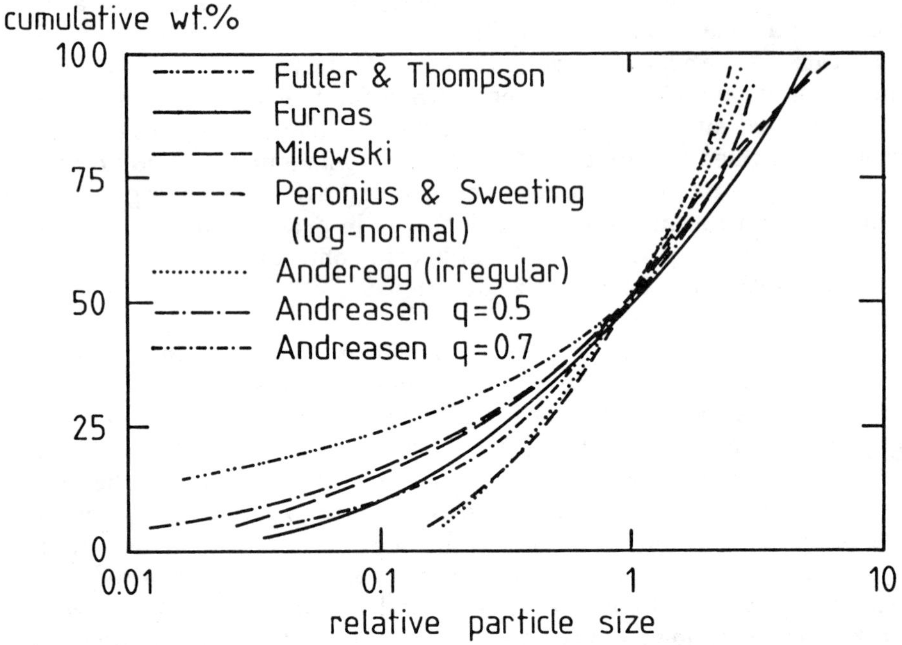

Figure 8.21 The similarity of optimized particle size distributions as empirically determined for several systems [2,29-31,57-60].

particles. There is a characteristic wide range of small particles in these optimized systems.

The early research of Fuller and Thompson [25] is still relevant. They found an optimal particle size distribution for maximum density that was independent of the mean size. They concluded that the highest packing density and highest strength concrete was obtained by

1) use of the largest mean particle size,
2) use of round particles as opposed to angular particles,
3) a high proportion by volume of large particles, and
4) a low proportion of intermediate sized particles.

References

1. C. C. Furnas, *The Relations Between Specific Volume, Voids and Size Composition in Systems of Broken Solids of Mixed Sizes*, Serial 2894, Report of Investigations, U.S. Bureau of Mines, North Central Experiment Station, Minneapolis, MN, October 1928.
2. C. C. Furnas, "Grading Aggregates I - Mathematical Relations for Beds of Broken Solids of Maximum Density," *Industrial and Engineering Chemistry*, 1931, vol. 23, pp. 1052-1058.
3. R. K. McGeary, "Mechanical Packing of Spherical Particles," *Journal of the American Ceramic Society*, 1961, vol. 44, pp. 513-522.
4. D. I. Lee, "Packing of Spheres and Its Effect on the Viscosity of Suspensions," *Journal of Paint Technology*, 1970, vol. 42, pp. 579-587.
5. K. E. Meiners, R. F. Geisendorfer, B. C. Mutsuddy and D. M. Bigg, *First Progress Report on Complex Metal Powder Shapes by Injection Molding*, Battelle Columbus Laboratories, Columbus, OH, August 1980.
6. D. J. Cumberland and R. J. Crawford, *The Packing of Particles*, Elsevier Science, Amsterdam, The Netherlands, 1987.
7. A. E. R. Westman and H. R. Hugill, "The Packing of Particles," *Journal of the American Ceramic Society*, 1930, vol. 13, pp. 767-779.
8. A. R. Dexter and D. W. Tanner, "Packing Density of Ternary Mixtures of Spheres," *Nature Physical Science*, 1971, vol. 230, pp. 177-179.
9. M. Cross, "Prediction of the Voidage of Packed Beds," *Proceedings Fourth International Symposium on Agglomeration*, C. E. Capes (ed.), Iron and Steel Society, Warrendale, PA, 1985, pp. 713-717.
10. M. Cross, W. H. Douglas and R. P. Fields, "Optimal Design Methodology for Composite Materials with Particulate Fillers," *Powder Technology*, 1985, vol. 43, pp. 27-36.
11. M. Suzuki and T. Oshima, "Verification of a Model for Estimating the Void Fraction in a Three Component Randomly Packed Bed," *Powder Technology*, 1985, vol. 43, pp. 147-153.
12. A. Marmur, "A Thermodynamic Approach to the Packing of Particle Mixtures," *Powder Technology*, 1985, vol. 44, pp. 249-253.

13. J. A. Dodds, "The Porosity and Contact Points in Multicomponent Random Packings Calculated by a Simple Statistical Geometric Model," *Journal of Colloid and Interface Science*, 1980, vol. 77, pp. 317-327.
14. J. Ayer and F. E. Soppet, "Vibratory Compaction: I, Compaction of Spherical Shapes," *Journal of the American Ceramic Society*, 1965, vol. 48, pp. 180-183.
15. T. Stovall, F. De Larrard and M. Buil, "Linear Packing Density Model of Grain Mixtures," *Powder Technology*, 1986, vol. 48, pp. 1-12.
16. M. Suzuki and T. Oshima, "Comparison Between the Computer Simulated Results and the Model for Estimating the Coordination Number in a Three Component Random Mixture of Spheres," *Powder Technology*, 1985, vol. 43, pp. 19-25.
17. N. Standish and D. E. Borger, "The Porosity of Particulate Mixtures," *Powder Technology*, 1979, vol. 22, pp. 121-125.
18. P. M. Worthington, "An Experiment on the Theory of Voids in Granular Materials," *Magazine of Concrete Research*, 1953, vol. 4, no. 12, pp. 121-126.
19. A. B. Yu and N. Standish, "Porosity Calculations of Multi-Component Mixtures of Spherical Particles," *Powder Technology*, 1987, vol. 52, pp. 233-241.
20. N. Standish and D. G. Mellor, "The Permeability of Ternary Coke Mixtures," *Powder Technology*, 1980, vol. 27, pp. 61-68.
21. H. T. Horsfield, "The Strength of Asphalt Mixtures," *Journal of the Society of Chemical Industry*, 1934, vol. 53, pp. 107T-115T.
22. H. E. White and S. F. Walton, "Particle Packing and Particle Shape," *Journal of the American Ceramic Society*, 1937, vol. 20, pp. 155-166.
23. D. R. Hudson, "Density and Packing in an Aggregate of Mixed Spheres," *Journal of Applied Physics*, 1949, vol. 20, pp. 154-162.
24. R. F. Fedors and R. F. Landel, "An Empirical Method of Estimating the Void Fraction in Mixtures of Uniform Particles of Different Size," *Powder Technology*, 1979, vol. 23, pp. 225-231.
25. W. B. Fuller and S. E. Thompson, "The Laws of Proportioning Concrete," *American Society of Civil Engineers Transactions*, 1907, vol. 59, pp. 67-143.
26. H. Y. Sohn and C. Moreland, "The Effect of Particle Size Distribution on Packing Density," *Canadian Journal of Chemical Engineering*, 1968, vol. 46, pp. 162-167.

27. H. J. Frost and R. Raj, "Limiting Densities for Dense Random Packing of Spheres," *Journal of the American Ceramic Society*, 1982, vol. 65, pp. C19-C21.
28. M. J. Powell, "Computer Simulated Random Packing of Spheres," *Powder Technology*, 1980, vol. 25, pp. 45-52.
29. F. O. Anderegg, "Grading Aggregates II - The Application of Mathematical Formulas to Mortars," *Industrial and Engineering Chemistry*, 1931, vol. 23, pp. 1058-1064.
30. W. A. Gray, *The Packing of Solid Particles*, Chapman and Hall, London, UK, 1968.
31. N. Peronius and T. J. Sweeting, "On the Correlation of Minimum Porosity with Particle Size Distribution," *Powder Technology*, 1985, vol. 42, pp. 113-121.
32. G. P. Bierwagen and T. E. Saunders, "Studies of the Effects of Particle Size Distribution on the Packing Efficiency of Particles," *Powder Technology*, 1974, vol. 10, pp. 111-119.
33. M. Suzuki and T. Oshima, "Coordination Number of a Multiple Component Randomly Packed Bed of Spheres with Size Distribution," *Powder Technology*, 1985, vol. 44, pp. 213-218.
34. R. K. Eckhoff and G. Enstad, "Derivation of Particle Size Distributions from Measurements on Plane Sections Through Particle Beds. Effect of Regularity of Particle Packing," *Powder Technology*, 1975, vol. 11, pp. 1-10.
35. A. D. Randolph and M. A. Larson, *Theory of Particulate Processes*, Academic Press, New York, NY, 1971.
36. R. M. German, *Powder Metallurgy Science*, Metal Powder Industries Federation, Princeton, NJ, 1984, pp. 27-36.
37. D. F. Heath, "Normal or Log-Normal: Appropriate Distributions," *Nature*, 1967, vol. 213, pp. 1159-1160.
38. M. Strauss, T. Ring, A. Bleier and H. K. Bowen, "The Effect of Particle Size Distribution on the Coagulation of Ceramic Suspensions," *Particulate and Multiphase Processes*, vol. 3, T. Ariman and T. N. Veziroglu (eds.), Hemisphere, New York, NY, 1987, pp. 57-67.
39. J. E. Smith and M. L. Jordan, "Mathematical and Graphical Interpretation of the Log-Normal Law for Particle Size Distribution Analysis," *Journal of Colloid Science*, 1964, vol. 19, pp. 549-559.
40. A. R. Dexter and D. W. Tanner, "Packing Densities of Mixtures of Spheres with Log-Normal Size Distributions," *Nature Physical Science*, 1972, vol. 238, pp. 31-32.
41. R. J. Wakeman, "Packing Densities of Particles with Log-

Normal Size Distributions," *Powder Technology*, 1975, vol. 11, pp. 297-299.
42. G. P. Korfiatis and C. M. Manikopoulos, "Correlation of Maximum Dry Density and Grain Size, *Journal of the Geotechnical Engineering Division, Proceedings of the American Society of Civil Engineers*, 1982, vol. 108, pp. 1171-1176.
43. T. W. Seaton, "Density - Sand Grain Distribution Effect on Physical Properties," *Transactions of the American Foundrymen's Society*, 1960, vol. 68, pp. 45-48.
44. A. S. Goldman, H. D. Lewis and W. M. Visscher, "Random Packing of Particles: Simulation With a One Dimensional Parking Model," *Technometrics*, 1974, vol. 16, pp. 301-309.
45. G. P. Korfiatis and C. M. Manikopoulos, "Maximum Dry Density for a Continuous Grain Size Distribution," *Civil Engineering for Practicing and Design Engineers*, 1984, vol. 3, pp. 1139-1158.
46. R. M. Koerner, "Limiting Density Behavior of Quartz Powders," *Powder Technology*, 1969, vol. 3, pp. 208-212.
47. F. Sigulinski and I. Stamenkovic, "Packing and Consolidation of Ceramic Particles of Very Different Size," *Journal of Powder and Bulk Solids Technology*, 1977, vol. 3, pp. 61-63.
48. T. Itoh, Y. Wanibe and H. Sakao, "Relation Between Packing Density and Particle Size Distribution in Random Packing Models of Spheres," *Journal of the Japan Institute of Metals*, 1986, vol. 50, pp. 740-746.
49. H. D. Lewis and A. Goldman, "Theorems for Calculation of Weight Ratios to Produce Maximum Packing Density of Powder Mixtures," *Journal of the American Ceramic Society*, 1966, vol. 49, pp. 323-327.
50. R. C. Burk and P. S. Apte, "A Packing Scheme for Real Size Distributions," *Ceramic Bulletin*, 1987, vol. 66, pp. 1389-1392.
51. N. Ouchiyama and T. Tanaka, "Estimation of the Average Number of Contacts Between Randomly Mixed Solid Particles," *Industrial and Engineering Chemistry Fundamentals*, 1980, vol. 19, pp. 338-340.
52. T. Tanaka and N. Ouchiyama, "Coordination Numbers and Porosity of Packing Calculated from Particle Size Distribution," *Proceedings of the Fourth International Symposium on Agglomeration*, C. E. Capes (ed.), Iron and Steel Society, Warrendale, PA, 1985, pp. 9-16.
53. M. J. Powell, "Distribution of Near Neighbours in Randomly Packed Hard Spheres," *Powder Technology*, 1980, vol. 26, pp.

221-223.

54. F. Spunick, "On the Dense Packing of Spheres," *Transactions of the American Mathematical Society*, 1949, vol. 65, pp. 14-26.
55. M. E. Wise, "Dense Random Packing of Unequal Spheres," *Philips Research Reports*, 1952, vol. 7, pp. 321-343.
56. I. G. Shatalova, N. S. Gorbunov and V. I. Likhtman, "Physicochemical Principles of Vibratory Compacting," *Vibratory Compacting Principles and Methods*, H. H. Hausner, K. H. Roll and P. K. Johnson (eds.), Plenum Press, New York, NY, 1967, pp. 1-206.
57. J. V. Milewski, "Packing Concepts in the Use of Filler and Reinforcement Combinations," *Handbook of Reinforcements for Plastics*, J. V. Milewski and H. S. Katz (eds.), Van Nostrand Reinhold, New York, NY, 1987, pp. 14-33.
58. C. Ludwig, "Principles of Agglomerating Processes," *Transactions of the American Foundrymen's Society*, 1955, vol. 63, pp. 466-469.
59. J. Parisi, O. C. Nutter and C. Michalowski, "Sieve Ratios and Processing for Strong Molding Sands," *Transactions of the American Foundrymen's Society*, 1958, vol. 66, pp. 278-283.
60. A. H. M. Andreasen, "Ueber die Beziehung Zwischen Kornabstufung und Zwischenraum in Produkten aus losen Kornern (mit einigen Experimenten)," *Kolloid Zeitschrift*, 1930, vol. 50, pp. 217-228.

Chapter Nine

Particle Deformation and Compaction

Chapter Outline

Introduction

Many uses of particles involve the application of pressure to increase density and strength. The external pressure repacks and deforms the particles into a higher density mass. Simultaneously, compaction imparts shape to the powder mass, which is useful in manufacturing products ranging from pills to automotive gears [1-3]. Thus, interest in particle compaction spans several fields, including powder metallurgy, ceramics, soils, pharmaceuticals, and geology.

The study of particle compaction focuses largely on the relation between density and the applied pressure or load. Besides a density increase with compaction, there is an increase in the packing coordination as the contacts between particles are deformed and enlarged. Compaction is not a reversible process since the particle shape is modified by the pressure. There is a dramatic increase in interparticle bond strength associated with compaction because of shearing along the contact zones. Actually, the initial particle contacts are composed of asperities bridging between particles. As the load is increased, the number of contacting asperities increases, and they flatten and grow to form a planar contact surface [4]. These asperities eventually merge to form bonding surfaces between particles.

In this presentation, the fundamentals of particle deformation and densification are emphasized. The conceptual stages of densification are introduced along with a discussion of the differences between compaction of ductile and brittle particles. Subsequently, the models and equations of density versus applied pressure are discussed. Influences from particle size and shape are considered later in the chapter. The process of compaction involves simultaneous changes in the packing coordination and particle shape. These changes are important to understanding the packing of many natural substances, such as the cellular structure of both animal and plant tissue. Finally, the chapter will discuss the technological implications of compaction, including the important subject of load transmission in compaction tooling.

Stages of Compaction

Figure 9.1 is a simplified diagram of the change in density versus applied pressure during particle compaction. The diagram is nominally divided into four zones, which are not distinctly separate. The initial density without an applied pressure corresponds to the loose packing condition. The densification rate (first derivative) continuously declines as the compact density increases. As illustrated in Figure 9.2, the porosity, coordination number (number of contacts), and contact area all vary with the pressure [5]. These observations were made on spherical bronze particles during compaction. Based on such observations a generalized concept of particle compaction has emerged [4-21]. Various

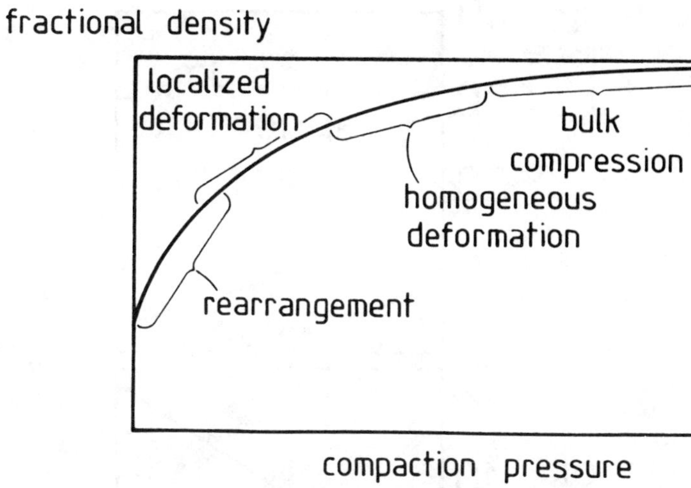

Figure 9.1 An idealized plot of fractional density versus pressure for particle compaction showing the four overlapping stages.

measurements and tools have been used to monitor the changes in packing structure during compaction. These have included scanning and transmission electron microscopy, optical metallography, x-ray diffraction, hardness, microhardness, resistivity, surface area, mercury porosimetry, permeability, and small-angle neutron scattering. The results of these studies allow a nominal division of compaction into four stages. Initially, elastic deformation occurs at point contacts. With further pressurization the number of contacts grows as particle rearrangement and sliding occur. Concomitantly, the contact areas expand and the zone of plastic deformation around each contact spreads. At high compaction pressures, rearrangement ceases and densification is by contact enlargement. The concentrated pressure at the contact points causes mass flow into the neighboring pores, reducing the pore size and porosity. Along with contact enlargement, strain hardening occurs; both factors increase the stress needed for continued densification. At very high densities the actual particle character is lost as massive deformation takes place.

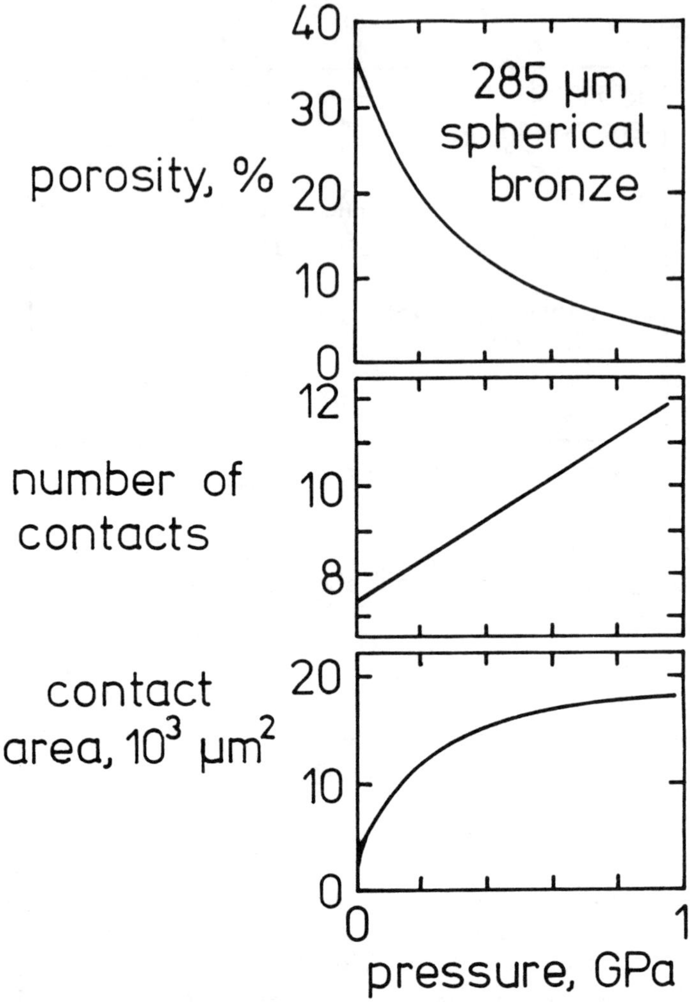

Figure 9.2 The compaction behavior of spherical bronze particles as measured by the compact porosity, number of contacts (coordination number), and area of each contact [5].

In the case of spherical particles, the contacts deform and the spheres are transformed into polyhedra as seen in Figure 9.3. This scanning electron micrograph shows the fracture surface of a spherical steel powder after extensive compaction. The initially

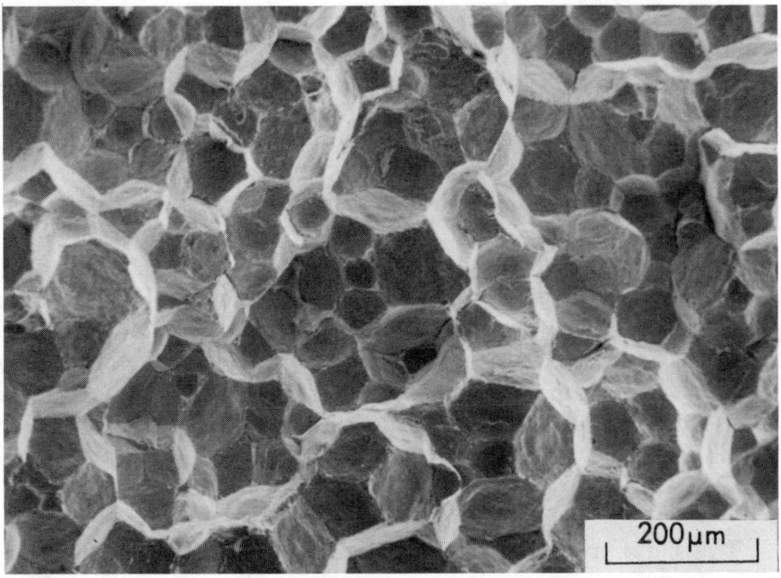

Figure 9.3 A scanning electron micrograph of a fracture surface showing the polyhedral shape of spherical steel particles pressed to a fractional density of 1.00.

spherical shape has been replaced by an extensively deformed, planar surface structure. During such deformation, shear occurs along the particle contacts by sliding. A consequence of this shear is that the contacts bond together by cold welding, giving an increase in compact strength. Additionally, the initially random structure takes on a more ordered character as density increases. There is danger in generalizing compaction behavior because of the several variables that can influence the process: the details of the compaction stages depend on particle and material characteristics [6-9,20-25]. Although this makes the specific response of particles to an applied pressure dependent on the situation, there are still some general behavior patterns. This section describes the idealized stages of compaction. Actual powder compaction is characterized by an intermixing of the stages, without distinct steps.

The initial transition with pressurization is from a loose array of particles to a closer packing. This is termed rearrangement, where both the packing density and coordination number increase with small levels of pressurization. Generally, rearrangement occurs at pressures less than 0.05 MPa [9]. The particles start with a packing density characteristic of random packing. From this starting condition, the amount of rearrangement is inversely dependent on the initial packing density. At low loads, the point contacts between particles undergo elastic deformation [26]. Additionally, large packing voids due to particle bridging are collapsed by particle motion under small stresses. Increasing pressure enlarges the contacts and provides closer packing with lower porosity and smaller pores [27]. The more irregular the particle shape, the lower the initial packing density. Accordingly, the density change by rearrangement is larger for powders of lower packing density.

The random dense packing is often cited as the upper limit to the rearrangement stage of compaction. However, other densification processes become active prior to this limit. For irregular particles, fracture is observed at pressures as low as 30 kPa [21]. Nonspherical particles undergo a substantial reduction in volume by rearrangement (up to a 55% decrease in volume for flakes). In contrast, spherical particles undergo as little as a 1% density increase by rearrangement.

The second stage of compaction involves localized deformation starting at the point contacts between particles [28]. As pressure is increased, the localized deformation flattens the asperities and the deformation spreads throughout the contact region. At the contact, the stress distribution is not uniform [29,30]. The maximum stress occurs at the center of the contact and the minimum compressive stress is at the edge of the contact. For a compressive force F applied over a circular contact of diameter X, the maximum stress σ_m is given as

$$\sigma_m = 6\, F/(\pi\, X^2)\,. \qquad (9.1)$$

In contrast, the maximum shear stress occurs away from the contact zone with a magnitude of approximately $0.3\sigma_m$. Along the contact face, the shear stress is dependent on the amount of slip between the particles.

For brittle particles, the onset of plastic deformation can lead to fracture, giving fragmentation of the original particles and densification by repacking of the fragments. In experiments with glass spheres of varying sizes, fracture was observed at a fractional density of 0.63, independent of the particle size [31]. Fracture initiates at the point of maximum tensile stress which occurs on the contact boundary as a radial stress. The onset of fracture can be detected by an increase in the surface area and a decrease in the mean particle size. Figure 9.4 illustrates the difference in surface area versus density for coal and salt [16]. Coal contains micropores that lead to

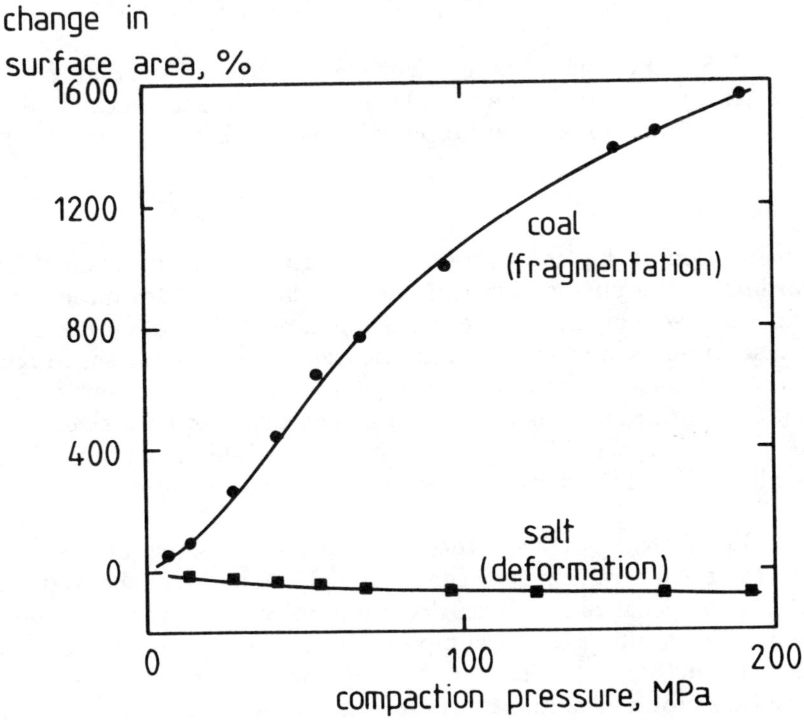

Figure 9.4 The change in specific surface area versus compaction pressure for coal and salt; the different behaviors reflect the difference in compaction events [16].

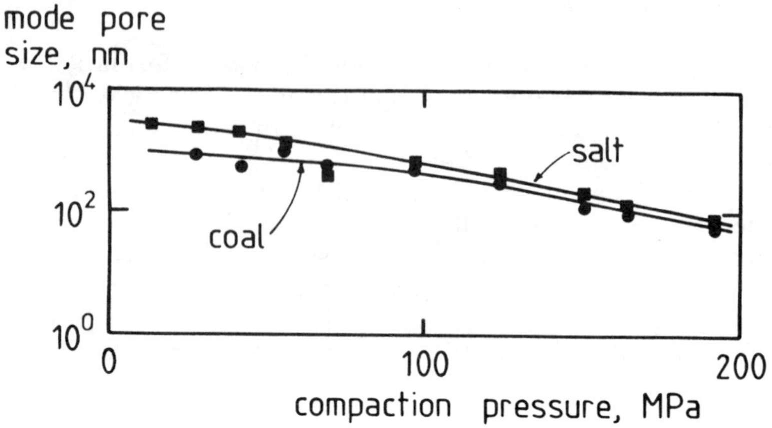

Figure 9.5 The mode pore size shown as a function of the compaction pressure for salt and coal. These data were collected under the same experimental conditions as used in Figure 9.4 [16].

fragmentation during compaction, while salt deforms by plastic flow. Accordingly, the surface area of the coal increases dramatically as fragments are generated. In contrast, plastic deformation leads to a decrease in surface area with densification. For comparison, Figure 9.5 plots the mode pore size for the compaction of these same particles [16]. Both materials show similar decreases in pore size; in the case of salt the pore filling is by plastic flow while in the case of coal pore filling is by fragment generation and flow into pores.

For ductile particles, the concentrated pressure at the particle contacts will eventually exceed the yield strength of the material. This causes localized plastic deformation at the contact points. Material flows into the unoccupied regions neighboring the contacts, leading to further densification. During compaction, there is a preferential elimination of the larger pores and a shift of the pore size distribution to smaller pores. Since the particle centers move closer together, new contacts are formed continuously as densification takes place. Indeed, sliding along the contact faces is observed well into the compaction process [5]. The zone of plastic deformation spreads as the pressure is increased until the entire particle deforms.

For a brittle substance, the deformation process is partially replaced by fragmentation. In contrast with pore filling by plastic flow, the fragments move into the voids, decreasing the porosity and mean pore size. When fragmentation begins there is a decrease in the compaction rate. The degree of fragmentation versus deformation during compaction is dependent on the material hardness and plasticity. Fragmentation reduces the particle size, increases the surface area, alters the particle shape, and increases the number of contacts per unit volume. This latter event leads to a stress reduction at each contact point as the number of contact points increases. As a generalization, plastic deformation gives faster densification than fragmentation.

At low pressures the stress is concentrated at the contact points between particles. With higher pressures, homogeneous deformation occurs throughout the compact [25,32]. The third stage of compaction is characterized by deformation throughout the particle mass, as can be seen by increases in the bulk properties like hardness. Particles able to undergo plastic deformation become strain hardened, increasing their resistance to further deformation. The combination of increased particle resistance to deformation, increased number of contact points, and enlarged contact zones act to resist further deformation. Materials with rapid strain hardening or low fracture toughness densify in the third stage of compaction by fragmentation. In this instance, the number of particles and the population of contact points continuously increase during densification.

For spheres, the onset of the third stage of compaction occurs at a fractional packing density of approximately 0.85. In this stage, independent of the densification process, further elimination of pores requires ever increasing expenditures of energy. The densification rate decreases as the density increases, which means that the energy input is inversely related to the remaining porosity.

At very high compaction pressures, massive deformation occurs, leaving small pores at the junction of three or more particles. This typically starts at fractional densities over 0.95 and compaction pressures in excess of 1 GPa. The fourth stage of compaction corresponds to a fully hardened compact. Plastic deformation is exhausted and the compact behaves as a full density elastic solid undergoing hydrostatic bulk deformation. An increase in pressure leads to a corresponding decrease in volume, yet no significant permanent volume change is obtained upon relaxation of the pressure. The fourth stage of compaction is rarely observed in practice.

After compaction, the external stress is removed and the compact undergoes elastic relaxation; this is referred to as springback. The elastic relaxation is very evident from the failure of a compact to fit back into rigid tooling after ejection. At low compaction pressures the degree of springback is low. For high compaction pressures, springback increases with approximately the square of the compaction pressure [33]. As the particle hardness increases, the degree of springback likewise increases. One consequence of springback is that the shear stresses associated with compact ejection can cause compact failure. This is typically evident as delamination cracks perpendicular to the direction of pressurization.

Density-Pressure Relations

The relation between density and the applied pressure is an important aspect of compaction. As shown in Figure 9.6 for 2 mm silver bromide particles, an increase in the applied pressure results in an increase in the fractional packing density [34]. The response of a particle mass to pressure is termed the compressibility, and is often defined as the density attained at a preset compaction pressure. Various theories of densification during compaction note that the change in fractional density f with pressure P can be expressed as follows [23,32,34-37]:

$$df/dP = K\,(1 - f) \tag{9.2}$$

where K is a proportionality constant and $1 - f$ is the porosity. Rearranging and integrating Equation 9.2 gives

$$\ln\,[(1 - f)/(1 - f_o)] = -\,K\,P \tag{9.3}$$

where f_o equals the fractional density without an applied pressure (the apparent density). An alternative form suggested by Heckel [7,32] incorporates a term A for initial rearrangement,

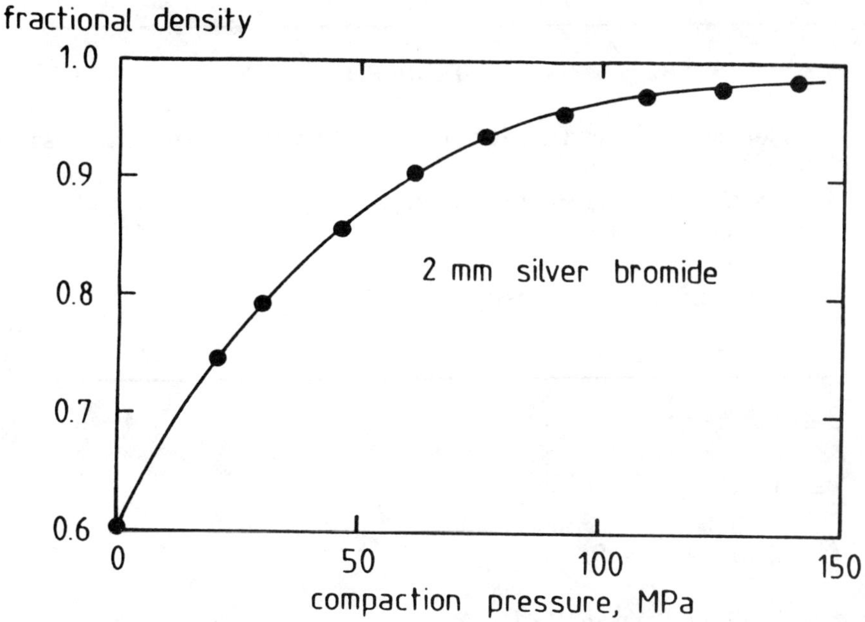

Figure 9.6 The fractional density for silver bromide versus the compaction pressure [34].

$$1 - f = (1 - f_o) \exp(-K P + A) \tag{9.4}$$

which is widely applied to powder compaction, especially in the intermediate pressure range of 100 to 700 MPa.

Equations 9.3 and 9.4 ignore the mechanism of compaction, yet they provide empirical starting points for many of the compaction models. Numerous models have been presented for the dependence of density on compaction pressure that incorporate terms for rearrangement, plastic deformation, fracture, hardness, and strain hardening. Table 9.1 summarizes several models [20,23,25,26,32, 35,38-42], but most of these are unsatisfactory for several reasons, not the least of which is their empirical origin [31,43,44].

Table 9.1

Models of Compact Density as a Function of the Applied Pressure

A, B, K, K_1, K_2, and m are constants
P = applied pressure
f = fractional density
f_o = initial fractional density

$f = A P^{2/3} - B$	[38]
$f = A P^{m} - B$	[38]
$f = (1 - f_o) \exp(-K P)$	[35,39]
$f = 1 - (1 - f_o) \exp(-K P - B)$	[39]
$f = f_o (1 + A P)/(1 + B P)$	[31,39]
$f = (A + B \log P)^{-1}$	[9]
$1/(1 - f) = A_1 \exp(-K_1 P) + A_2 \exp(-K_2 P)$	[25]
$f = f_o + A P^{1/3}$	[40]
$f = f_o + A P^{3/2}$	[41]
$f = (1 - f_o) \exp(-K P + A)$	[26,32]

Although all models adequately predict a decrease in porosity with an increase in pressure, they often lack a physical basis. Generally, the models oversimplify the complexity of compaction and do not adequately include the stages outlined above. One problem is the number of adjustable parameters. The transition between stages is often poorly defined. For example, most models fail to include rearrangement as a low pressure component of densification. Additionally, very high pressure compaction appears to be poorly treated by the existing models. Some of the variations have attempted to incorporate packing density, yield strength, strain hardening, fracture strength, or rearrangement corrections [22,26, 42,45,46]. Various comparisons of the models to experimental measurements do not clarify the matter. There are differing levels of success depending on the powder and material characteristics. In spite of the shortfallings, empirical equations that predict the density as a function of the compaction pressure are important in generalizing behavior and comparing different particles. Most probably the largest problem is that compaction is not a simple combination of stages and mechanisms; thus, the mathematical treatments need to better reflect the complexity of the actual process. Accordingly, simple models involving only the pressure and a few adjustable constants are nominal acknowledgments of the underlying complexity.

It is recognized that rearrangement is larger but overall densification is inferior for particles with a low inherent packing density. However, attempts to link compaction behavior to packing density or coordination number have been unsuccessful [24,47]. Furthermore, the success of the various models is dependent on differences in material properties. Equations that are suitable for materials that plastically deform are unsuitable for materials that fracture.

Changes in Particle Shape and Size

During compaction, the particles move closer together and the coordination number increases as new contacts form. Figure 9.7 demonstrates such an increase in coordination number for spherical bronze powder [5]. The initial packing coordination was near seven, while the full density value was near 14. Simple equations can be used to relate the the fractional density f to the mean coordination number $\bar{N}_c$ [48-50]; for fractional densities below 0.85,

$$\bar{N}_c = N_{c\text{-}o} + 9.5\,(f - f_o) \tag{9.5}$$

while for fractional densities above 0.85,

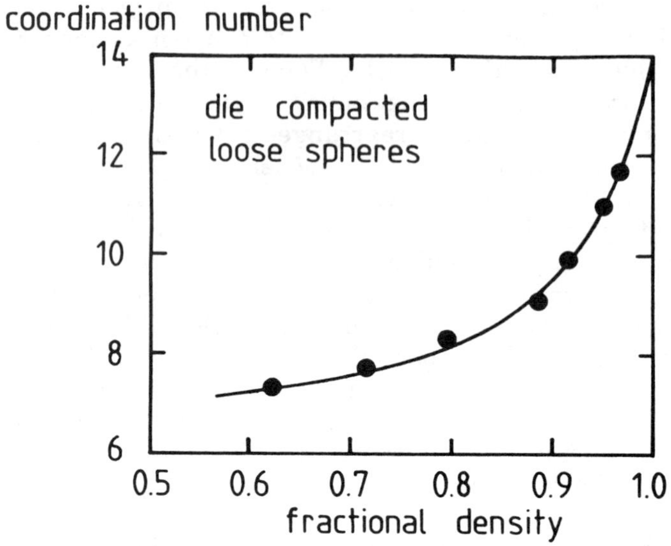

Figure 9.7 The increase in coordination number with fractional density during compaction of spherical bronze particles [5].

$$\bar{N}_c = N_{c\text{-}o} + 2 + 9.5\,(f - 0.85) + 881\,(f - 0.85)^3\,. \qquad (9.6)$$

The parameter $N_{c\text{-}o}$ is the packing coordination at the loose packed density f_o. For monosized spheres in a random dense packing, the initial density would be approximately 0.64 and the initial coordination number would be six. Thus, at full density Equation 9.6 predicts a coordination number of 12.4, slightly below the expected value of 14. Polynomial relations have been used to express the dependence of coordination number on packing density [49,51]. One simple form that provides an accurate fit to the experimental data is

$$\bar{N}_c = 14 - 10.4\,(1 - f)^{0.38}\,. \qquad (9.7)$$

The study of particle shape at full density is a problem of importance to cell biologists, metallurgists, physicists, and many others [52]. In 1727, Stephen Hales reported that fresh peas poured into a pot and compressed gave a coordination number of 12 with a dodecahedron shape. Plateau in 1873 and Kelvin in 1887 both reported the possible coordinations and polyhedra shapes expected at full density [53]. Plateau's experiments were on foams, and based on these Kelvin suggested the 14-sided tetrakaidecahedron as the most economical (lowest surface area to volume ratio) space filling polyhedron. Since that time, several experimental determinations of the packing coordination for compacted monosized spheres have been made. Compaction of randomly packed spheres results in a packing coordination near 14 at full density. Coordination numbers reported in the literature (typically fabricated from soft materials like lead compacted to full density) range from 13.4 to 14.2, with an average of 13.9 [5,17,19,53-63]. Alternatively, if an ordered packing of monosized spheres is compacted to full density, the final shape is a 12-faced dodecahedron [55,61]. Likewise, if the particles are allowed to swell to fill space (without shear motion), then the final grain shape is also 12-faced. The difference in final grain shape reflects the greater relative particle motion during compaction for the random packing. Additionally, there is nonuniform distortion of the particle when compaction is performed uniaxially (in a cylindrical die). The major deformation is in particle height, yielding as much as a factor of two difference in contact size in the axial and radial directions [7,63].

Sketches of the model full density grain shapes of a tetrakaidecahedron, rhombic dodecahedron, and pentagonal dodecahedron are shown in Figure 9.8. These structures obey the Euler relation, which says that for a convex polyhedron the number of corners plus faces equals the number of edges plus two. The rhombic dodecahedron is composed of 12 faces, each with four edges shared by two faces, giving a total of 24 edges and 14 corners per particle. The tetrakaidecahedron is composed of eight hexagons and six squares, giving 14 faces, 36 edges, and 24 corners. If L is the length of an edge, then the volume V, surface area S, and grain size G are given as follows:

$$V = (128)^{1/2} L^3 = 11.31 L^3 \tag{9.8}$$

$$S = (432^{1/2} + 6) L^2 = 26.78 L^2 \tag{9.9}$$

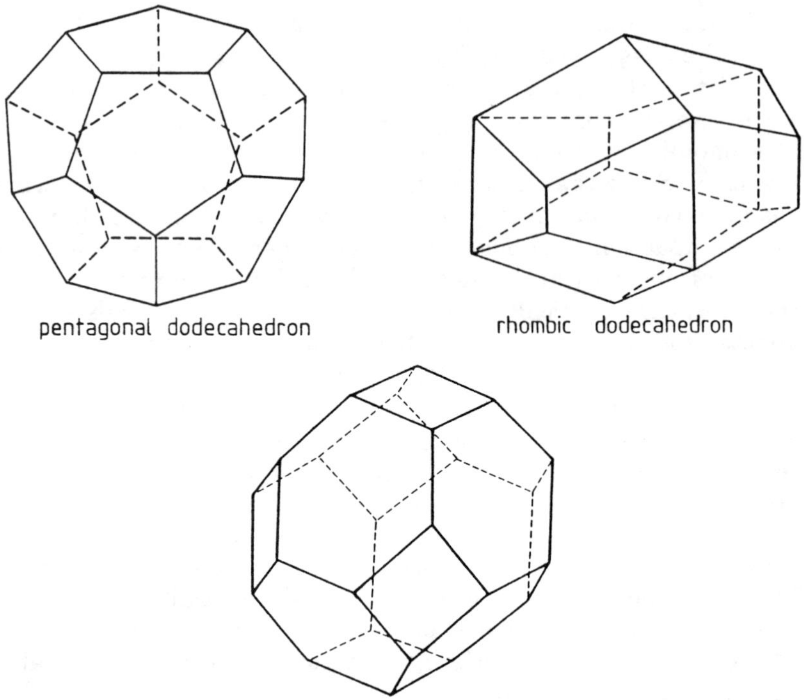

Figure 9.8 Three typical particle shapes after densification; the pentagonal dodecahedron, rhombic dodecahedron, and tetrakaidecahedron.

$$G = 8^{1/2}\,L = 2.83\,L\,. \tag{9.10}$$

For the pentagonal dodecahedron with an edge length L,

$$V = 7.663\,L^3 \tag{9.11}$$

and

$$S = 20.65\,L^2\,. \tag{9.12}$$

The final shape of randomly packed spheres compacted to full density is typically not a true tetrakaidecahedron. Although the final particle is a 14-faced polyhedron, there are distributions to the number of faces per particle and number of edges per face. Figure 9.9 shows the frequency distribution for the coordination number (number of faces per polyhedron) at two points in the densification process for compressed lead spheres [19]. At full density the mode is 14 faces, but some polyhedra are detected with up to 19 faces. In these compacted structures, the 14-sided particles are composed of a mixture of faces with four, five, and six edges (squares, pentagons, and hexagons). Approximately 50% of the faces have five edges. The pentagonal shape is not predicted by any of the models, although such a shape is noted in many foams, biological structures, and polycrystalline materials. Gradients in the initial packing density also affect the observed polyhedron shape after compaction.

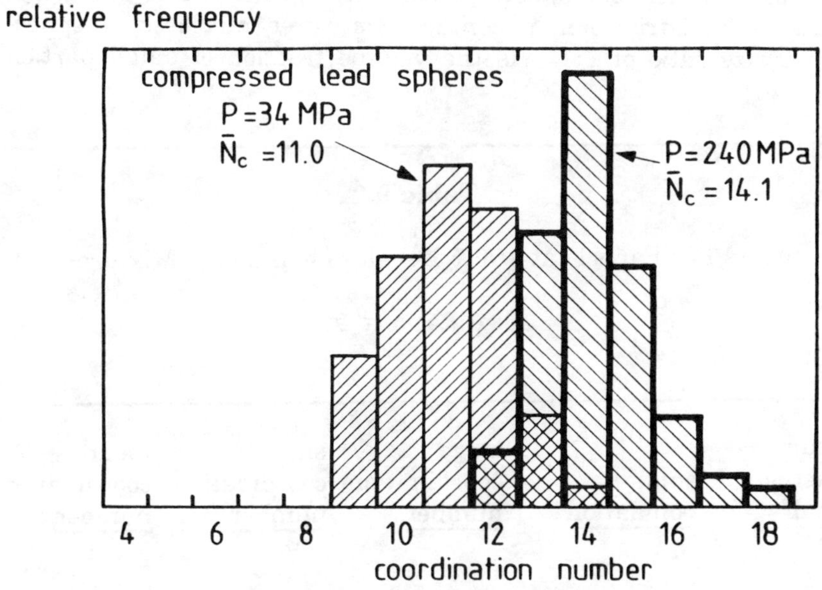

Figure 9.9 The coordination number distribution for compressed lead spheres at two stages of compaction [19].

Thus, the low density associated with the wall effect translates to a lower packing coordination for particles located near a container wall after compaction.

According to the model of particle compaction presented in Figure 9.1, the effect of an increasing compaction pressure is to first increase the size of initial point contacts. Subsequently, new contacts form as the separation distance between particles decreases. The first contacts tend to have more edges per face while the later contacts have fewer edges per face. Furthermore, there is a direct correlation between the contact size and the point in the compression cycle where the contact formed. Those particles that are initially far apart will form small contacts with fewer edges. For compacts formed from mixtures of spheres with differing sizes, the mean coordination number approaches a limiting value of 14 contacts at full density. However, the larger spheres have more faces and the smaller spheres have fewer faces. Table 9.2 demonstrates the distribution in contacts for three mixtures of spheres with a size ratio of 2:1. As the volume fraction of small particles

Table 9.2

Contacts Per Particle at 100% Density for Bimodal Mixtures [53].

Size Ratio = 2:1

volume fraction small	small population	large coordination number	small coordination number	mean coordination number
0.20	0.67	19.98	9.50	13.0
0.50	0.89	25.62	12.04	13.6
0.80	0.97	30.18	13.30	13.8

increases, the large particles are engulfed and exhibit a large number of contacts. This behavior is analogous to that observed in the coordination number of randomly mixed spheres. Even though there are wide differences in the number of contacts per particle, the mode number of edges remains at five (pentagonal) for most all cases.

During compaction of ductile particles the contact area between particles will vary with the degree of densification. A usual assumption is that the area can be approximated by a circle of diameter X and normalized by the particle diameter D. Observations of particles at intermediate levels of deformation confirm that the contacts are largely circular until the contact zones merge at a fractional density of approximately 0.75. The resulting relation between the fractional density and contact size is estimated as [64]

$$X/D = [1 - (f_o/f)^{2/3}]^{1/2} \tag{9.13}$$

or alternatively as [48]

$$X/D = [3/(4\pi)(f - f_o)]^{1/2} \tag{9.14}$$

where f_o is the initial packing density ($X = 0$) and f is the packing density after deformation. Another derivation expresses the relation between the contact size and density in an exponential form [58]. Based on experimental observations, the mean coordination number can be estimated from the density and contact size as

$$\bar{N}_c = -6.32 + 23.4\,f - 24.4\,f\,(X/D)^2\,. \tag{9.15}$$

Because of nonuniform deformation, Equations 9.13 and 9.14 tend to underestimate the contact zone size. On a more empirical level, the following relation is reasonable [16]:

$$X/D = A\,ln(P) - B \tag{9.16}$$

with A and B equal to adjustable constants.

In summary, the compaction of spherical particles involves a transition in shape from spheres to polyhedra as the pressure is increased. Both the coordination number and contact size increase as the particles are compacted, while the free surface area decreases. For randomly packed spheres the eventual shape is a 14-sided polyhedron with approximately 50% of the faces having a pentagonal shape. If the initial packing is an ordered close-packed structure, then particle sliding is inhibited and the coordination number remains constant at 12. In mixtures of differing sized spheres, the mean coordination number increases toward a value of 14, although the larger spheres will have more than 14 contacts and the smaller spheres will have less than 14 contacts. Furthermore, in the case of constrained uniaxial compaction, the contacts perpendicular to the compaction axis are larger than those parallel to the direction of compaction. There has not been a similar level of study for nonspherical particle shapes, but for equiaxed particles the same principles hold.

Effects of Material and Particle Characteristics

Several efforts have been made to link the dominant material and particle characteristics to the compaction process. Compactability depends directly on material properties like strain hardening, yield strength, and hardness [7,22,23,45,65-68]. For example, Figure 9.10 demonstrates a universal behavior for metal particles. This figure gives the fractional density versus the compaction pressure divided by the bulk material yield strength for nickel, copper, tungsten, and iron particles [26]. Generally, a lower hardness, strength, or strain hardening rate allows a higher degree of compaction.

An obvious concern is the difference between ductile and brittle particles; deformation versus fragmentation. From a phenomenological view, such mixed behavior is best represented by an equation involving two or more exponential pressure terms. The effect of particle hardness has been studied for oxide materials, where a high particle hardness hinders compaction as illustrated in Figure 9.11 [69]. This figure plots the degree of densification during compaction for four ionic materials of varying hardness but equivalent particle sizes. The higher the hardness, the lower the density

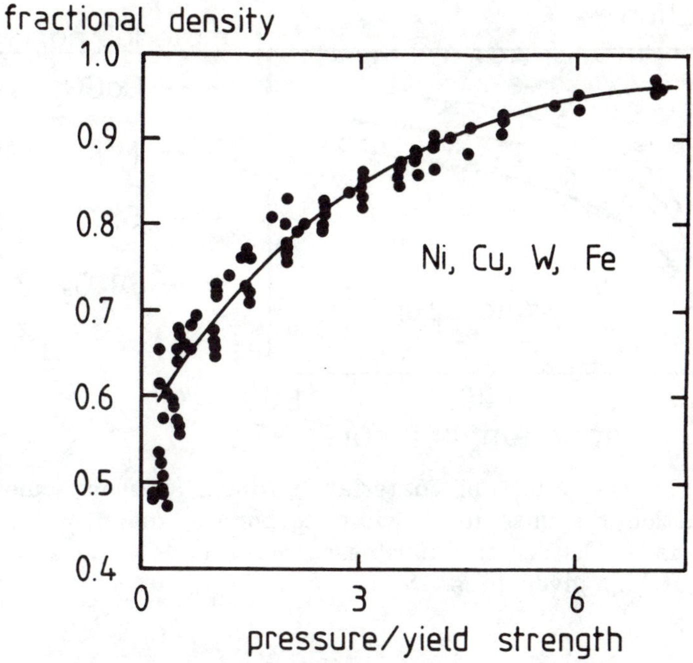

Figure 9.10 Fractional density versus the normalized compaction pressure (pressure divided by bulk material yield strength) for materials that deform during compaction [26].

for any compaction pressure. For the harder particles, both fracture and deformation are observed. The fragments formed during compaction have a wide range of sizes. Because of the size range and limited fragment mobility, the rate of pore filling by fragmentation is relatively inefficient. Consequently, hard and brittle materials are difficult to compact and high pressures are required in practice. The surface condition of the powder has an influence on compaction, especially for irregular particles [16,70]. Hybrid behavior occurs when a mixture of hard and soft particles is compacted. The compressibility is dominated by the connectivity of the hard phase; when it is interconnected the compactability is poor [71].

Particle size and shape control the interparticle friction, packing density, and pore size. Larger particles give an initially higher packing density and exhibit higher densities at all compaction

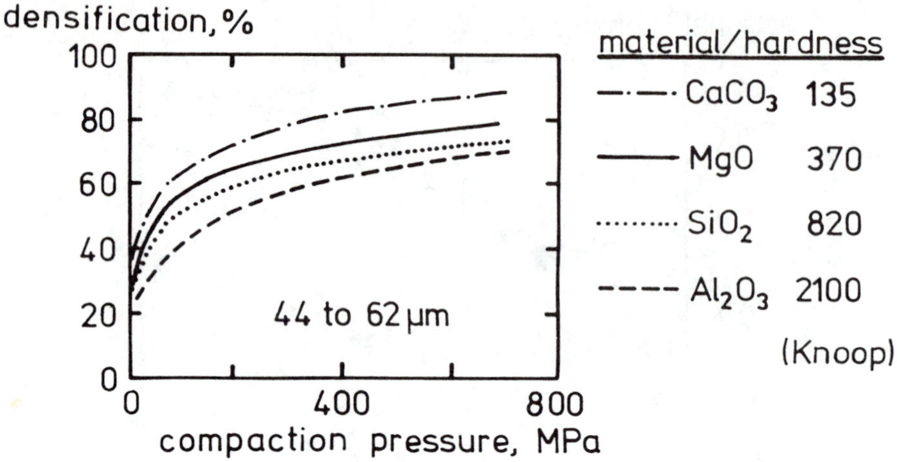

Figure 9.11 The effect of material hardness (Knoop scale) on the compaction response for calcium carbonate, magnesia, silica, and alumina [69]. As the hardness increases the densification decreases at any given pressure.

pressures. The smaller particles are more difficult to compact because of the greater number of point contacts, higher strain hardening rate, and higher stress concentration at the small pores [24,33]. Since pore size scales with particle size, compaction of small particles is more difficult. Also, for small particles the number of particle contacts is larger, and the deformation zones within each particle merge at lower pressures. Hence, densification is easier for large particles. A wide particle size distribution increases the density at any compaction pressure [72].

Particle shape has a large effect on the initial packing density and rearrangement behavior. For spherical particles the rearrangement contribution to densification is relatively small and isotropic. In contrast, irregular particles undergo considerable rearrangement at low compaction pressures. Consequently, there is a very strong sensitivity to particle shape [73], with the larger particles exhibiting the greatest level of orientation. This behavior can be used to create oriented structures with specific crystallographic texture [11,74]. This has been exploited in the fabrication of improved superconducting ceramic structures.

Particles with internal porosity, including agglomerated particles, are difficult to compact. Such particles exhibit two stages of compaction. The first stage involves the collapse of the larger interparticle pores, while the higher pressure stage involves elimination of the intraparticle pores. Consequently, porous and agglomerated particles exhibit high initial compressibilities, but resist compaction to high final densities [75,76].

The equation linking density and pressure must account for these various factors to adequately model the observed behavior. For ductile powders that do not fragment, the compaction behavior is adequately represented by terms dependent on the yield strength and rate of strain hardening. At the higher pressures, the rate of strain hardening controls compaction. Those factors which impair compressibility include small particle sizes, irregular particle shapes, high interparticle friction, high hardness, high yield strength, rapid strain hardening, internal porosity, agglomeration, and low initial packing fractions.

Load Transmission in Packed Particles

Beyond the particle and material characteristics, there are effects from the compaction geometry to be considered. Most cases of particle compaction involve uniaxial loading in rigid tooling where most of the particle motion is along the compact axis. The friction between the die wall and the powder causes a decrease in the pressure with depth in the particle bed. Particles under load exhibit a pseudo-fluid character that converts the uniaxial stress into a radial stress. Consequently, the redistribution in stress leads to a decrease in pressure with depth in a powder bed [23,36,37,77-82];

$$P = P_o \exp(-4uzx/D_c) \qquad (9.17)$$

where P is the pressure at a point x below the top surface, P_o is the applied pressure, u is the coefficient of friction between the powder and die wall, z is the ratio of radial to axial stress, and D_c is the diameter of the compact. The parameter z is a characteristic related to the powder friction. Equation 9.17 is applicable to single action pressing, where the load is applied from only one direction. This equation shows that the pressure decreases with depth in the

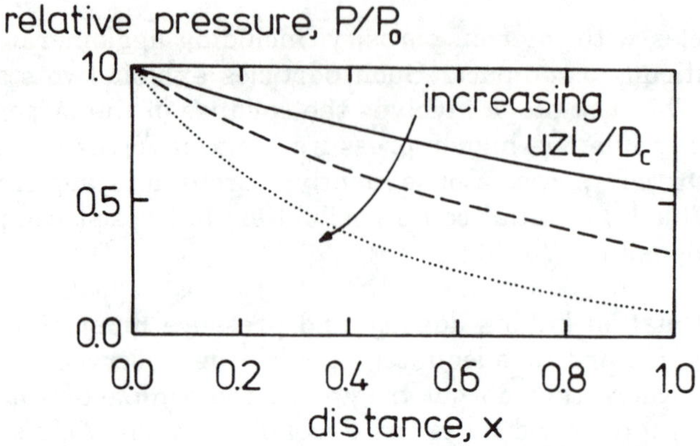

Figure 9.12 Pressure variation with depth in a particle compact for three possible values of the parameter uzL/D_c, where L is the compact thickness.

particle bed. A sketch of this behavior is given in Figure 9.12. This plot shows the ratio P/P_o versus the distance x for three hypothetical combinations of parameters. As the dimensionless group zuL/D_c increases, the inhomogeneities in pressure (and density) increase. An alternative plot of the this behavior is given in Figure

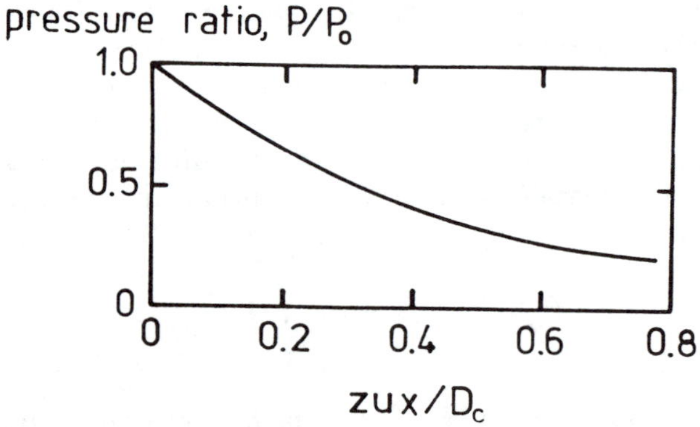

Figure 9.13 The pressure divided by the applied pressure shown as a function of the generalized compaction parameter zux/D_c where x is the depth into the compact.

9.13, where the pressure ratio is shown versus the parameter group zux/D_c. This form demonstrates the interaction of compact depth, diameter, and friction on the pressure distribution.

The decline in pressure between the top and bottom of the compact depends on the parameter uzL/D_c, where L is the compact thickness. Figure 9.14 illustrates the effects of the compact thickness-to-diameter ratio by plotting the bottom to top pressure ratio versus the thickness-to-diameter ratio [80]. This plot is for copper compacted with upper punch pressures of 690 and 276 MPa

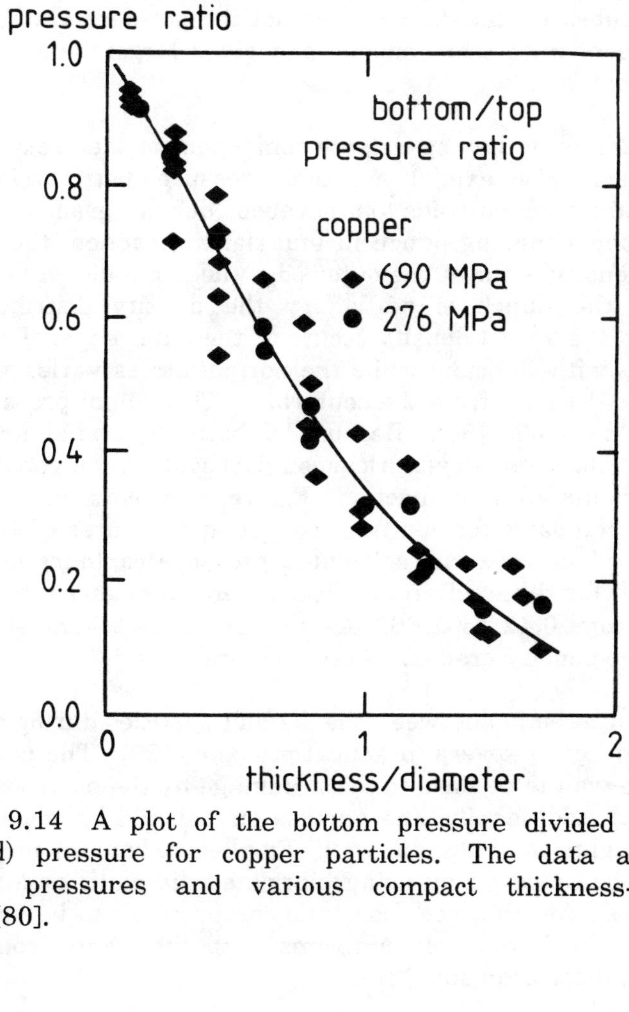

Figure 9.14 A plot of the bottom pressure divided by the top (applied) pressure for copper particles. The data are for two applied pressures and various compact thickness-to-diameter ratios [80].

[83]. This illustrates the detrimental effects of long, thin compacts upon density uniformity.

As the friction between particles is varied, differences in pressure and density gradients are possible. Experiments by Kamm et al. [78,84] demonstrated these gradients by using lead grids embedded in the powder compacts. X-ray photographs show the uneven densification and die wall friction effect. Thus, a main adjustable parameter in compaction is the die wall friction. Lubricants lower the coefficient of friction, but when mixed directly with the particles, a lubricant also increases the parameter z. Thus, it is best to lubricate the die wall and not the powder. Unlubricated compaction consumes more energy and gives larger density gradients in the compact.

In addition to having a depth component of the pressure gradient, compacts also exhibit a radial pressure distribution. This radial pressure gradient leads to a parabolic density gradient [78,80, 83-86]. Under a moving punch in uniaxial compaction, the highest stress and density occur at the outer edge (near the die wall). Alternatively, if the punch is stationary the density distribution is reversed and the lowest density occurs at the outer edge. The shear stress varies with the cube while the normal stress varies with the square of the distance from the centerline. This radial pressure distribution is parabolic [80]. Because of both the radial and axial pressure distributions, a cylindrical compact will exhibit considerable density gradients after compaction. Figure 9.15 plots the isodensity contours for compacts formed from copper at pressures of 700 MPa [83]. These plots show the calculated pressure contours in half of each compact for three different thickness-to-diameter ratios (L/D_c), increasing from 0.42 to 1.66. As the thickness-to-diameter ratio increases, the density gradients become more severe.

The interaction between the die and particles during compaction is the source of several practical problems [20]. The preference is for the die wall to be hard to avoid damage by the particles during deformation. Additionally, the particle size should be much larger than the roughness of the die wall. Further reductions in die wall friction can be achieved by using lubricated die walls or agglomerated particles. Alternatively, isostatic compaction can be used with a flexible die wall to avoid problems with lubricants, component shapes, or density gradients [6].

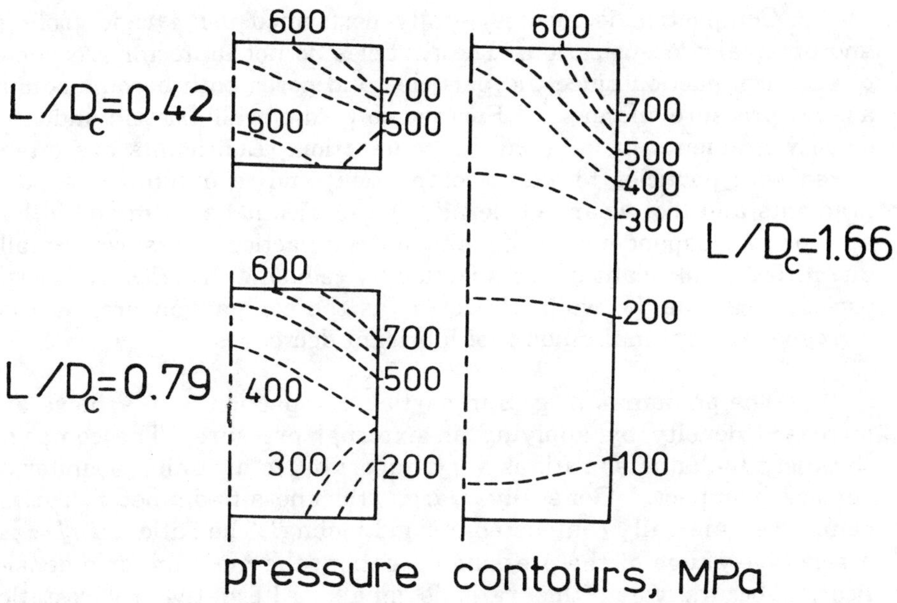

Figure 9.15 The calculated pressure distributions in split compacts of copper pressed at 700 MPa. The effect of an increase in the thickness-to-diameter ratio is to create greater inhomogeneity [83].

Summary and Technological Implications

Pressure is used routinely to form loose particles into controlled bulk shapes with prescribed dimensions and increased packing density. The compaction process involves both rearrangement and deformation of the particles, with resulting particle fracture in some cases. The generalized sequence of stages involves particle rearrangement followed by localized deformation. As the compaction pressure is increased the zone of deformation expands to eventually encompass the entire particle. Because of shear during compaction, interparticle bonds develop which give an increased strength to the particle mass. The densification mechanism is dependent on the specific powder and material characteristics. These factors include the material hardness, yield strength, strain hardening, ductility, fracture toughness, particle size, and particle shape.

Compaction is most typically performed using rigid tooling and uniaxial pressurization. The particles do not uniformly respond to such compaction since the particles and pores both become squat as the pressure increases. Furthermore, die wall friction induces density gradients in uniaxial die compaction. Lubricants are often mixed with particles to aid in compaction, and to minimize density gradients and tool wear. Generally the lubricant has a mixed influence on the dependence of density on compaction pressure. Small quantities of lubricant aid compaction by reducing the friction forces, especially at the die wall. However, as the compaction pressure is increased the optimal amount of lubricant decreases.

The predominant goal in particle compaction is to achieve an increased density by applying an external pressure. The compact thickness-to-diameter ratio is very important to attaining a uniform density compact. Generally, large thickness-to-diameter ratios cannot be uniaxially compacted in rigid tooling. The ratio uzL/D_c is a sensitive gauge of the pressing operation; the most uniform densification occurs when this ratio is small. Alternative hydrostatic compaction technologies provide isotropic pressurization and result in homogeneous compacts. Although uniaxial compaction is the most common process, there are also several other compaction processes. These include rolling, isostatic compaction, forging, shock wave consolidation, and various processes at elevated temperatures where particles are softer.

Various literature compilations and reviews are available for those interested in further details on compaction [2,8,9,14,20,87]. Generalization in compaction is difficult since there are three categories of variables that must be considered: powder characteristics (size, shape), material properties (hardness, plasticity), and operation variables (lubrication, strain rate). All of these have a measurable role in particle compaction.

References

1. F. V. Lenel, *Powder Metallurgy Principles and Applications*, Metal Powder Industries Federation, Princeton, NJ, 1980.
2. K. Shinohara, "Rheological Property of Particulate Solids," *Handbook of Powder Science and Technology*, M. E. Fayed and L. Otten (eds.), Van Nostrand Reinhold, New York, NY, 1984,

pp. 129-169.
3. E. Klar (ed.), *Powder Metallurgy Applications, Advantages and Limitations*, American Society for Metals, Metals Park, OH, 1983.
4. J. A. Greenwood and J. H. Tripp, "The Elastic Contact of Rough Spheres," *Journal of Applied Mechanics*, 1967, vol. 89, pp. 153-159.
5. H. F. Fischmeister, E. Arzt and L. R. Olsson, "Particle Deformation and Sliding During Compaction of Spherical Powders: A Study by Quantitative Metallography," *Powder Metallurgy*, 1978, vol. 21, pp. 179-187.
6. P. Popper, *Isostatic Pressing*, Heyden, London, UK, 1976.
7. R. W. Heckel, "An Analysis of Powder Compaction Phenomena," *Transactions of the Metallurgical Society of AIME*, 1961, vol. 221, pp. 1001-1008.
8. R. P. Seelig, "Introduction to Seminar - Review of Literature on Pressing of Metal Powders," *Transactions of AIME*, 1947, vol. 171, pp. 506-518.
9. W. D. Jones, *Fundamental Principles of Powder Metallurgy*, Edward Arnold Publishers, London, UK, 1960, pp. 270-319.
10. J. D. Bernal, K. R. Knight and I. Cherry, "Growth of Crystals from Random Close Packing," *Nature*, 1964, vol. 202, pp. 852-854.
11. J. R. F. Arthur and T. Dunstan, "Radiological Techniques Developed to Describe Particle Packing," *Powder Technology*, 1969, vol. 3, pp. 195-207.
12. P. J. James, "Fundamental Aspects of the Consolidation of Powders," *Powder Metallurgy International*, 1972, vol. 4, pp. 82-85, 145-149, 193-198.
13. B. O. Sundstrom and H. F. Fischmeister, "A Continuum Mechanical Model for Hot and Cold Compaction," *Powder Metallurgy International*, 1973, vol. 5, pp. 171-174.
14. H. F. Fischmeister, "Powder Compaction: Fundamentals and Recent Developments," *Proceedings of the Institution of Mechanical Engineers*, 1982, vol. 196, pp. 105-121.
15. H. Kuno and M. Tsuchiya, "The Use of Cumulative Pore Volume Difference in the Study of Pore Changes Occurring During Compaction or Sintering," *Powder Technology*, 1987, vol. 52, pp. 187-192,
16. J. S. Hardman and B. A. Lilley, "Mechanisms of Compaction of Powdered Materials," *Proceedings of the Royal Society of London A*, 1973, vol. A333, pp. 183-199.

17. J. D. Bernal, S. V. King and J. L. Finney, "Random Close-Packed Hard-Sphere Model," *Discussions of the Faraday Society*, 1967, vol. 43, pp. 60-69.
18. J. W. Marvin, "Cell Shape Phenomena Interpreted in Terms of Compressed Lead Shot," *Science*, 1937, vol. 26, pp. 493-494.
19. J. W. Marvin, "The Shape of Compressed Lead Shot and its Relation to Cell Shape," *American Journal of Botany*, 1939, vol. 26, pp. 280-288.
20. J. T. A. M. Welzen, "Die Compaction of Ceramic Powders," *Journal of Materials Education*, 1986, vol. 8, pp. 187-238.
21. M. C. Kostelnik, F. H. Kludt and J. K. Beddow, "The Initial Stage of Compaction of Metal Powders in a Die," *International Journal of Powder Metallurgy*, 1968, vol. 4, no. 4, pp. 19-27.
22. G. G. Brown, *Unit Operations*, John Wiley & Sons, New York, NY, 1950.
23. R. M. German, *Powder Metallurgy Science*, Metal Powder Industries Federation, Princeton, NJ, 1984.
24. E. Klar, "Relationships Between Pore Characteristics and Compacting Properties of Copper Powders," *Journal of Materials*, 1972, vol. 7, pp. 418-424.
25. R. L. Hewitt, W. Wallace and M. C. de Malherbe, "Plastic Deformation in Metal Powder Compaction," *Powder Metallurgy*, 1974, vol. 17, pp. 1-12.
26. R. W. Heckel, "A Normalized Density-Pressure Curve for Powder Compaction," *Transactions of the Metallurgical Society of AIME*, 1962, vol. 224, pp. 1073-1074.
27. G. Bockstiegel, "The Porosity - Pressure Curve and its Relation to the Pore Size Distribution in Iron Powder Compacts," *Modern Developments in Powder Metallurgy*, vol. 1, H. H. Hausner (ed.), Plenum Press, New York, NY, 1966, pp. 155-187.
28. L. A. Mitchell and M. D. Rowe, "Influence of Asperity Deformation Mode on Gas Leakage Between Contacting Surfaces," *Journal of Mechanical Engineering Science*, 1969, vol. 11, pp. 534-545.
29. S. Timoshenko and J. N. Goodierder, *Theory of Elasticity*, McGraw-Hill Book Co., New York, NY, 1951, pp. 372-384.
30. H. Deresiewicz, "Mechanics of Granular Matter," *Advances in Applied Mechanics*, vol. 5, H. L. Dryden and T. von Karman (eds.), Academic Press, New York, NY, 1958, pp. 233-306.
31. K. Kawakita and K. H. Ludde, "Some Considerations on Powder Compression Equations," *Powder Technology*, 1970, vol. 4, pp. 61-68.

32. R. W. Heckel, "Density-Pressure Relationships in Powder Compaction," *Transactions of the Metallurgical Society of AIME*, 1961, vol. 221, pp. 671-675.
33. C. L. Huffine and C. F. Bonilla, "Particle Size Effects in the Compression of Powders," *American Institute of Chemical Engineers Journal*, 1962, vol. 8, pp. 490-493.
34. I. Shapiro and I. M. Kolthoff, "Studies on the Aging of Precipitates and Coprecipitation. XXXVIII, The Compressibility of Silver Bromide Powders," *Journal of Physical Chemistry*, 1947, vol. 51, pp. 483-493.
35. K. Kim, "Hot Compaction Equations for Metal Powders and Porous Preforms," *International Journal of Powder Metallurgy*, 1988, vol. 24, pp. 31-37.
36. J. H. Shaxby and J. C. Evans, "On the Properties of Powders. The Variation of Pressure with Depth in Columns of Powders," *Transactions of the Faraday Society*, 1923, vol. 19, pp. 60-72.
37. E. E. Walker, "The Properties of Powders. Part VI. The Distribution of Densities in Columns of Compressed Powder," *Transactions of the Faraday Society*, 1923, vol. 19, pp. 83-86.
38. I. Fatt, "Compressibility of a Sphere Pack - Comparison of Theory and Experiment," *Journal of Applied Mechanics*, 1957, vol. 79, pp. 148-149.
39. G. W. Halldin, K. T. Yung and T. H. Tsai, "A New Analysis of Compressibility and Green Strength of Metal Powders," *Progress in Powder Metallurgy*, 1981, vol. 37, pp. 383-391.
40. D. Duzevic, V. Petrovic and M. M. Ristic, "Compacting Phenomenology of Coarse Titanium Powders," *Science of Sintering*, 1984, vol. 16, pp. 21-28.
41. G. B. Smith, "Compressibility Factor Development of a General Formula," *Metal Industry (London)*, 1948, vol. 72, pp. 427-428.
42. J. K. Beddow, *Particulate Science and Technology*, Chemical Publishing Co., New York, NY, 1980.
43. G. Bockstiegel, "Relations Between Pore Structure and Densification Mechanism in the Compacting of Iron Powders," *Iron Powder Metallurgy*, H. H. Hausner, K. H. Roll and P. K. Johnson (eds.), Plenum Press, New York, NY, 1968, pp. 54-71.
44. M. J. Donachie and M. F. Burr, "Effects of Pressing on Metal Powders," *Journal of Metals*, 1963, vol. 15, pp. 849-854.
45. A. K. Kakar and A. C. D. Chaklader, "Deformation Theory of Hot Pressing-Yield Criterion," *Transactions of the Metallurgical Society of AIME*, 1968, vol. 242, pp. 1117-1120.
46. M. M. Carroll and K. T. Kim, "Pressure - Density Equations for

Porous Metals and Metal Powders," *Powder Metallurgy*, 1984, vol. 27, pp. 153-159.
47. R. A. Davis and H. Deresiewicz, "A Discrete Probabilistic Model for Mechanical Response of a Granular Medium," *Acta Mechanica*, 1977, vol. 27, pp. 69-89.
48. J. V. Kumar, *Physics of Powder Consolidation*, Report MDN R&D 10 1987, Mishra Dhatu Nigam Limited, Hyderabad, India, 1987.
49. E. Arzt, "The Influence of an Increasing Particle Coordination on the Densification of Spherical Powders," *Acta Metallurgica*, 1982, vol. 30, pp. 1883-1890.
50. D. Bouvard and E. Ouerdraogo, "A Simulation Material for Modelling the Hot-Forming of Metal Powders," *Powder Technology*, 1986, vol. 46, pp. 255-262.
51. R. E. Collins, *Flow of Fluids Through Porous Media*, Reinhold Publishing Corp., New York, NY, 1961.
52. C. S. Smith, "Further Notes on the Shape of Metal Grains: Space-Filling Polyhedra with Unlimited Sharing of Corners and Faces," *Acta Metallurgica*, 1953, vol. 1, pp. 295-300.
53. E. B. Matzke, "Volume-Shape Relationships in Lead Shot and Their Bearing on Cell Shapes," *American Journal of Botany*, 1939, vol. 26, pp. 280-295.
54. F. T. Lewis, "The Shape of Compressed Spheres," *Science*, 1937, vol. 86, pp. 609-611.
55. M. Gardner, "Reflections on the Packing of Spheres," *Scientific American*, 1960, vol. 202, no. 5, pp. 174-184.
56. L. Oger, J. P. Troadec, D. Bideau, J. A. Dodds and M. J. Powell, "Properties of Disordered Sphere Packings I. Geometric Structure: Statistical Model, Numerical Simulations and Experimental Results," *Powder Technology*, 1986, vol. 46, pp. 121-131.
57. P. S. Clarke, J. W. Orton and A. J. Guest, "Electrical Conductivity and Hall Effect Measurements in Semiconducting Powders. Study of Percolation Effects," *Physical Review B*, 1978, vol. 18, pp. 1813-1817.
58. A. K. Kakar and A. C. D. Chaklader, "Deformation Theory of Hot-Pressing," *Journal of Applied Physics*, 1967, vol. 38, pp. 3223-3230.
59. J. A. Dodds, "The Porosity and Contact Points in Multicomponent Random Packings Calculated by a Simple Statistical Geometric Model," *Journal of Colloid and Interface Science*, 1980, vol. 77, pp. 317-327.

60. D. P. Haughey and G. S. G. Beveridge, "Structural Properties of Packed Beds - A Review," *The Canadian Journal of Chemical Engineering*, 1969, vol. 47, pp. 130-140.
61. S. Melmore, "Densest Packing of Equal Spheres," *Nature*, 1947, vol. 159, p.817.
62. R. G. Carlson and F. E. Westermann, "Hot Pressing of Lead Spheres," *Planseeberichte fuer Pulvermetallugie*, 1962, vol. 10, pp. 15-23.
63. A. K. Kakar and A. C. D. Chaklader, "Deformation Theory of Hot Pressing (Density Equation for Hot Pressing Conditions)," *Journal of Applied Physics*, 1968, vol. 39, pp. 2486-2488.
64. I. H. Moon and J. S. Choi, "Dependence of Green Strength on Contact Area Between Powder Particles for Spherical Copper Powder Compacts," *Powder Metallurgy*, 1985, vol. 28, pp. 21-26.
65. C. G. Goetzel, *Treatise on Powder Metallurgy*, volume 1, Interscience Publishers, New York, NY, 1949, pp. 259-312.
66. P. J. James, "Particle Deformation During Cold Isostatic Pressing of Metal Powders," *Powder Metallurgy*, 1977, vol. 20, pp. 199-204.
67. R. L. Hewitt, W. Wallace and M. C. deMalherbe, "The Effects of Strain Hardening in Powder Compaction," *Powder Metallurgy*, 1973, vol. 16, pp. 88-106.
68. A. Lawley, "Powder Consolidation," *Advances in Powder Technology*, G. Y. Chin (ed.), American Society for Metals, Metals Park, OH, 1982, pp. 75-97.
69. A. R. Cooper and L. E. Eaton, "Compaction Behavior of Several Ceramic Powders," *Journal of the American Ceramic Society*, 1962, vol. 45, pp. 97-101.
70. K. E. Easterling and A. R. Tholen, "The Role of Surface Energy and Powder Geometry in Powder Compaction," *Powder Metallurgy*, 1973, vol. 16, pp. 112-118.
71. N. B. Singh, "Pressure-Densification of Powder Composites," *Powder Metallurgy*, 1971, vol. 14, pp. 277-288.
72. R. M. Koerner, "Effects of Particle Size and Size Distribution on the Compaction of Iron and Nickel Powders," *Modern Developments in Powder Metallurgy*, vol. 4, H. H. Hausner (ed.), Plenum Press, New York, NY, 1971, pp. 115-126.
73. Paramanand and P. Ramakrishnan, "Effect of Powder Characteristics on Compaction Parameters and Ejection Pressure of Compacts," *Powder Metallurgy*, 1984, vol. 27, pp. 163-168.

74. N. S. Lidorecko, A. S. Kagan, S. P. Chizhik, A. E. Kovalshii, S. M. Prokoshina and O. V. Sokolov, "Packing Texture of Brittle Materials," *Soviet Powder Metallurgy and Metal Ceramics*, 1969, vol. 8, pp. 840-841.
75. F. W. Dynys and J. W. Halloran, "Compaction of Aggregated Alumina Powder," *Journal of the American Ceramic Society*, 1983, vol. 66, pp. 655-659.
76. D. E. Niesz, R. B. Bennett and M. J. Snyder, "Strength Characterization of Powder Aggregates," *Ceramic Bulletin*, 1972, vol. 51, pp. 677-680.
77. L. A. Shofman, "Dependence of the Density of Metal Powder Compacts on Compaction Pressure," *Soviet Powder Metallurgy Metal Ceramics*, 1968, vol. 7, pp. 596-603.
78. R. Kamm, M. Steinberg and J. Wulff, "Plastic Deformation in Metal Powder Compacts," *Transactions of AIME, Institute of Metals Division*, 1947, vol. 171, pp. 439-456.
79. D. Train, "Transmission of Forces Through a Powder Mass During the Process of Pelleting," *Transactions of the Institution of Chemical Engineers*, 1957, vol. 35, pp. 258-266.
80. R. A. Thompson, "Mechanics of Powder Pressing: I, Model for Powder Densification," *Ceramic Bulletin*, 1981, vol. 60, pp. 237-243.
81. E. G. Schwartz and A. S. Weinstein, "Model for Compaction of Ceramic Powders," *Journal of the American Ceramic Society*, 1965, vol. 48, pp. 346-350.
82. A. Stewart, "The Pelleting of Granular Materials," *Engineering*, 1950, vol. 169, pp. 175-177 and 203-204.
83. P. Duwez and L. Zwell, "Pressure Distribution in Compacting Metal Powders," *Transactions of the Metallurgical Society of AIME*, 1949, vol. 185, pp. 137-144.
84. R. Kamm, M. A. Steinberg and J. Wulff, "Lead-Grid Study of Metal Powder Compaction," *Transactions of the AIME*, 1949, vol. 180, pp. 694-706.
85. M. Y. Nazmy and M. S. Abdel-Azim, "Investigation of the Dimensional Changes in Soft Metal Powder Compacts," *Powder Metallurgy*, 1974, vol. 17, pp. 13-20.
86. G. C. Kuczynski and I. Zaplatynskj, "Density Distribution in Metal Powder Compacts," *Transactions of the Metallurgical Society of AIME*, 1956, vol. 206, p. 215.
87. H. H. Hausner, *Bibliography on the Compaction of Metal Powders*, Hoeganaes Corp., Riverton, NJ, 1967.

Chapter Ten

Connectivity in Mixtures of Two Phases

Chapter Outline

Introduction and Measurement of Connectivity

In some applications involving mixed particles of differing compositions, the connectivity is an important attribute. Connectivity is a measure of the degree of contact of one phase in a multiple phase mixture. It is expressed as the mean number of contacts one particle makes with similar composition particles. The connectivity differs from the coordination number which is a measure of the total number of contacts, independent of the composition of the contact. A fully connected structure would conceivably allow transport of heat or electricity throughout the compact via one phase without involving the other phase. This would be analogous to a set of interlinked caves, where it is possible to travel from any one cave to any other cave by a tortuous set of interconnections, without the need to dig any new passageways.

Maxwell was the first to consider the connectivity of a mixture. In 1881 he examined dispersed conducting particles in an insulating matrix in a desire to explain the conductivity variation with composition. Subsequently, the problem has been linked with percolation theory and its application to conductivity of filled polymers, antistatic coatings, and thin film resistors, as well as to disease control in orchards, liquid flow in soil, gel formation, permeability in porous structures, and magnetic behavior of ferrites [1-12]. Percolation models are composed of networks that involve nodes and connections between the nodes. In an ordered packing, the nodes are the lattice sites and the connections are the particle contacts. Node problems concern the proportion of occupied sites, while connection problems treat property variations with the number of broken interconnections between nodes. This chapter emphasizes the site problem, but the reader should realize that there are parallel treatments of problems that deal with missing connections between sites. In both cases, there is a sharp transition in behavior over a narrow composition range (composition measured by the fraction of completed connections or occupied sites). At a critical composition, the percolation network becomes connected over long-range distances. A characteristic feature of such events is a large change in properties with a small change in connectivity at the critical point. Zallen [13] lists several examples of such behavior.

The connectivity of mixed powders is simply the mean number of contacts with similar composition particles per particle. Figure 10.1 is an optical micrograph of a polished cross section through a composite material where the rounded particles have a mean connectivity of 1.2 contacts per particle in two dimensions. Figure 10.2 illustrates the basic concept of connectivity in two dimensions. In the disconnected case, the dark disks do not form a continuous chain. As the concentration of dark disks increases, a level is eventually attained where a continuous pathway between the dark disks exists throughout the structure. For a two-dimensional ordered lattice, it is observed that the number of disks needed for conduction varies with the coordination number [9,14].

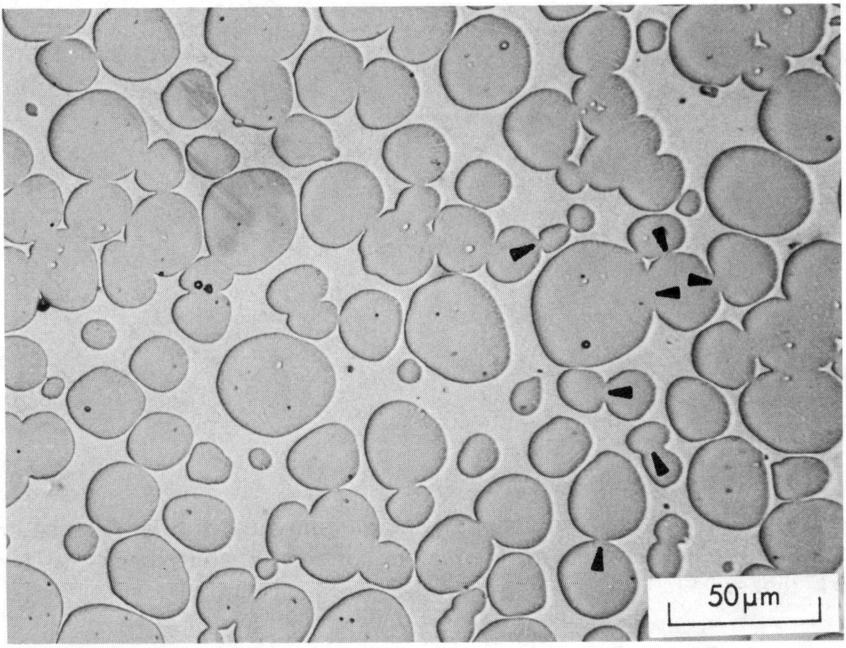

Figure 10.1 An optical micrograph of a liquid phase sintered composite with a connected microstructure of tungsten dispersed in a nickel-iron-tungsten matrix. A few connections are marked by arrowheads.

As an example, for a two-dimensional structure with a coordination of three, at least 70% of the disks need to be conducting to allow current flow through the network. When the two-dimensional packing is dense, with a coordination number of six, then 50% of the disks need to be conducting to have conductivity in the network. At the intermediate coordination of four, the required proportion of conducting disks is 59%. Unlike the three-dimensional case, a two-dimensional structure does not exhibit mutual interconnectivity. Thus, if the dark phase is connected over long ranges, then the light phase is disconnected. At intermediate compositions in three dimensions both phases can be interconnected. In two dimensions at least 50% of one phase is needed to establish connectivity, thereby denying connectivity to the other phase.

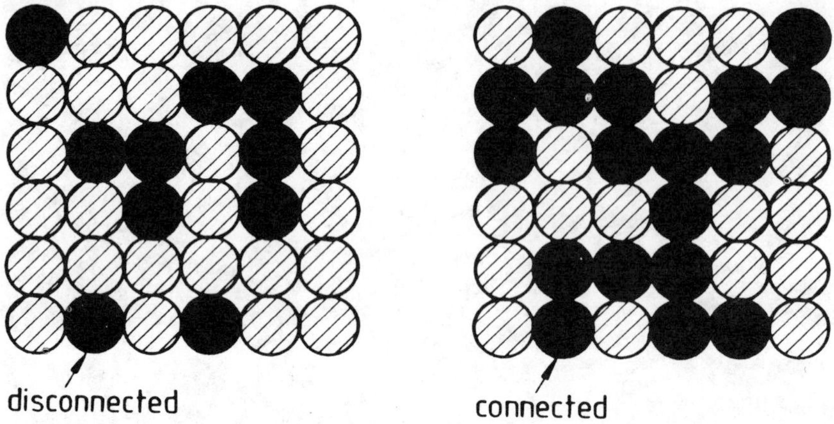

Figure 10.2 A two-dimensional sketch showing the difference in structure between disconnected and connected structures for the dark disks.

In three dimensions, the connectivity must often be measured indirectly. A typical method is to measure conductivity when the two phases differ greatly in conductivities [2,15]. In this case, the behavior shown in Figure 10.3 occurs [16]. This figure plots the calculated relative conductivity of the composite (normalized to the conductivity of phase 2), versus composition for an ordered simple cubic structure. The particles have been randomly substituted on lattice sites with conductivities of either σ_1 or σ_2. Three cases are shown in Figure 10.3, representing relative conductivities for phase 1 equal to 10^{-5}, 0.1, or 0.5, that of phase 2. Initially, at low concentrations of conducting phase, there is insufficient contact between conducting particles to alter the bulk conduction. When there is a large difference in conductivities, there is a threshold point at which the network becomes conductive. As seen for the case of $\sigma_1 / \sigma_2 = 10^{-5}$, a critical state occurs where the conductivity changes rapidly with a slight increase in concentration of the conducting phase. This critical concentration corresponds to the formation of a continuous network of conducting phase. In addition to measuring conductivity, connectivity can also be determined using thermal transport, selective leaching, metallography, corrosion, thermal expansion, or Hall effect

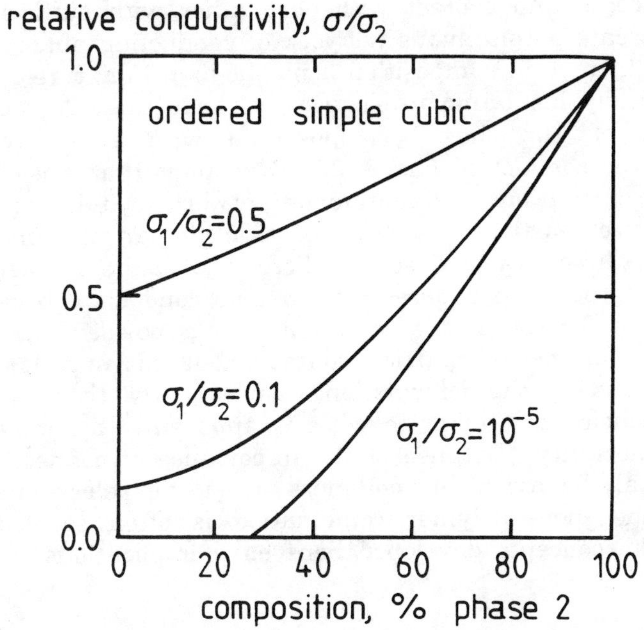

Figure 10.3 The relative conductivity of a three-dimensional ordered simple cubic lattice versus composition for three different conductivity ratios of the component phases.

measurements [2,15,17-21]. Optical microscopic observations on a polished cross section is a common approach, but this is only a two-dimensional view of the three-dimensional structure. Consequently, the level of connection between particles will appear low.

A primary application of connectivity concepts is the electrical conductivity of a plastic-metal composites. In contrast with electrical behavior, the thermal conductivity shows less variation with composition, but is also dependent on connectivity [22]. This is expected based on Figure 10.3, since thermal conductivities generally vary less between materials than electrical conductivities. As a simple example, consider the fabrication of a polymer filled with spherical metal particles. In a random mixture the composite will be nonconducting at low concentrations of metal. However, as the metal concentration is increased, the metallic spheres will eventually come into contact with each other with sufficient frequency to create a continuous network of conducting phase. Thus, over a small range in composition a plastic-metal mixture converts from electrically insulating to electrically conducting. A change in conductivity of 10^{20} has been observed with a 5% (volume) increase in the conducting phase [23]. Questions that arise in considering the formation of a continuous network include the effects of particle size, particle size ratio, particle shape, and amount of each phase. In a practical sense, there is also the problem that point contacts between spheres may not be conductive because of surface films. Pressure is often applied to a powder mixture to rupture films at the interparticle contacts, but this may form new contacts because of the deformation. Consequently, there is opportunity for considerable error in what is apparently a simple measure of connectivity. Because of the importance of connectivity to the electrical behavior of polymers used for electromagnetic shielding, the metal-polymer combination is often studied [24]. However, the concepts developed here have implications in many fields.

In this treatment of connectivity, the problem is divided into several segments. The first problem considered is the formation of a connected structure of monosized spheres in a fluid-like matrix. This represents a random mixture of spheres similar to that which exists in metal-filled polymers. Next, the problem of a mixture of equal sized spheres is treated. Here the emphasis is on the packing density and the volume concentration of each phase. This creates a need to discuss the role of interparticle porosity on the overall connectivity. As will be shown subsequently, mixtures of unequal size spheres alter the critical volume fraction for connectivity based on the particle size ratio. Finally, the effect of particle shape is discussed, although there is relatively little known about this area.

Monosized Spheres in a Matrix Phase

For the first treatment, a monosized spherical powder will be assumed to be randomly dispersed in a matrix phase. This avoids the complication induced by segregation of the particles. Additionally, it is assumed that the particles do not have any surface layers that inhibit contact formation. For such a situation it is estimated that the metal particle network will be fully connected and electrically conducting at a volume fraction near 0.3 and higher [4,25-28]. There is an upper limit for the metal content of this structure equal to the random dense packing fraction for the powder; for monosized spheres this is a fractional content of 0.64. Thus, electrically conductive metal-filled plastics can be fabricated from monosized spheres over the approximate range from 0.30 to 0.64.

To demonstrate these principles, Gurland [18] performed experiments with silver spheres dispersed in a phenolic matrix. Figure 10.4 shows the results of his experiments in a plot of resistivity on a logarithmic scale versus volume fraction of silver. A critical point for the formation of an interconnected network, and therefore a reduced resistivity, occurred at approximately 1.5

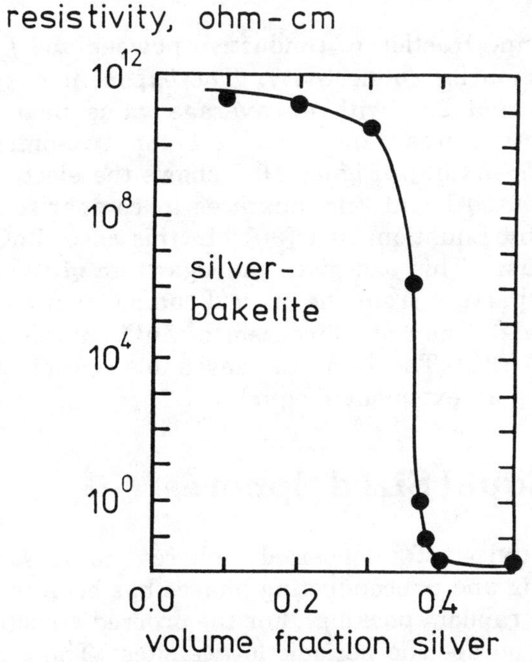

Figure 10.4 The resistivity of mixtures of silver powder in bakelite as a function of the composition of the mixture [18].

contacts per particle. This corresponds to a fractional density between 0.36 and 0.38 in Gurland's experiment. Subsequent analysis suggests that the silver spheres were not truly randomly packed; thus the experimental value for the critical packing fraction is slightly higher than the theoretical value.

Other treatments of this problem have assumed some form of ordered packing for the monosized spheres. In these cases, the critical volume fraction of solid phase randomly dispersed in the matrix is estimated to be near 0.3 [4,29]. Near the critical point, the conductivity of the mixture σ follows Equation 10.1 [1,4,27,30-32],

$$\sigma = \sigma_o (f - f_c)^s \tag{10.1}$$

where f is the volume fraction of conductive powder and f_c is the critical volume fraction for connectivity. The exponent s typically varies between 1.4 and 2.0, with an average value near 1.7 for three-dimensional structures (and near 1.4 for two-dimensional structures) [1,4,16,28-31,33]. Figure 10.5 shows the electrical conductivity behavior of ZnO and ZnS mixtures in comparison to the behavior predicted by Equation 10.1 [30]. In this case, ZnO is the more conductive phase. This plot gives the logarithm of the conductivity versus the departure from the critical composition $f - f_c$, also on a logarithmic scale. The critical content of ZnO for this mixture was estimated at 0.135. The fit is in reasonable agreement with Equation 10.1, giving an exponent s equal to 1.7.

Mixtures of Equal Sized Spheres

The connectivity of monosized spheres composed of a mixture of conducting and nonconducting phases has been treated as either an ordered or random packing. For the ordered structure, the spheres are located on specific periodic lattice sites. The calculated fraction of the spheres (randomly selected) that are needed to form a connected structure depends on the coordination number for the lattice. According to percolation theory, the critical point of network formation is near 1.5 contacts per particle. Denoting this critical number of contacts per particle as C_p allows calculation of the fraction of similar phase neighbors needed for long range connectivity in

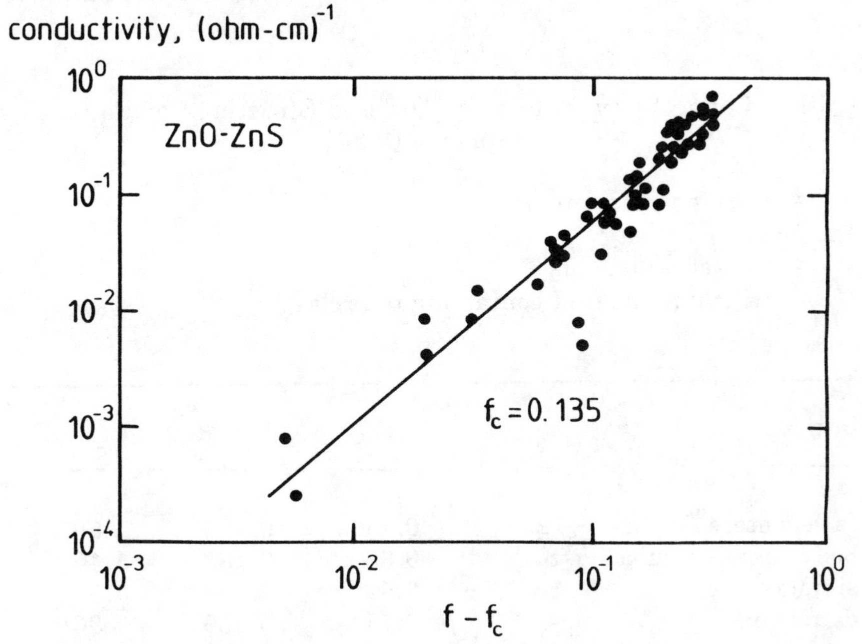

Figure 10.5 Conductivity of mixtures of ZnO and ZnS versus composition, expressed in terms of the departure from the critical composition of 13.5% ZnO [30].

terms of the packing coordination number N_c as follows:

$$C_p = N_c P_c \tag{10.2}$$

where P_c is the probability that a neighboring particle will be of the same composition at the critical condition. Table 10.1 lists the behavior for four ordered lattices, showing the coordination number N_c, packing density f, probability of similar neighbors at the critical condition P_c, and the critical number of contacts per particle C_p. Since the probability varies inversely with the coordination number, the value of C_p appears to be insensitive to the structure [29,34]. Other calculations for a connected structure give slightly different values

Table 10.1

Critical Connectivity Values for Ordered Structures of Equal Sized Spheres [2,34]

N_c = coordination number
f = packing density
P_c = contact probability
C_p = critical number of contacting particles

lattice	N_c	f	P_c	C_p
face-centered cubic	12	0.74	0.125	1.50
body-centered cubic	8	0.68	0.183	1.46
simple cubic	6	0.52	0.254	1.52
diamond	4	0.34	0.389	1.56

for P_c [3,28,35]. For the coordinations numbers of 12, 8, 6, and 4, the typical values are 0.20, 0.25, 0.31, and 0.43, respectively [5,9]. Other calculations place the value of C_p in the range of 1.7 to 2.4 contacts per particle. However, the critical number of contacts per particle is relatively insensitive to the structure for spherical particles, and roughly equals 1.5 [2]. The product of the fractional density and the probability of a similar composition neighbor is nearly constant. In all cases of ordered equal sized spheres, the fraction of conducting phase must exceed approximately 15% of the total volume for the composite to be conducting.

There is a complication associated with porosity in a random packing of spheres. As noted earlier, a mixture of equal sized spheres will not totally fill space. Indeed, at least 36% porosity can be expected with randomly mixed equal sized spheres. The pores then represent an insulating phase that decreases the connectivity of the conducting phase.

The critical number of contacts per particle needed for a connected structure in a random packing is also 1.5 [2,36]. Thus, for an average coordination number of 6 and a packing density of 0.64, the estimated critical probability of a contact is 0.25. This means that one out of four spheres must be conducting for the random mixture to be conducting. Assuming a mixture of plastic and metallic spheres of equal size, of the total volume 64% is solid, composed of at least three parts plastic spheres and one part metallic spheres. In terms of the total volume, this 25% metallic component constitutes a volume fraction of 0.16 at the critical point. Other estimates have placed the critical ratio of conducting particles at values ranging from approximately 1 in 3 to 1 in 6 (18 to 36%) for the case of randomly mixed equal sized spheres [26-29,36,37]. Part of the discrepancy arises from the problems of making good point contacts between the spheres. Not all contact points are good conductors, since contact resistance inhibits conduction. The application of pressure to the structure helps to break down the surface films that interfere with conduction. However, compaction also alters the structure since many close contacts can be completed, giving a higher conduction than should really be observed. This will increase the overall connectivity and will decrease the critical volume fraction at which connectivity first occurs.

In the case of a compacted mixture of equal size spheres, the coordination number will approach 14. In this case, both phases will be interconnected for compositions between 18 and 82% by volume [38]. Outside of this range the major phase will be connected and the minor phase will be disconnected. The maximum interconnectivity will occur in the composition range between 40 and 60% of each phase. Indeed, for a 50-50 composition with a coordination number of 14, there will be only 1 particle in 10,000 that is isolated from neighbors of similar composition.

Mixtures of Unequal Sized Spheres

In the previous section it was demonstrated that a substantial fraction of conducting phase was needed to make the mixture of equal sized spheres conducting. It is often desired to obtain conduction in the mixture using lower concentrations. This is best attained by using unequal particle sizes, where the smaller particles are the conducting phase. The small particles effectively coat the large particles, giving conduction at low concentrations because of segregation to the interparticle regions. Figure 10.6 illustrates this concept for four situations involving two particle size ratios and two concentrations of particles. Assuming the dark particles are conducting, then the mixture is conducting at a lower concentration for the case where the dark particles (phase 1) are much smaller than the light particles (phase 2). This is because of the formation of networks along the surfaces of the large particles. An extension of this concept to other particle shapes leads to the use of metallic fibers or flakes as ideal low concentration fillers for promoting conduction in an otherwise insulating material.

If the conducting particle size is small compared to the nonconducting particle size, then the network is treated as a segregated structure. For spheres the critical volume fraction of

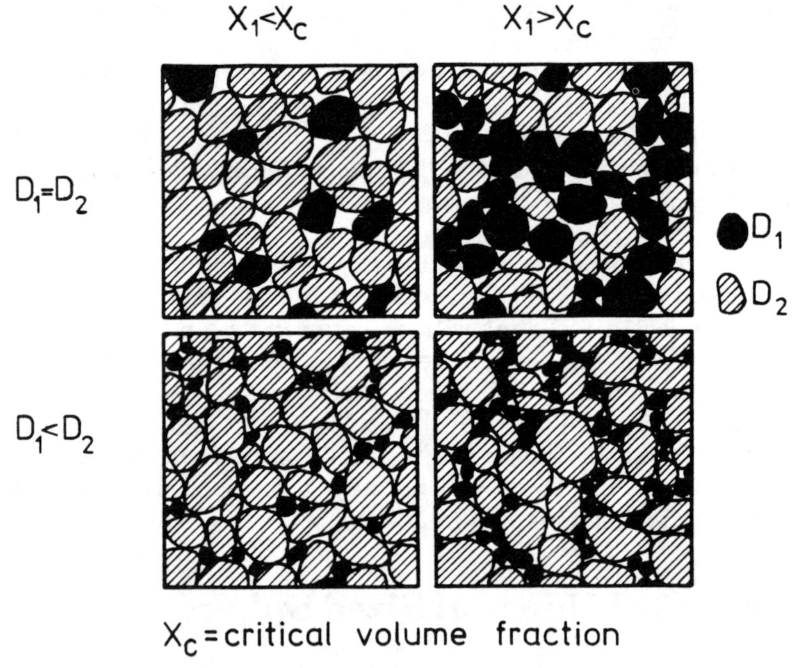

Figure 10.6 A two-dimensional sketch of the role of composition and particle size ratio on connectivity. Less of the dark phase is needed to form a connected network as the relative particle size of the phase decreases.

conducting phase X_c can be estimated in terms of the particle size ratio,

$$(1 - X_c)/X_c = 2.99\, D_i\ /D_c \qquad (10.3)$$

where D_i and D_c are the particle sizes for the insulating and conducting phases, respectively [15]. The first signs of conduction will occur at concentrations as low as 32.5% of the critical value calculated using Equation 10.3. There is a conductivity transition region near the critical composition, as described by Equation 10.1. Figure 10.7 plots the resistivity measurements of polymer-metal mixtures, with resistivity on a logarithmic scale and composition on a volume fraction scale [39]. The smaller the metal particle size with respect to the polymer, the lower the concentration needed for conduction. Equation 10.3 predicts that relatively small metallic particle sizes

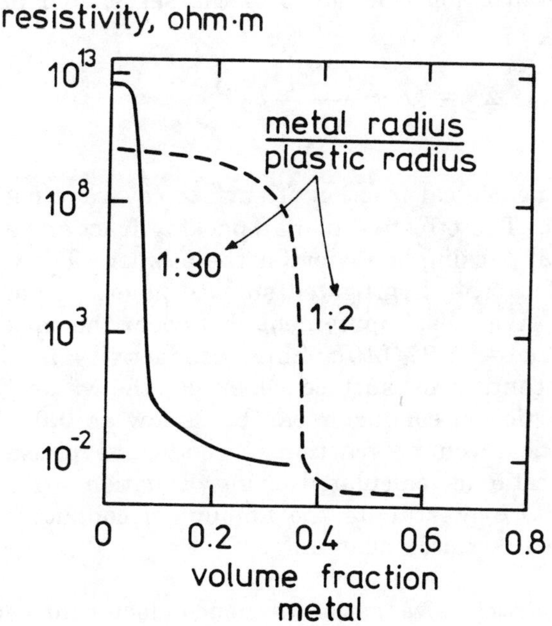

Figure 10.7 The resistivity of mixtures of plastic and metal powders versus the composition for two particle size ratios [39]. Conduction occurs at a lower concentration of metal for the smaller particle size ratio.

will provide conductivity to the mixture at low volume fractions. Indeed, with large particle size differences, volume fractions of 0.05 to 0.06 are used in practice.

Various models have been used to predict the relation between composition, particle size, and composite conductivity for spheres of unequal size. One can assume the particles are either randomly distributed or segregated [40]. Most of this work assumes the particle mixtures have been compacted to eliminate pores. In the case of randomly distributed particles, the critical amount of conducting phase is estimated from simple statistical arguments. However, this overestimates the critical amount of conducting phase for most structures. Alternatively, if the conducting particles are smaller than the nonconducting particles, then an assumption of segregation to the nonconductor surface proves valid. In this case, the critical volume fraction of conducting phase X_c is estimated from the effective planar packing fraction ω of the segregated phase as follows:

$$X_c = \left[1 + \frac{\omega D_i}{4 P_c D_c}\right]^{-1} \tag{10.4}$$

with P_c equal to the critical fraction of surface coverage needed for conductivity [2,40]. The effective planar packing fraction relates to the two-dimensional packing behavior for the powder. The value for ω ranges from 1.1 to 1.4. Segregated surface layers of small conducting particles give best agreement between the model and experiment when $\omega = 1.27$ [40]. Since conductivity is observed with less than a continuous surface coverage, the values for the critical volume fraction of conductor can be as low as 0.05. Figure 10.8 plots the critical volume fraction of conductive phase versus the particle size ratio as calculated using Equation 10.4 [2,40]. This figure tends to overestimate the amount of conductive phase needed for particle size ratios near unity.

At large particle size ratios of nonconductor to conductor (matrix to filler) and reasonably high volume fractions, both phases are interconnected. However, at small particle size ratios and low volume fractions of conductor phase, there is poor conduction. Between these two conditions is the transition region. Clearly, the volume fraction of conductive phase needed for a connected structure depends on the particle size ratio. For example, a size ratio of

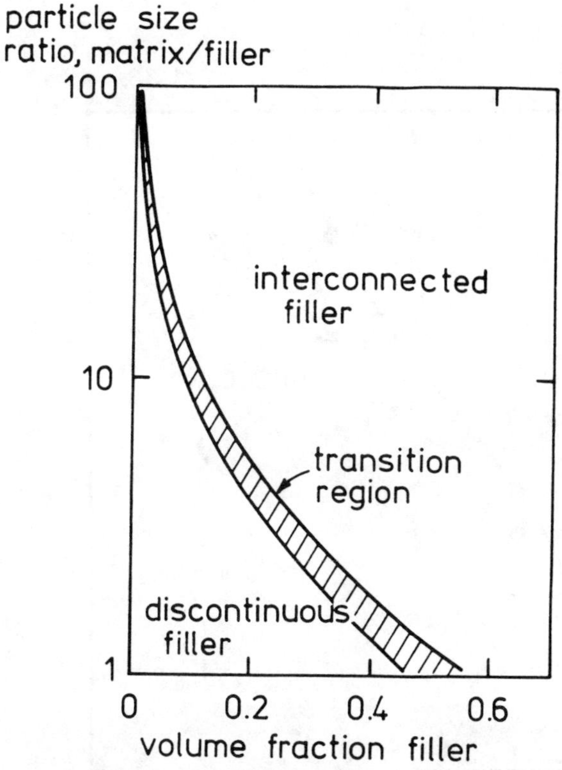

Figure 10.8 The boundary between a connected filler and disconnected filler as a function of the volume fraction and particle size ratio [2].

16:1 gives 0.07 as the critical volume fraction of conducting phase [41]. As the randomness of the structure increases there is less segregation of the conducting phase and a greater concentration is needed to establish connectivity.

A more general problem is the binary mixture of two differing particle sizes, where the conducting spheres can be either larger or smaller than the nonconducting spheres. For this general case, the larger the average coordination of a phase, the more continuous the chains of particle contacts and the greater its connectivity. Figure 10.9 shows the results for this general case [26]. The number fraction of the conductive phase is plotted versus

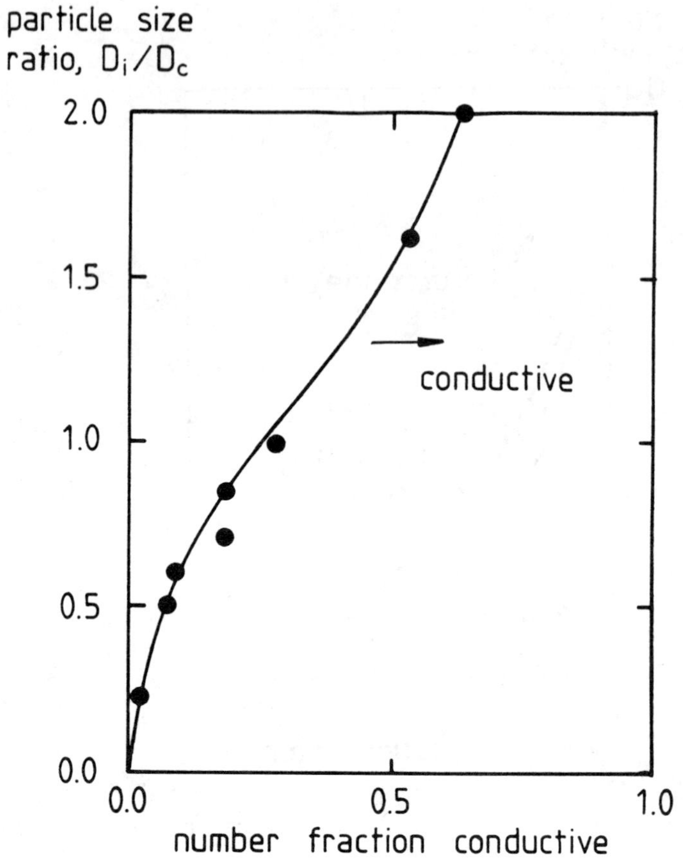

Figure 10.9 A plot of the conduction threshold based on number fraction versus the particle size ratio [26].

the particle size ratio. This plot shows that connectivity requires a larger number fraction as the particle size ratio is decreased, but that the volume fraction behaves in the opposite manner.

Finally, for the case where a particle size distribution is used rather than monosized spheres, the general effect is a decrease in connectivity for the conducting phase as the distribution widens. This relates to a lower probability of contact at a given volume fraction for the wider size distributions, because the smaller particles have lower coordination numbers than the larger particles. Since connectivity depends on a specific frequency of contact, a broad particle size distribution requires a higher concentration of conducting phase to attain a connected structure.

Particle Shape Effects

As already noted, particle shape has an important effect on connectivity. Generally, the focus is on techniques to improve conductivity of a metal-filled polymer by using nonspherical metallic particles [42]. Thus, for particles other than spheres, most of the attention has been on flake or fiber particle shapes.

The critical volume fraction for connectivity X_c depends on the aspect ratio of the conducting particle. As the conducting particles become elongated, the shape of the conductivity versus volume fraction curve remains similar to that shown earlier; however, the onset of conduction (connectivity) occurs at lower concentrations. One important case is a conducting fiber dispersed in a polymer matrix, where the polymer particle size is given as D_i. For a conducting fiber of circular cross section with diameter D_c and length L, the needed volume fraction for connectivity is shown in Figure 10.10 [2,22]. The needed volume fraction of fibers decreases as the length-to-diameter ratio increases. This is because an average of only two contacts per fiber are needed to maintain connectivity. With long randomly oriented fibers, considerable distances exist between contacts; hence, low concentrations are acceptable. In an extreme case of a length-to-diameter ratio of 1000:1, the needed volume fraction is only 0.005.

For a conducting fibrous shape with an elliptical cross section, let the length-to-maximum-diameter ratio be given by F_1 and let the maximum-to-minimum-diameter ratio of the elliptical shape be given by F_2. Then for this shape the critical volume fraction of fibers for connectivity is estimated as [15]

$$\frac{X_c}{1 - X_c} = \frac{2.3\, D_c F_1 F_2}{D_i (F_1 F_2 + F_1 + 1)} \qquad (10.5)$$

where D_c is the minimum diameter of the elliptical cross section. Equation 10.5 can be applied to most any nonspherical particle

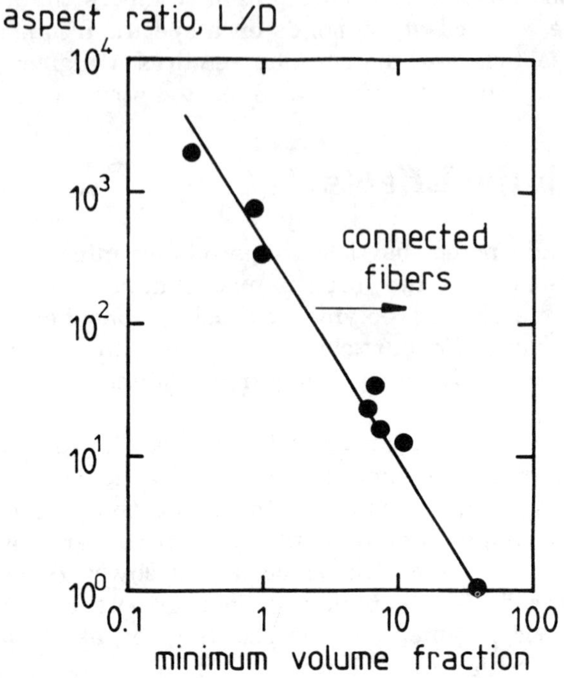

Figure 10.10 Relation between the fiber aspect ratio and minimum conductive filler concentration required to produce an electrically conductive mixture [2,22].

shape, where F_1 and F_2 are characteristic aspect ratios for the particles.

As illustrated by Equation 10.5, the volume fraction needed for fibers to form a connected structure depends on the fiber aspect ratio. Hence, long fibers are most beneficial to maintaining connectivity. Likewise, if the particles are flakes, the critical volume fraction is low. In all of these models there is a need to qualify the differences between a random structure and one where the conducting phase is segregated.

Summary

The connectivity of mixed particles is important when the phases are dissimilar in properties. Concern over connectivity is dominated by electrical conductivity in metal-filled polymers. Low concentrations of metallic particles are needed when the particle is a fiber or flake. Other applications of connectivity concepts are found in problems dealing with fluid flow, disease spreading, rheocasting, sintering, polymer structure, coatings, and alloy corrosion.

For the general case of mixed spheres of equal size, the minor phase will be connected when it constitutes roughly 30% of the particles. The actual value depends on the packing density and homogeneity. The critical point where the minor phase becomes connected throughout a packed structure is highly dependent on the homogeneity of the structure. A segregated structure becomes more probable as the particle size of the minor phase becomes smaller than the particle size of the major phase. In this case, the critical volume fraction decreases. In practical experiences with mixed spheres of unequal size, volume fractions as low as 0.05 have provided a conductive composite of metal particles mixed with nonconductive particles. This allows use of low concentrations of expensive materials in fabricating various devices such as electromagnetic shields and conducting pastes. Lower critical volume fractions are possible using nonspherical filler particle shapes. Connectivity depends on attaining a critical mean number of contacts per particle. With fibers this allows low concentrations of a phase while connectivity is still attained on a large scale.

References

1. T. J. Coutts, "Percolation Conduction in Mixed Insulator-Conductor Systems," *Thin Solid Films*, 1976, vol. 38, pp. 313-321.
2. D. M. Bigg, "Electrical Properties of Metal-Filled Polymer Composites," *Metal-Filled Polymers*, S. K. Bhattacharya (ed.), Marcel Dekker, New York, NY, 1986, pp. 165-226.
3. R. H. Wang and L. T. Fan, "Critical Contact Number for Percolation Processes," *Powder Technology*, 1979, vol. 23, pp. 139-142.

4. S. Kirkpatrick, "Percolation and Conduction," *Reviews of Modern Physics*, 1973, vol. 45, pp. 574-588.
5. D. Stauffer, *Introduction to Percolation Theory*, Taylor and Francis, London, UK, 1985.
6. G. Deutscher, R. Zallen and J. Adler (eds.), *Percolation Structures and Processes*, Adam Hilger, Briston, UK, 1983.
7. R. T. Smythe and J. C. Wierman, *First-Passage Percolation on the Square Lattice*, Springer-Verlag, Berlin, Federal Republic of Germany, 1978.
8. B. D. Hughes and B. W. Ninham (eds.), *The Mathematics and Physics of Disordered Media*, Springer-Verlag, Berlin, Federal Republic of Germany, 1983.
9. A. L. Efros, *Physics and Geometry of Disorder, Percolation Theory*, Mir Publishers, Moscow, USSR, 1986.
10. A. R. Brown, *Optimum Packing and Depletion*, Macdonald, London, UK, 1971.
11. R. Blanc and E. Guyon, "Transport Properties of an Assembly of Spheres," *Percolation Structures and Processes*, G. Deutscher, R. Zallen and J. Adler (eds.), Adam Hilger, Bristol, UK, 1983, pp. 229-250.
12. J. M. Hammersley, "Origins of Percolation Theory," *Percolation Structures and Processes*, G. Deutscher, R. Zallen and J. Adler (eds.), Adam Hilger, Bristol, UK, 1983, pp. 47-57.
13. R. Zallen, "Introduction to Percolation: A Model for All Seasons," *Percolation Structures and Processes*, G. Deutscher, R. Zallen and J. Adler (eds.), Adam Hilger, Bristol, UK, 1983, pp. 4-16.
14. J. C. Wierman, "Critical Probabilities in Percolation Models," *The Mathematics and Physics of Disordered Media*, B. D. Hughes and B. W. Ninham (eds.), Springer Verlag, Berlin, Federal Republic of Germany, 1983, pp. 300-313.
15. S. K. Bhattacharyya, S. Basu and S. K. De, "Effect of Size, Shape and Oxide Content of Metal Particles on the Formation of Segregated Networks in PVC Composites," *Composites*, 1978, vol. 9, pp. 177-183.
16. S. Kirkpatrick, "Classical Transport in Disordered Media: Scaling and Effective-Medium Theories," *Physical Review Letters*, 1971, vol. 27, pp. 1722-1725.
17. R. M. German, "The Two-Dimensional Connectivity of Liquid Phase Sintered Microstructures," *Metallurgical Transactions A*, 1987, vol. 18A, pp. 909-914.
18. J. Gurland, "An Estimate of Contact and Continuity of

Dispersions in Opaque Samples," *Transactions of the Metallurgical Society of AIME*, 1966, vol. 236, pp. 642-646.
19. M. G. Mathew and K. S. Mendelson, "Hall Effect in the 'Composite Sphere' Material," *Journal of Applied Physics*, 1974, vol. 45, pp. 4370-4372.
20. L. I. Tuchinskii, "Thermal Expansion of Composites with a Skeletal Structure," *Soviet Powder Metallurgy Metal Ceramics*, 1983, vol. 22, pp. 659-664.
21. Z. Hashin, "Analysis of Composite Materials - A Survey," *Journal of Applied Mechanics*, 1983, vol. 50, pp. 481-505.
22. J. V. Milewski and D. E. Davenport, "Short Metal Fibers and Flakes," *Handbook of Reinforcements for Plastics*, J. V. Milewski and H. S. Katz (eds.), Van Nostrand Reinhold, New York, NY, 1987, pp. 103-114.
23. A. Malliaris and D. T. Turner, "Influence of Particle Size on the Electrical Resistivity of Compacted Mixtures of Polymeric and Metallic Powders," *Journal of Applied Physics*, 1971, vol. 42, pp. 614-618.
24. S. K. Bhattacharya, *Metal-Filled Polymers*, Marcel Dekker, New York, NY, 1986.
25. R. Kusy and D. T. Turner, "Effects of Nickel on Mechanical Properties of PVC," *Society of Plastics Engineers Journal*, 1973, vol. 29, pp. 56-61.
26. L. Oger, J. P. Troadec, D. Bideau, J. A. Dodds and M. J. Powell, "Properties of Disordered Sphere Packings II. Electrical Properties of Mixtures of Conducting and Insulating Spheres of Different Sizes," *Powder Technology*, 1986, vol. 46, pp. 133-140.
27. M. J. Powell, "Site Percolation in Randomly Packed Spheres," *Physical Review B*, 1979, vol. 20, pp. 4194-4198.
28. H. Ottavi, J. Clerc, G. Giraud, J. Roussenq, E. Guyon and C. D. Mitescu, "Electrical Conductivity of a Mixture of Conducting and Insulating Spheres: An Application of Some Percolation Concepts," *Journal of Physics C: Solid State Physics*, 1978, vol. 11, pp. 1311-1328.
29. M. J. Powell, "Site Percolation in Random Networks," *Physical Review B*, 1980, vol. 21, pp. 3725-3728.
30. P. S. Clarke, J. W. Orton and A. J. Guest, "Electrical Conductivity and Hall Effect Measurements in Semiconducting Powders. Study of Percolation Effects," *Physical Review B*, 1978, vol. 18, pp. 1813-1817.
31. J. P. Straley, "Critical Experiments for the Conductivity of

Random Resistor Lattices," *Physical Review B*, 1977, vol. 15, pp. 5733-5737.
32. R. M. Scarisbrick, "Electrically Conducting Mixtures," *Journal of Physics D: Applied Physics*, 1973, vol. 6, pp. 2098-2110.
33. G. Deutscher, A. Kapitulnik and M. Rappaport, "Percolation in Metal - Insulator Systems," *Percolation Structures and Processes*, G. Deutscher, R. Zallen and J. Adler (eds.), Adam Hilger, Bristol, UK, 1983, pp. 207-228.
34. V. K. S. Shante and S. Kirkpatrick, "An Introduction to Percolation Theory," *Advances in Physics*, 1970, vol. 20, pp. 325-356.
35. H. Scher and R. Zallen, "Critical Density in Percolation Processes," *Journal of Chemical Physics*, 1970, vol. 53, pp. 3759-3761.
36. J. P. Fitzpatrick, R. B. Malt and F. Spaepen, "Percolation Theory and the Conductivity of Random Close Packed Mixtures of Hard Spheres," *Physics Letters*, 1974, vol. 47A, pp. 207-208.
37. A. Sur, J. L. Lebowitz, J. Marro, M. H. Kalos and S. Kirkpatrick, "Monte Carlo Studies of Percolation Phenomena for a Simple Cubic Lattice," *Journal of Statistical Physics*, 1976, vol. 15, pp. 345-353.
38. J. W. Cahn, "A Model for Connectivity in Multiphase Structures," *Acta Metallurgica*, 1966, vol. 14, pp. 477-480.
39. R. P. Kusy, "Applications," *Metal-Filled Polymers*, S. K. Bhattacharya (ed.), Marcel Dekker, New York, NY, 1986, pp. 1-142.
40. R. P. Kusy, "Influence of Particle Size Ratio on the Continuity of Aggregates," *Journal of Applied Physics*, 1978, vol. 48, pp. 5301-5305.
41. R. P. Kusy and D. T. Turner, "Electrical Resistivity of a Polymeric Insulator Containing Segregated Metallic Particles," *Nature*, 1971, vol. 229, pp. 58-59.
42. J. Delmonte, *Metal-Filled Plastics*, Reinhold Publishing, New York, NY, 1961.

Chapter Eleven

Pore Size and Shape

Chapter Outline

- **Introduction**
- **Pore Characterization Techniques**
- **Surface Area Measurement**
- **Pores in Ordered Sphere Packings**
- **Pores in Random Packings**
- **Summary**
- **References**

Introduction

The packing density is not the only property of importance to particulate structures. Pores also constitute a useful aspect of packed particles, since they control properties such as filtration, permeability, electrode activity, fluid trapping, drainage, wicking, elastic modulus, thermal conductivity, and surface area. Pore structures are the subject of study in investigations of several materials including rocks, ceramics, wood, textiles, charcoal, and catalysts. Porosity is a gross measure of the packing characteristics, and more detail is usually needed on the packing structure. Thus, other important pore characteristics beyond porosity are measured: pore size, shape, surface area, and interconnectivity.

In a packing of particles the pores are angular and irregular in cross section, with an inherent distribution in size that reflects both the random particle packing and the irregular pore cross section. Pore sizes range from the molecular level up to macropores that are seen with the unaided eye. Generally, the pore size depends on the particle size and the porosity. However, the pore shape is highly variable from point to point within a compact and between compacts. In packed particles the pores have a high level of interconnectivity. With compaction this connectivity is degraded with a loss of permeability. The pore channels are also tortuous, meaning that fluid passing through the pores travels a distance much greater than the bulk thickness. Actually, the pore surface consists of an adsorbed film that dictates the interfacial properties of packed particles. Control over the surface state can influence the interaction of the pore structure with the environment. For example, the wetting behavior can be modified by the presence of oxides, organic molecules, or other adsorbed films. Consequently, in practice surface treatments are used to create dramatic shifts in wetting or other surface related properties.

Since pore structure is complex, it cannot be characterized by a single parameter [1,2]. Several features may be needed to fully specify the pore structure, including porosity, pore size, pore shape, connectivity between pores, tortuosity, and the distributions in all of these. Various models have been developed to link particle size to pore size. These are only rough guides to pore structure since there is a further sensitivity to the porosity and particle shape. [3]. The goal is to model the performance of a porous material based on the primary characteristics. With proper models, attributes such as diffusivity, permeability, capillarity, and strength can be predicted from the basic structural attributes.

A useful measure of the pore structure is its resistance to fluid flow. Because of its analogy to fluid flow through pipes, a measure of flow resistance is given by the hydraulic radius R_h. There is uncertainty as to the appropriate technique for estimating the hydraulic radius for packed particles [1,4-10]. Table 11.1 lists some of the mathematical expressions for the hydraulic radius offered in the literature. Inherently, the hydraulic radius relates to the flow resistance, with a basic dependence on the pore volume-to-surface area ratio. For a tube this means that the hydraulic radius is directly proportional to the tube diameter, but this relation is less

Table 11.1

The Hydraulic Radius of Porous Structures

R_h = hydraulic radius
S_v = specific surface area of packing per unit volume
f = fractional packing density
D = sphere diameter
n = exponent between 1.2 and 1.9, depending on particle shape
A = constant with a typical value near 0.6

$R_h = (1 - f)/(f S_v)$	*spheres [4]*
$R_h = 3\,(1 - f)/(f S_v)$	*spheres [1]*
$R_h = (1 - f)\,D/(3\,f)$	*spheres [5,7]*
$R_h = A\,(1 - f)\,D^n$	*irregular [3,9]*
$R_h = A\,(1 - f)\,D$	*irregular [10]*

direct for packed particles.

This chapter focuses on pore structure and its relation to particle size and porosity. The first topic is a discussion of the techniques for characterizing pore structure. Although there are several techniques, mercury porosimetry is the most widespread. The subsequent section examines measurement of surface area by vapor adsorption processes. Next, the characteristics of pores in ordered packings of spheres is presented. Pore size, connectivity, and shape will be emphasized at this stage. This treatment provides a basis for discussing the pores in random structures. Obviously, this is a less exact, yet very important area. Later in this book the discussion of capillary and permeability characteristics will build on the concepts introduced in this chapter.

Pore Characterization Techniques

There is a wide variety of pore characterization techniques, including x-ray scattering, microscopic sectioning, mercury porosimetry, small-angle neutron scattering, radiographic analysis, as well as measurements of gas adsorption, gas permeability, liquid expulsion, liquid infiltration, electrical conductivity, and electrode activity [1,2,11-25]. In addition, because several of the properties of the powder system will vary with the pore characteristics, for example the thermal conductivity, elastic modulus, or electrical conductivity, these characteristics provide information on the pore structure. Accordingly, on occasion, indirect measures of the pore structure are used to characterize the packing structure. Each of the pore characteristics is a distributed parameter within the packing, meaning that there is a natural variation from region to region. Beyond these parameters, it is also necessary to know the tortuosity of the pore structure. Tortuosity is a measure of the actual pore length divided by the compact thickness.

For an ordered sphere packing, the pore size can be related to the insphere diameter which is the constriction size between the spheres; it is the diameter of a sphere capable of just passing through the void between the spheres. For a close-packed structure the size of the insphere is 0.1547 times the diameter of the close-packed sphere. However, with a random particle packing and irregular particle shape the relation between the insphere size and the actual pore size is less specific. With mixtures of particles of various sizes, the distribution in pore sizes can be quite large, leading to difficulties in assessing the pore structure. The narrowest part of the flow path between the connecting units determines the pressure associated with flow. However, the curvature may be highly variable in comparison with the pore size, although certain minimum energy configurations are often assumed [26]. Because many of the techniques for measuring pore size assume specific relations between the insphere size and pore size, there is often considerable error in the resulting pore size measurement. The typical assumption is that fluid flow into or out of the pores relates to the curvature at the pore constriction. Accordingly, the volume assigned to the pore is poorly estimated using the insphere size.

Pore size and pore shape measurements by quantitative stereology are accurate but difficult to perform. The pore structure can be described in terms of polyhedral units that inscribe the pores [27]. The size and connectivity of the polyhedral units can be related to observable features ranging from the porosity to the flow behavior. This approach requires careful serial sectioning to construct representative pore models. Random cross sections are preferred in practice, but these will not capture the largest dimension of the pores; the pores will appear smaller than their actual size. Even so, random two-dimensional cross sections allow accurate deduction of the mean pore size and spacing [28-30]. Furthermore, under appropriate conditions it is possible to deduce the true pore size distribution from the distribution measured in two dimensions.

For monosized spheres the mean separation between pores is directly proportional to the particle size [31]. Likewise, the pore size as measured in two dimensions is an accurate guide to the three-dimensional size if a simple pore shape is assumed. Microscopy has the disadvantage of requiring tedious individual pore measurements. Unfortunately, in most cases the pores are not simple cylindrical shapes and the relations between two-dimensional data and true pore characteristics are unknown. Figure 11.1 shows an example of the pore structure attained using epoxy infiltration into the pores between packed particles [32,33]. For this micrograph, the solid structure was dissolved after the pores were impregnated with epoxy, thereby leaving a replica of the pore structure. From such micrographs, the level of connectivity, shape variation, size, and curvature of real pore structures is apparent. By contrast, the models used in pore size analysis assume a much simpler pore structure.

Most common techniques for measuring the size of pores in packed powders rely on the capillary behavior. This has the advantage that simple measurements can be used to give information on a large number of pores. A wetting fluid will preferentially fill a pore structure, and work must be performed to remove the fluid. Alternatively, work must be performed to force a nonwetting fluid into a pore structure. In each case, the work performed is a gauge of the pore size. The rate at which a pore is infiltrated with a wetting fluid provides one measure of the pore size [34-36]. Large pores fill rapidly compared to small pores. Measuring the weight gain versus time can give a mean pore size, since the depth (or weight) of

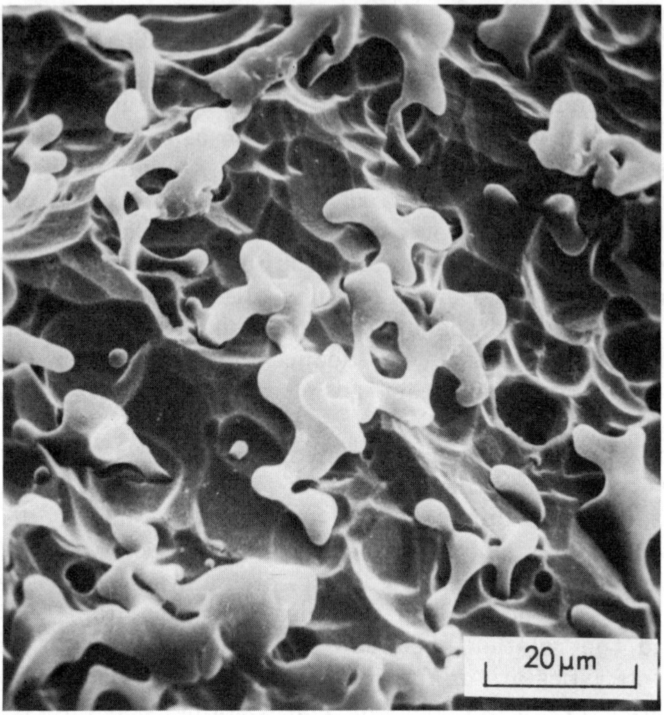

Figure 11.1 Scanning electron micrograph of a pore structure replica formed by plastic infiltration of pores with subsequent removal of the solid structure.

infiltration varies approximately with the pore size as follows [37,38]:

$$L = (1/\pi)\,[d\,t\;\gamma\cos(\theta)/\mu]^{1/2} \tag{11.1}$$

where L is the depth of infiltration, d is the pore diameter, t is the infiltration time, θ is the wetting angle, γ is the liquid-vapor surface energy, and μ is the liquid viscosity. The weight gain can be related to the infiltration depth by determining the fluid density and the compact porosity. This type of measurement assumes cylindrical, monosized pores. If the fluid filling the pores is electrically conducting (for example, a salt solution), then the pore size distribution can be calculated from measurements of bulk sample conductivity versus time [36].

When a pore structure is saturated with a wetting liquid, an external gas pressure is needed to expulse the liquid from the pores. The pressure needed for the gas to break into a pore is inversely related to the insphere size [2]. Mercury porosimetry provides the antithesis situation. In this case a high pressure is used to drive mercury into the pore structure from the outer surface of the compact. Since mercury is nonwetting with respect to most materials, the amount of mercury entering the pores depends on the applied pressure and the pore size. From a scan of intruded mercury volume versus applied pressure, a rough calculation of the pore size distribution is possible [6,20,39-44]. This assumes that the pores have a simple shape, usually cylindrical.

The resistance to mercury intrusion varies inversely with the pore size. For a cylindrical pore, the applied pressure P needed for intrusion relates to the contact angle between mercury and the test material θ, pore diameter d, and mercury-vapor surface energy γ as follows:

$$P = -4 \gamma \cos(\theta)/d . \qquad (11.2)$$

This is known as the Washburn equation.

For mercury, the contact angle with most materials is typically between 130 and 140° while the surface energy ranges near 0.485 J/m^2 [23]. A mercury porosimetry experiment is carried out by surrounding the material with mercury after the pore structure is evacuated to remove trapped gases. Pressure is then applied to the mercury to force intrusion into the pores connected to the external surface of the compact. Since the pores are irregular in cross section, the filling of a pore occurs over a range of pressures [40]. Figure 11.2 illustrates this point. Initially, the mercury fills the pore when the applied pressure corresponds to the insphere diameter in Equation 11.2. Hence, pore filling is controlled by the size of the largest constrictions between pores. As the pressure is increased, the curvature of the mercury in the pore increases and fills the corners of the pore. Subsequently, higher pressures lead to

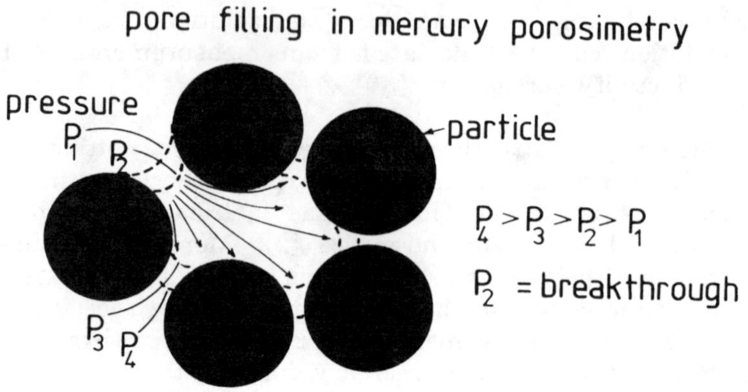

Figure 11.2 A sketch of the process of mercury intrusion into a pore between spheres, showing progressive pore filling as the applied pressure is increased.

more complete filling of the pore. Note that pore filling occurs over a range of pressures. Furthermore, preferential filling occurs in the surface connected pores, so that actual behavior depends on the sample size [45,46].

A record of the volume of mercury intruded versus the applied pressure is used to assess the pore size distribution. Figure 11.3 shows how the void space in an ordered monosized sphere packing fills with increasing pressure under ideal conditions [42]. The void space begins filling at the breakthrough pressure associated with the insphere diameter. Continued pressurization increases the degree of pore filling, but the process continues over a considerable range of pressures. Therefore, models that assume filling at a single pressure are in considerable error when applied to packed particles. The breakthrough pressure approximately corresponds to the pressure required to fill the largest 25% of the pores [47]. The actual value depends on the connectivity of the pore structure (see Chapter Ten). Figure 11.4 plots the results of measurements of pore filling in monosized sphere packings [12]. The fractional porosity intruded is shown versus the normalized pressure for four randomly packed particle sizes. These plots demonstrate the progressive filling behavior that is characteristic of

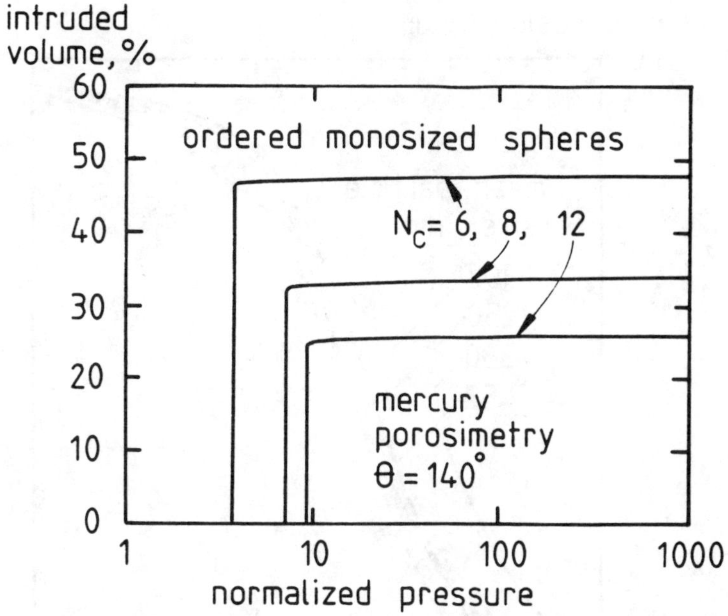

Figure 11.3 The volume of mercury intruded in pores in an ordered packing of monosized spheres [42].

mercury porosimetry.

Since the intruded pore size varies with the inverse of the applied pressure, very high pressures allow access to pores as small as 1 nm. Beyond the size distribution, pore shape and surface area can be estimated from mercury porosimetry. A hysteresis occurs on depressurization which is partly due to the nonuniform pore shape and the difference in contact angle between an advancing and receding liquid. The greater the departure from a cylindrical pore shape, the greater the hysteresis [47-50]. This results from the nonuniform pore cross section, which limits the flow of mercury both in and out of the pore based on the constrictions between particles.

On depressurization, a retention force is exerted on the mercury. Figure 11.5 plots the penetration hysteresis on intrusion

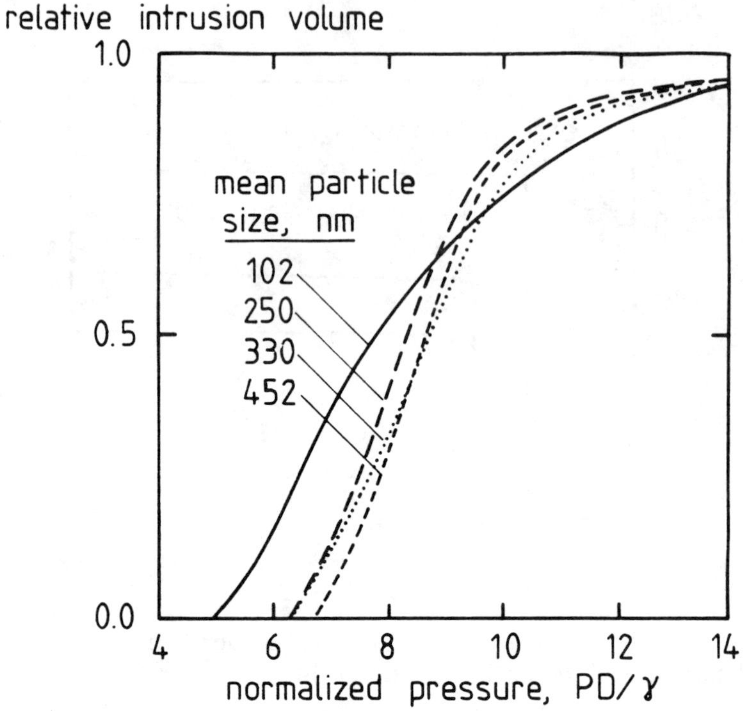

Figure 11.4 The normalized intrusion volume versus the normalized pressure during mercury porosimetry filling of pores between four monosized sphere packings [12].

and extrusion for glass spheres [6]. The differing path for the pressurization and depressurization cycles is due to the retention of mercury in the pores. This hysteresis increases as the pore shape becomes more irregular and as the pore size distribution becomes wider. The smaller the ratio of the insphere size to the pore size the greater the hysteresis during depressurization. In contrast to intrusion, which is controlled by the size of the connections between pores, the extrusion behavior is dictated by the pore size and its release of mercury [51]. Accurate interpretation of mercury porosimetry data can be difficult because of these pore shape complexities. However, mercury porosimetry is useful in comparing pore structures. It provides information on the pore size, shape, and surface area in a fairly automated analysis. Accordingly,

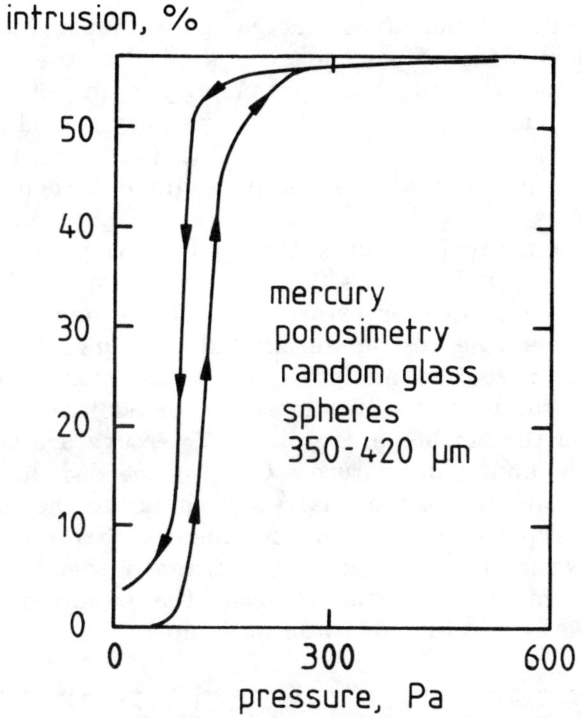

Figure 11.5 A plot of volume versus applied pressure illustrating the penetration hysteresis for mercury porosimetry of a random packing of nearly monosized glass spheres [6].

mercury porosimetry is widely applied to the characterization of pore characteristics [52]. One review cites 998 references in this area prior to 1981 [24].

Various assumptions concerning pore shape and interconnectivity are possible for packed particles. These assumptions influence the analysis of mercury porosimetry data as well as the relative errors in the data [39]. Furthermore, the mercury porosimetry technique has several potential sources of error. Thus, data from this technique must be treated judiciously [23,25,39,40,46,53]. For one thing, the pores must be open to the compact surface to be filled by the intruding mercury. A frequent problem relates to the model selected for the pore shape. The model must have some realistic relation to the actual pore shape. Most typically, a cylindrical model of pore shape is applied to the intrusion data, meaning that the size measurements are in considerable error. With a cylindrical pore, the retained mercury should be near zero, yet in practice up to 98% of the intruded mercury may remain trapped upon depressurization in a highly irregular pore structure. There is also a time dependent rate of pore filling that can lead to a measurement error if the analysis is performed too rapidly. Equation 11.1 illustrates the relation between infiltration time and compact size. Other errors arise from time dependent interactions between the test material and mercury. These are often overlooked in routine analysis. They are most noticeable when comparing samples of differing sizes. In addition, both the contact angle and the surface energy of the mercury are sensitive to contamination and temperature. Care is needed to ensure isothermal conditions during the analysis to minimize the contributions of thermal expansion to the measurements. Errors also arise from the compressibility of the mercury, trapped gas, test material, and measurement apparatus. Many of these sources of error can be eliminated by careful calibration procedures.

A nondestructive alternative to mercury porosimetry relies on water (or other wetting fluid) expulsion from a saturated pore structure [54]. This is also known as the bubble point procedure. Gas pressure is applied to the exterior of a compact with the pores saturated by the wetting fluid. There is no flow of gas through the compact until the pressure exceeds that needed to expel the liquid from the largest continuous pore. At this point a gas bubble passes through the pore structure and emerges at the surface, leading to the bubble point concept. The size of the pore is estimated using Equation 11.2. With subsequently higher gas pressures smaller pores will be opened and a greater quantity of gas will flow through the structure. Figure 11.6 illustrates how the flow rate versus pressure behavior will differ between a saturated and dry specimen. This difference is characteristic of the pore structure. Thus, a measurement of flow rate versus the applied pressure provides a rough means of characterizing the pore size distribution. Assuming laminar flow behavior for the gas permeation through the pores and a cylindrical pore shape gives a calculation of the number of pores n_i of a given size d_i as follows:

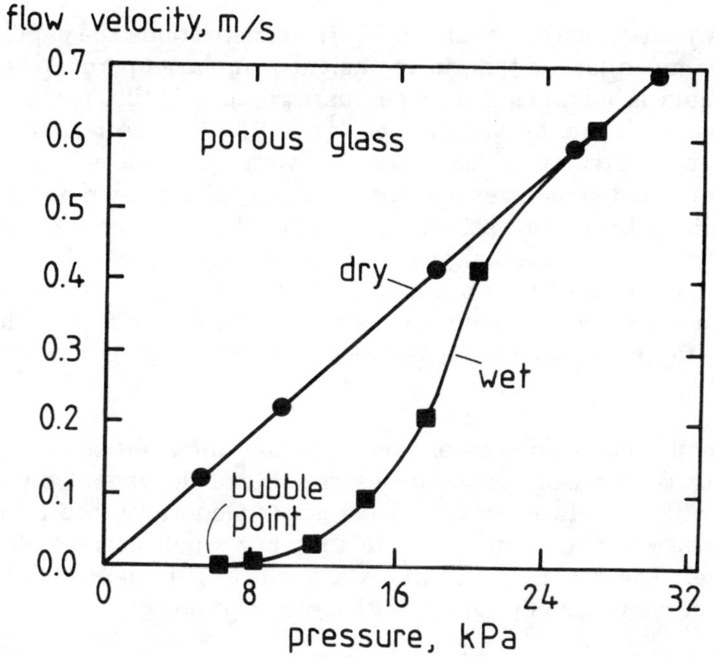

Figure 11.6 A comparison of the gas flow rate versus applied pressure for a dry powder compact and one saturated with a wetting liquid where the applied pressure must exceed the bubble point for the largest pore.

$$n_i = (41 \ \mu L/d_i) \, dQ/dP \tag{11.3}$$

where μ is the gas viscosity, L is the compact thickness, and dQ/dP is the change in gas flow rate with pressure. Using this equation, the pore size distribution can be estimated from the flow rate versus pressure curve such as shown in Figure 11.6. Future chapters will treat aspects of this behavior in greater detail. In Chapter Twelve the relations between drainage and pore size will be described and in Chapter Thirteen the permeability of packed powders will be covered.

As noted already, a liquid behaves differently when it is advancing than when it is receding within a porous structure. In part, this difference relates to the constrictions and the capillary pressure variation with curvature. At the constrictions the capillary pressure is much larger than in the neighboring large pore. When a wetting liquid is advancing in a porous structure it fills pores at a lower pressure than at which they drain [34,51]. Alternatively, drying a pore structure filled with a wetting fluid will give a sequence of liquid structures just the reverse of that shown in Figure 11.2 for filling by a nonwetting fluid. Not all of the wetting liquid can be displaced by external gas pressurization, because of the formation of isolated pendular bonds at particle contacts. Yet the measured pore sizes seem reasonable using this bubble test procedure, especially for large spherical particles with low packing densities [55].

Condensation processes also provide information on pore sizes [19]. Both adsorption and desorption are dependent on pore size, curvature, surface area, and porosity. Generally, the adsorption processes are more sensitive to the very small pores, below 1 μm in size. The mean curvature of a surface ρ is defined by the two radii of curvature on any point by the relation

$$1/\rho = 0.5\,(1/R_1 + 1/R_2) \tag{11.4}$$

where R_1 and R_2 are the principal radii of curvature for the pore (perpendicular to each other). The vapor pressure of a condensed liquid that occupies a curved surface is directly dependent on the curvature according to

$$k\,T\,\ln(P/P_o) = -2\,\gamma\,\Omega/\rho \tag{11.5}$$

where k is Boltzmann's constant, T is the absolute temperature, P_o is the equilibrium vapor pressure, P is the vapor pressure over the curved surface, γ is the liquid-vapor surface energy, and Ω is the atomic volume. Equation 11.5 assumes that the pore is concave. Because of the vapor pressure dependence on curvature, measurements of adsorption or desorption versus external vapor pressure provide information on the distribution of pore surface curvature. On this basis, if an adequate pore shape model exists, then pore size information can be generated from measurements of vapor pressure [6].

Comparisons have been made between these various techniques for measuring pore size [12,13,19,23,35,51,56,57]. Generally, the techniques do not give equivalent mean pore sizes nor even similar distributions. Typically, mercury porosimetry gives a large error in the pore size and a distribution skewed toward the smaller pores. On the basis of such comparisons, it is felt that microscopy coupled with mercury porosimetry provides the best means for characterizing pore structures. All of the techniques have their problems, but the large amount of research on the mercury porosimetry technique provides an excellent basis for data treatments.

Surface Area Measurement

All surfaces have an atomic scale texture that includes steps and voids. Even flat objects like plate glass have a higher actual surface area than is evident from simple bulk geometric measurements; for plate glass this amounts to an excess of only 5% [21]. However, for powders the excess surface area can be up to ten times that perceived from the particle size. This surface area influences the adhesion, friction, packing, and agglomeration of powders. Thus, in linking the model sphere packings to actual particle packings, new tools are needed to assess the surface structure on the atomic scale.

For monosized spheres the surface area can be calculated directly from the particle size. In working with powders, the specific surface area is usually the basis for examination; it is defined as the ratio of surface area per unit powder mass. In the case of monosized spheres, the specific surface area S_m depends on the particle diameter D and the theoretical density for the material ρ_t as

$$S_m = 6/(\rho_t D) . \tag{11.6}$$

Note that this formula is exactly the same for cubic particles if the size parameter D is the cube edge length. If there is a known distribution in particle sizes, then the specific surface area can be estimated by summing the surface area and mass contributions from each size class of the population distribution. For a distribution with

n size classes, this procedure gives

$$S_m = (6/\rho_t)\left(\sum_{i=1}^{n} D_i^2 N_i\right) / \left(\sum_{i=1}^{n} D_i^3 N_i\right) \qquad (11.7)$$

where N_i is the number of particles in each size class. To accurately calculate specific surface area by Equation 11.7 one needs a precise knowledge of the distribution function that links the particle size and population of particles. The actual surface areas can vary considerably from that calculated using the particle size [58]. This leads to the need to directly measure the surface area for small particles.

The permeation rate for a gas in a porous structure gives some information on the surface area. During permeation the gas molecules adsorb on the powder surface for a molecular vibrational cycle and then quickly desorb. After desorption, the molecular velocity is random. Consequently, the adsorption event removes energy from the flowing gas near the wall, leading to a drag effect. As the surface area to volume ratio increases there will be more drag on the permeating gas. As long as the pore structure is open to the flowing gas and the flow rate is reasonably low, the permeability provides a gauge of the powder surface area. Usually, these relations assume a spherical particle and give only nominal surface area assessments.

The surface area can also be calculated from mercury porosimetry measurements [23,41,44,59]. The work performed in intruding the nonwetting mercury into the pore structure between packed particles depends on the covered surface area as follows:

$$dE/dS = \gamma \cos(\theta) \qquad (11.8)$$

where dE/dS is the work performed per unit surface area. Since the pressure times the volume intruded equals the work performed in covering the pore surface,

$$S_m \, m \, \gamma \cos(\theta) = \int P \, dV \qquad (11.9)$$

with m equal to the sample mass. Thus, the mercury porosimetry data that give information on the pore size and shape also provide surface area information.

Gas adsorption is a common approach to determining the surface area and pore size for particles [59-61]. When a liquid evaporates from a pore surface a thin film remains on the pore. The thickness of the film depends on the relative saturation pressure of the vapor as illustrated in Figure 11.7 [21]. This plot shows the thickness of nitrogen versus the relative vapor pressure (a ratio of the vapor pressure to the saturation vapor pressure). Because the partial pressure for a curved surface differs from that of a flat surface, the pore size can be estimated by the desorption behavior. Likewise the total amount of gas adsorbed gives a basis for determining the surface area for a porous structure. The use of gas adsorption and desorption measurements to characterize pore structures requires that the particle size be very small so that the quantities of gas involved are measurable.

At a given vapor pressure, the gas molecules continually strike the surface and maintain an equilibrium. The number of molecules that strike the surface per unit time depends on the gas pressure,

$$N = N_o P/(2 \pi M k T)^{1/2} \tag{11.10}$$

where N is the number of molecules striking the surface per unit time and per unit surface area, N_o is Avogadro's number, P is the pressure, M is the molecular weight of the gas, k is Boltzmann's constant, and T is the absolute temperature. From this rate of arrival, there will be a fraction that adsorb on the powder surface. This fraction depends on the fraction of surface sites that are occupied θ and a parameter called the sticking probability α:

$$N_a = N \theta \alpha \tag{11.11}$$

where N_a is the number that adsorb per unit area and time. Simultaneously with adsorption, desorption will also occur. The rate of adsorption depends on the coverage of molecules on the powder surface (1 - θ), vibrational frequency η, and the energy available for evaporation. The energy enters as an Arrhenius equation,

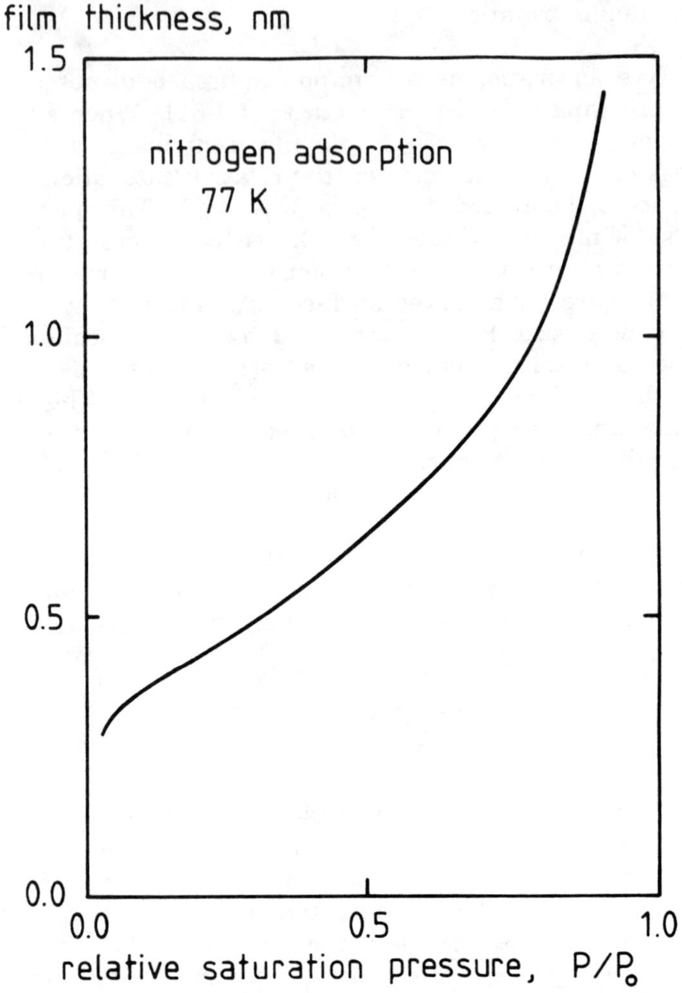

Figure 11.7 The adsorbed film thickness at 77 K as a function of the vapor pressure of nitrogen gas in equilibrium with the surface.

$$N_d = N_m (1 - \theta)\ \eta\ exp(-Q/kT) \qquad (11.12)$$

with N_d being the number of molecules that desorb per unit time and area, N_m the number of molecular monolayers adsorbed on the surface, and Q the evaporation energy. At equilibrium the rates of adsorption and desorption will be equal. Then, assuming uniformity to the surface structure allows one to calculate the specific surface area from the adsorption behavior:

$$S_m = N_o W A/M . \tag{11.13}$$

In this equation, W is the weight of gas adsorbed on the powder to form one monolayer, and A is the area occupied by an adsorbed molecule. The value of W is actually measured through the adsorption changes with vapor pressure. The adsorbing gas is brought into contact with the powder after its surface has been cleaned. The weight, vapor pressure change, or thermal conductivity of the gas is measured as equilibrium adsorption occurs, allowing determination of w, the actual weight adsorbed, as a function of P. This allows extrapolation of W, from the BET equation named after Brunauer, Emmett and Teller [60]:

$$P/w = 1/(KW) + P/W \tag{11.14}$$

where w is the adsorbed gas weight and K is a constant equal to

$$K = N_a(1 - \theta)/(P N_d \theta) . \tag{11.15}$$

In turn, the surface area of the powder is then calculated from Equation 11.13. Figure 11.8 shows the adsorption and desorption isotherms for a small particle size alumina [61]. For this powder the calculated specific surface area is 162 m^2/g, and the mean pore size is near 10 nm.

Although the process may appear complex mathematically, surface area measurements by gas adsorption are routinely applied to small particle size powders. Evaluation of the weight adsorbed versus pressure at pressures between 5 and 35% of the saturation vapor pressure gives a rapid and accurate measure of surface area. In turn, gas adsorption also provides information on the pore structure. The corrections for pore shape effects are small for gas adsorption in comparison with mercury porosimetry.

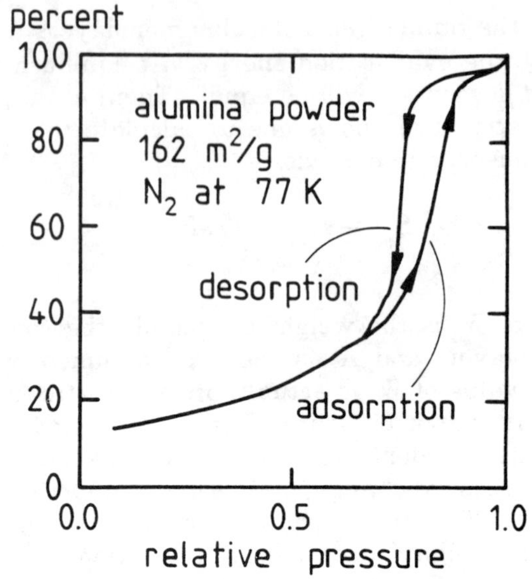

Figure 11.8 The isothermal adsorption and desorption curves for a fine alumina powder [61].

Gas adsorption measurements have been used for extracting information on the packing coordination of powders [62]. A wetting fluid is adsorbed into the powder, where it forms preferential pendular bonds at the particle contacts. A subsequent analysis of surface area by gas adsorption will show a decrement in the surface area in proportion to the number of contact bonds. For packed 15 nm alumina particles tested at a fractional packing density of 0.47, this technique gives a coordination number near 8. This is slightly larger than the expected value, probably because the technique cannot separate near contacts from actual contacts. Even so, such an approach is useful in extending packing concepts developed on large scale model packings to a smaller size range.

Pores in Ordered Sphere Packings

The pores in ordered packings of monosized spheres would seem to be the simplest to study. However, even here the pore structure is complex [63-65]. Graton and Fraser [64] recognized the complexity of pores in ordered packings and based their analysis of random packings on this complexity. Figure 11.9 sketches the pore shapes in unit cells with packing coordinations of 6, 8, and 12. To further illustrate the complexity, Figure 11.10 shows the structure of the pore in a simple cubic packing at five different cross sections. As can be seen, the pore shape alters rapidly over small distances in the unit cell. The cross sectioned shape varies from a square to a rounded diamond. Accordingly, random packing structures are even more complex and variable.

The insphere size for four ordered packings are shown in Table 11.2 [6]. The multiple entries represent the variations in

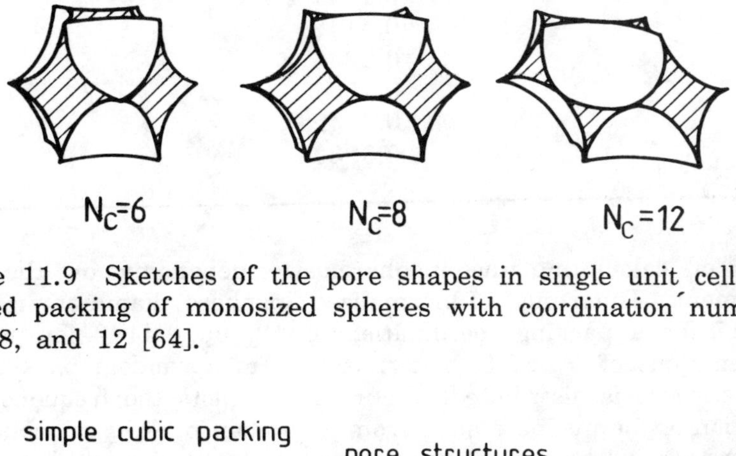

Figure 11.9 Sketches of the pore shapes in single unit cells for ordered packing of monosized spheres with coordination numbers of 6, 8, and 12 [64].

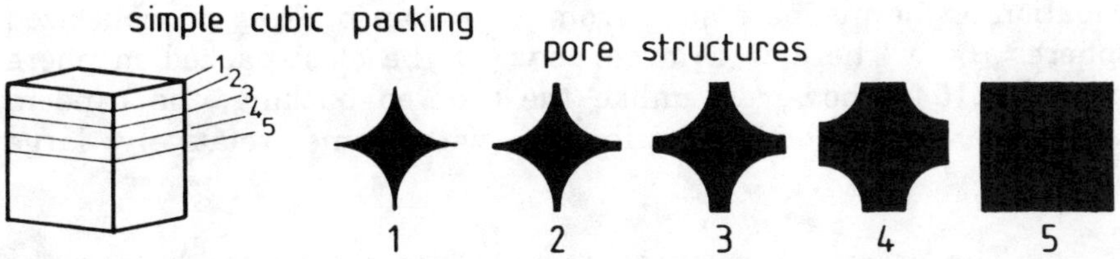

Figure 11.10 The variation in pore morphology with sectioning depth for a unit cell of the simple cubic packing [64].

Table 11.2

Insphere Sizes for Ordered Packings [6]

coordination number	relative insphere diameter
12	0.1547
10	0.2649
	0.1547
8	0.4142
	0.2247
	0.1547
6	0.4142
	0.3333

ways of calculating the insphere size depending on the pore placement. The range of normalized insphere diameters is from 0.1547 for a packing coordination of 12, up 0.4142 for packing coordinations of 6 and 8. Alternatively, in a random packing the insphere size is distributed. Figure 11.11 plots the frequency distribution of the insphere sizes from a random packing of monosized spheres [66]. There is a large peak at the close-packed insphere size of 0.1547; however, unlike the ordered packing, the random structure has a wide distribution with some relatively large openings.

The ordered structures provide one means for linking the pore size to capillary behavior. Various relations have been proposed between the capillary pressure and the particle size [12,63, 67,68]. These generally take on the form

$$P = G\ \gamma/D \tag{11.16}$$

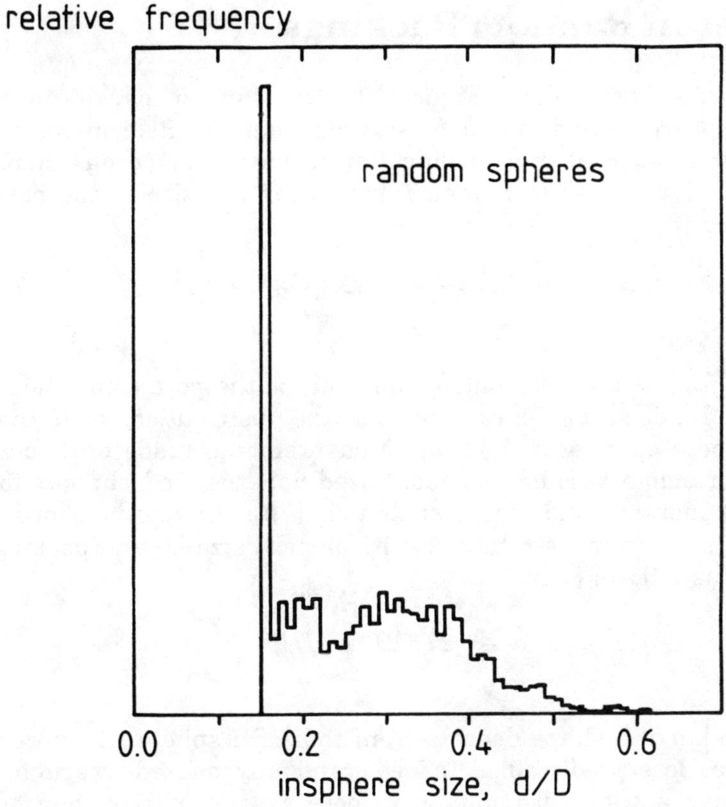

Figure 11.11 The insphere size distribution for randomly packed spheres [66].

where P is the capillary pressure associated with the pore, γ is the liquid-vapor surface energy, and D is the sphere diameter. The parameter G represents the geometric link to the packing structure; its value ranges between 9.6 and 25.8, depending on the structure.

Pores in Random Packings

As illustrated in Figure 11.11, a random packing of monosized spheres exhibits a pore size distribution. Reasonable success has been reported in modeling flow behavior in porous structures using a Rayleigh distribution function for the size of the pore connections [69],

$$f(d) = \alpha^2 d \exp(-\alpha^2 d^2/4) \tag{11.17}$$

where $f(d)$ is the distribution function for the pore connection sizes, α is an adjustable parameter for the distribution such that the mode pore opening is $1.414/\alpha$. Measurements made on two-dimensional random packings of monosized spheres lend support to such an exponential void size distribution [70]. In another model, the mean pore size $\bar{d}$ is related to the particle size D and packing density f as follows [71]:

$$\bar{d} = (2/3)D\,(1 - f)/f\,. \tag{11.18}$$

As the particle shape departs from that of a sphere, the pore shape becomes less predictable. Indeed, randomly packed irregular particles have a wider distribution in pore characteristics than ordered packings. Mathematical models have been offered for the pore shape, and an extensive review on this subject is given by van Brakel [1]. Basically, as the local porosity between particles increases, the relative pore size and population also increase. But there are no pores larger than 80% of the sphere size in random packings (unless bridging takes place).

Several possible shapes have been suggested for the pores in random packings [65,66,72-75]. Some of the most useful models treat the pore structure as a mixture of tetrahedral units of varying size. An example of such a distribution is given in Figure 4.15. Between 48 and 73% of the pores in a dense random packing of monosized spheres are in tetrahedral sphere clusters [76]. Figure 11.12 shows the calculated distribution in porosity for both the loose and dense random packings of monosized spheres [5,7,77]. The experimental determinations show a distribution that is bimodal.

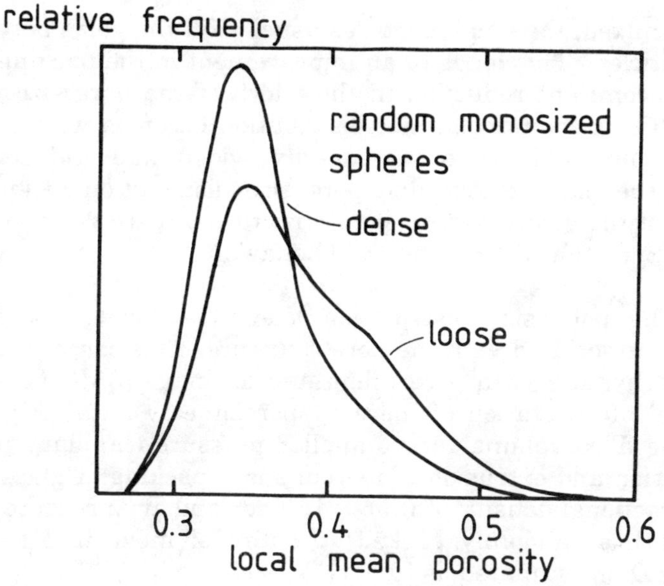

Figure 11.12 The distribution in porosity for unit tetrahedral cells in random loose and random dense packings of monosized spheres [7].

Each of the unit cells in a random packing has a characteristic relation between the cell coordination and the pore population. For the tetrahedral unit cell in a random packing there are 2.92 pores per sphere and these pores constitute 48.4% of the volume [65]. These models aid in understanding the relations between pore connectivity and permeability. As the porosity decreases the tortuosity increases. For example, the tortuosity of randomly packed spheres is larger than ordered spheres by roughly a factor of two [78]. The tortuosity of ordered packings gives a flow length that is approximately three times the thickness of the packing, while it is nearly seven times the thickness of a random packing.

There is merit in linking the pore size distribution to the particle size distribution for random sphere packings. For a narrow particle size distribution the pore size has a similar, if not wider, size distribution [79,80]. This is especially true for narrow log-normal particle size distributions where the pore size distribution has a similar slope. However, when large and small particles of very different sizes are mixed, the small particles usually fill the voids between the large particles. This leads to an improvement in the packing density and a concomitant reduction in the width of the pore size distribution. Thus, as the particle size distribution becomes wider, the pore size distribution does not necessarily widen and the correlation between the particle size and pore size distribution disintegrates. Because of the complexity in pore structures, there is no single distribution type that fits all observed behavior.

The pore size distribution leads to a range of capillary pressures associated with the pore structure. One consequence is an increased hysteresis in pore filling and drainage [66]. This complicates the interpretation of mercury porosimetry data. Figure 11.5 shows the filled volume versus applied pressure (capillary pressure) for intrusion and extrusion of mercury in a packing of glass spheres with a fractional density of 0.65 [6]. The capillary pressure P associated with a randomly packed structure of monosized spheres of diameter D is estimated as [9]

$$P = 10 \ \gamma \cos(\theta)/D \tag{11.19}$$

where γ is the liquid-vapor surface energy, and θ is the contact angle.

Summary

In model packings of large spheres, several principles have been isolated. Yet it is uncertain how applicable these principles are to the packing of small particles. In packings of very small particles there are many new variables that are not treated according to the model packings of large spheres. For example, the adhesion and capillary forces are much more important as the particle mass decreases. The surface area-to-volume ratio increases in inverse proportion to the particle size. Thus, surface frictional forces from oxides, adsorbed films, and atomic scale asperities can influence the density, bridging, strength, pore size, pore shape, coordination number, and other packing attributes. This is evident from the low packing densities of very small particles. With some submicron particles, packing fractional densities as low as 0.05 are observed. Obviously, the attributes of large monosized spheres (with a fractional density of 0.64) will not correlate well with the packing attributes of these small particles. The inability to examine the packing structure directly leads one to rely on the several measurement techniques introduced in this chapter.

Microscopic examination is very worthwhile in assessing the general pore structure, packing coordination, and porosity. Quantitative microscopy can be applied to a packing to measure the important features, and techniques exist for transforming two-dimensional size distributions to true three-dimensional distributions [29]. This is a tedious approach. Consequently, many alternative measurement techniques have been advanced for characterization of the pores and packing structure. These include gas permeability, fluid infiltration, fluid expulsion, mercury porosimetry, and gas adsorption measurements.

In the liquid wetting or intrusion techniques, the insphere diameter is an important parameter. Various relations have been provided for relating the hydraulic radius or capillary pressure to the actual particle size. These are most applicable to spherical particles. For random packings, especially of irregular particles, the models of pore size and shape are inadequate. Thus, efforts to relate the packing features to the particle size are prone to error [81]. The mean pore size and distribution width are both underestimated by characterization techniques based on capillary behavior. Combinations of techniques can be potent in characterizing packing structures [41,61,82]. As an example, mercury porosimetry and gas adsorption measurements can be cross-calibrated for pore volume, surface area, and pore size, giving

$$\bar{d}\, S_v = 1.28 + 2.72\, f - 4\, f^2 \qquad (11.20)$$

where $\bar{d}$ is the mean pore size, S_v is the surface area per unit volume, and f is the fractional density. Such relations provide an important basis for interrelating the packing characteristics of very small particles. Need for the techniques discussed in this chapter increases as the particle size decreases.

References

1. J. van Brakel, "Pore Space Models for Transport Phenomena in Porous Media, Review and Evaluation with Special Emphasis on Capillary Liquid Transport," *Powder Technology*, 1975, vol. 11, pp. 205-236.
2. R. M. German, *Powder Metallurgy Science*, Metal Powder Industries Federation, Princeton, NJ, 1984, pp. 225-255.
3. S. V. Belov, O. G. Kartuesov and V. M. Polyaev, "Mean Pore Size of Porous Metals," *Soviet Powder Metallurgy and Metal Ceramics*, 1972, vol. 11, pp. 733-734.
4. S. Debbas and H. Rumpf, "On the Randomness of Beds Packed with Spheres or Irregular Shaped Particles," *Chemical Engineering Science*, 1966, vol. 21, pp. 583-607.
5. D. P. Haughey and G. S. G. Beveridge, "Structural Properties of Packed Beds - A Review," *The Canadian Journal of Chemical Engineering*, 1969, vol. 47, pp. 130-140.
6. S. Kruyer, "The Penetration of Mercury and Capillary Condensation in Packed Spheres," *Transactions of the Faraday Society*, 1958, vol. 54, pp. 1758-1767.
7. D. P. Haughey and G. S. G. Beveridge, "Local Property Variations in a Randomly Packed Bed of Equal-Sized Spheres," *Chemical Engineering Science*, 1967, vol. 22, pp. 715-718.
8. S. V. Belov, O. G. Kartuesov and V. M. Polyaev, "Mean Pore Size of Porous Metals," *Soviet Powder Metallurgy and Metal Ceramics*, 1972, vol. 11, pp. 733-734.
9. R. M. German, "Theory of Thermal Debinding," *International Journal of Powder Metallurgy*, 1987, vol. 23, pp. 237-245.
10. C. Y. Liu, K. Murakami and T. Okamoto, "Effect of Capillary Pressure on Interdendritic Liquid Flow," *Acta Metallurgica*, 1986, vol. 34, pp. 159-166.
11. J. R. F. Arthur and T. Dunstan, "Radiological Techniques Developed to Describe Particle Packing," *Powder Technology*, 1969, vol. 3, pp. 195-207.
12. D. M. Smith and D. L. Stermer, "Mercury Porosimetry: Theoretical and Experimental Characterization of Random Microsphere Packings," *Journal of Colloid and Interface Science*, 1986, vol. 111, pp. 160-168.
13. K. G. Frase and K. Hardman-Rhyne, "Porosity in Spinel Compacts Using Small-Angle Neutron Scattering," *Journal of the American Ceramic Society*, 1988, vol. 71, pp. 1-6.

14. F. A. L. Dullien, "Wood's Metal Porosimetry and its Relation to Mercury Porosimetry," *Powder Technology*, 1981, vol. 29, pp. 109-116.
15. J. van Brakel and P. M. Heertjes, "Homogeneous Particle Packings and Porous Masses. Some Experimental Techniques," *Powder Technology*, 1974, vol. 9, pp. 263-271.
16. F. A. L. Dullien and V. K. Batra, "Determination of the Structure of Porous Media," *Flow Through Porous Media*, American Chemical Society, Washington, DC, 1969, pp. 2-30.
17. N. F. Astbury and J. Vyse, "A New Method for the Study of Pore Size Distribution," *Transactions of the British Ceramic Society*, 1971, vol. 70, pp. 77-96.
18. P. R. Marshall, "Pore Size Distribution and Permeability of Porous Metal Materials," *Proceedings of a Symposium on Powder Metallurgy*, Special Report number 58, Iron and Steel Institute, London, UK, 1956, pp. 71-80.
19. E. Robens, "The Measurement of Specific Surface Area and Pore Size Distribution of Powders," *Powder Metallurgy International*, 1986, vol. 18, pp. 12-16.
20. H. M. Rootare, "A Review of Mercury Porosimetry," *Advanced Experimental Techniques in Powder Metallurgy*, J. S. Hirschhorn and K. H. Roll (eds.), Plenum Press, New York, NY, 1970, pp. 225-252.
21. T. Allen, *Particle Size Measurement*, second edition, Chapman and Hall, London, UK, 1974.
22. T. Katan and E. A. Grens, "The Effect of Liquid Configuration in Porous Gas Electrodes," *Journal of the Electrochemical Society*, 1971, vol. 118, pp. 1881-1886.
23. J. van Brakel, S. Modry and M. Svata, "Mercury Porosimetry: State of the Art," *Powder Technology*, 1981, vol. 29, pp. 1-12.
24. S. Modry, M. Svata and J. van Brakel, "Thematic Bibliography of Mercury Porosimetry," *Powder Technology*, 1981, vol. 29, pp. 13-43.
25. L. Moscou and S. Lub, "Practical Use of Mercury Porosimetry in the Study of Porous Solids," *Powder Technology*, 1981, vol. 29, pp. 45-52.
26. J. M. Haynes, "Stability of a Fluid Cylinder," *Journal of Colloid and Interface Science*, 1970, vol. 32, pp. 652-654.
27. I. M. Storozhevskii, "Some Problems of Modeling a Porous Body," *Soviet Powder Metallurgy and Metal Ceramics*, 1968, vol. 7, pp. 409-413.
28. E. E. Underwood, *Quantitative Stereology*, Addison-Wesley,

Reading, MA, 1970.
29. F. A. L. Dullien, E. Rhodes and S. R. Schroeter, "Comparative Testing of Statistical Methods for Obtaining Particle Size Distributions," *Powder Technology*, 1969, vol. 3, pp. 124-135.
30. E. R. Weibel, *Stereological Methods*, vol. 1, Academic Press, New York, NY, 1979.
31. P. P. Bansal and A. J. Ardell, "Average Nearest Neighbor Distances Between Uniformly Distributed Finite Particles," *Metallography*, 1972, vol. 5, pp. 97-111.
32. R. M. German, "The Direct Observation of Open Porosity Networks," *Metallography*, 1972, vol. 5, pp. 462-465.
33. D. Kellam and P. S. Nicholson, "Pore Shape Changes During the Initial Stages of Sintering," *Journal of the American Ceramic Society*, 1971, vol. 54, pp. 127-128.
34. R. E. Collins, *Flow of Fluids Through Porous Media*, Reinhold Publishing Corp., New York, NY, 1961.
35. J. F. Clements and J. Vyse, "A Comparison of Methods of Measuring Pore Size Distribution in Refractories," *Transactions of the British Ceramic Society*, 1968, vol. 67, pp. 285-317.
36. D. G. Beechry and J. Vyse, "A New Method for the Study of Pore Size Distribution: Some Notes on Theory," *Transactions of the British Ceramic Society*, 1971, vol. 70, pp. 87-96.
37. J. van Brakel, S. Modry and M. Svata, "Mercury Porosimetry: State of the Art," *Powder Technology*, 1981, vol. 29, pp. 1-12.
38. R. M. German, *Liquid Phase Sintering*, Plenum Press, New York, NY, 1985.
39. B. P. Pandey and A. K. Singhal, "Evaluation of the Capillary Pressure Curve Techniques for Determining Pore Size Distribution - A Network Approach," *Powder Technology*, 1976, vol. 15, pp. 89-96.
40. L. K. Frevel and L. J. Kressley, "Modifications in Mercury Porosimetry," *Analytical Chemistry*, 1963, vol. 35, pp. 1492-1502.
41. H. M. Rootare and C. F. Prenzlow, "Surface Areas from Mercury Porosimeter Measurements," *Journal of Physical Chemistry*, 1967, vol. 71, pp. 2733-2736.
42. R. P. Mayer and R. A. Stowe, "Mercury Porosimetry: Filling of Toroidal Void Volume Following Breakthrough Between Packed Spheres," *Journal of Physical Chemistry*, 1966, vol. 70, pp. 3867-3873.
43. S. Kruyer, "The Penetration of Mercury and Capillary Condensation in Packed Spheres," *Transactions of the Faraday*

Society, 1958, vol. 54, pp. 1758-1767.
44. C. Orr, "Application of Mercury Penetration to Materials Analysis", *Powder Technology*, 1969, vol. 3, pp. 117-123.
45. I. Chatzis and F. A. L. Dullien, "Mercury Porosimetry Curves of Sandstones. Mechanisms of Mercury Penetration and Withdrawal," *Powder Technology*, 1981, vol. 29, pp. 117-125.
46. R. G. Larson and N. R. Morrow, "Effects of Sample Size on Capillary Pressures in Porous Media," *Powder Technology*, 1981, vol. 30, pp. 123-138.
47. N. C. Wardlaw and M. McKellar, "Mercury Porosimetry and the Interpretation of Pore Geometry in Sedimentary Rocks and Artificial Models," *Powder Technology*, 1981, vol. 29, pp. 127-143.
48. J. Kloubek, "Hysteresis in Porosimetry," *Powder Technology*, 1981, vol. 29, pp. 63-73.
49. W. K. Bell, J. van Brakel and P. M. Heertjes, "Mercury Penetration and Retraction Hysteresis in Closely Packed Spheres," *Powder Technology*, 1981, vol. 29, pp. 75-88.
50. G. P. Androutsopoulos and R. Mann, "Evaluation of Mercury Porosimeter Experiments Using a Network Pore Structure Model," *Chemical Engineering Science*, 1979, vol. 34, pp. 1203-1212.
51. G. R. Lapidus, A. M. Lane, K. M. Ng and W. C. Conner, "Interpretation of Mercury Porosimetry Data Using a Pore-Throat Network Model," *Chemical Engineering Communications*, 1985, vol. 38, pp. 33-56.
52. L. Zagar, "Methods of Determination of Pore Structure Permeability and Diffusion," *Proceedings of the International Symposium on Pore Structure and Properties of Materials*, RILEM-IUPAC, Prague, Czechoslovakia, September 1973, pp. C249-C277.
53. H. M. Rootare and A. C. Nyce, "The Use of Porosimetry in the Measurement of Pore Size Distribution in Porous Materials," *International Journal of Powder Metallurgy*, 1971, vol. 7, pp. 3-11.
54. C. Gelinas and R. Angers, "Improvement of the Dynamic Water Expulsion Method for Pore Size Distribution Measurements," *Ceramic Bulletin*, 1986, vol. 65, pp. 1297-1300.
55. D. I. W. Atkinson, "The Mechanism of the Displacement of Liquids from Porous Solids," *Transactions of the Institution of Chemical Engineers*, 1949, vol. 27, pp. 259-272.
56. V. M. Kaptsevich, V. K. Sheleg, A. N. Sorokina, V. V. Savich,

V. V. Mazyuk and T. A. Anishchik, "Investigation of Pore Size Distribution in Porous Powder Materials by Different Methods," *Powder Metallurgy and Metal Ceramics*, 1987, vol 16, pp. 234-236.

57. D. W. Smith, J. E. Smugeresky and B. A. Meyer, *The Dependence of Permeability and Filtration on Pore Morphology in Consolidated Particulate Media*, Technical Report SAND87-8227, Sandia National Laboratories, Livermore, CA, October 1987.
58. I. G. Shatalova, N. S. Gorbunov and V. I. Likhtman, "Physicochemical Principles of Vibratory Compacting," *Vibratory Compacting Principles and Methods*, H. H. Hausner, K. H. Roll and P. K. Johnson (eds.), Plenum Press, New York, NY, 1967, pp. 1-206.
59. S. Lowell, *Introduction to Powder Surface Area*, Wiley-Interscience, New York, NY, 1979.
60. S. Brunauer, P. H. Emmett and E. Teller, "Adsorption of Gases in Multimolecular Layers," *Journal of the American Chemical Society*, 1938, vol. 60, pp. 309-319.
61. C. S. O'Neil, "Computation of Nitrogen Sorption Pore Volume Distributions," *American Laboratory*, 1985, June, pp. 41-48.
62. W. H. Wade, "The Coordination Number of Small Spheres," *Journal of Physical Chemistry*, 1965, vol. 69, pp. 322-326.
63. N. H. Ceaglske and O. A. Hougen, "The Drying of Granular Solids," *Transactions of the Institute of Chemical Engineers*, 1937, vol. 33, pp. 283-314.
64. L. C. Graton and H. J. Fraser, "Systematic Packing of Spheres With Particular Relation to Porosity and Permeability," *Journal of Geology*, 1935, vol. 43, pp. 785-909.
65. H. J. Frost, "Cavities in Dense Random Packings," *Acta Metallurgica*, 1982, vol. 30, pp. 889-904.
66. G. Mason, "A Model of the Pore Space in a Random Packing of Equal Spheres," *Journal of Colloid and Interface Science*, 1971, vol. 35, pp. 279-287.
67. J. Dodds and P. J. Lloyd, "A Model for the Void Structure in Multi-Component Sphere Packs Applied to Capillary Pressure Curves," *Powder Technology*, 1971, vol. 5, pp. 69-76.
68. J. F. Pearse, T. R. Oliver and D. M. Newitt, "The Mechanism of Drying of Solids Part I -The Forces Giving Rise to Movement of Water in Granular Beds, During Drying," *Transactions of the Institution of Chemical Engineers*, 1949, vol. 27, pp. 1-8.
69. M. Sahimi, B. D. Hughes, L. E. Scriven and H. T. Davis,

"Dispersion in Flow Through Porous Media - I. One-Phase Flow," *Chemical Engineering Science*, 1986, vol. 41, pp. 2103-2122.

70. M. Shahinpoor, "Frequency Distribution of Voids in Randomly Packed Monogranular Layers," *Mechanics of Granular Materials: New Models and Constitutive Relations*, J. T. Jenkins and M. Satake (eds.), Elsevier Science, Amsterdam, Netherlands, 1983, pp. 173-186
71. R. M. German, "Porous Materials," *Advances in Powder Technology*, G. Y. Chin (ed.), American Society for Metals, Metals Park, OH, 1982, pp. 225-251.
72. M. Ahmadzadeh and B. Cantor, "Interstitial Structure and Diffusion in DRP Metallic Glasses," *Journal of Non-Crystalline Solids*, 1981, vol. 43, pp. 189-219.
73. J. L. Finney and J. Wallace, "Interstice Correlation Functions; A New, Sensitive Characterisation of Non Crystalline Packed Structures," *Journal of Non-Crystalline Solids*, 1981, vol. 43, pp. 165-187.
74. A. G. Kostornov and M. S. Shevchuk, "Effect of Porosity, Thickness and Differential Pressure on the Principal Properties of Permeable Fiber Materials," *Soviet Powder Metallurgy and Metal Ceramics*, 1973, vol. 12, pp. 752-756.
75. D. M. Smith, D. P. Gallegos and D. L. Stermer, "Mercury Porosimetry in Random Sphere Packings: Breakthrough Pressure and Pore Structure Determination," *Powder Technology*, 1987, vol. 53, pp. 11-22.
76. J. L. Finney, "Random Packings and the Structure of Simple Liquids I. The Geometry of Random Close Packing," *Proceedings of the Royal Society of London A*, 1970, vol. 319, pp. 479-493.
77. D. P. Haughey and G. S. G. Beveridge, "Local Voidage Variation in a Randomly Packed Bed of Equal-Sized Spheres," *Chemical Engineering Science*, 1966, vol. 21, pp. 905-916.
78. J. W. Evans, M. H. Abbasi and A. Sarin, "A Monte Carlo Simulation of the Diffusion of Gases in Porous Solids," *Journal of Chemical Physics*, 1980, vol. 72, pp. 2967-2973.
79. L. W. De Backer, "Characterization of Some Retention and Transfer Properties of Granular Materials in Relation with Geometrical and Liquid Properties," *Advances in the Mechanics and the Flow of Granular Materials*, vol. 1, M. Shahipoor (ed.), Trans Tech Publ., Clausthal, Federal Republic of Germany, 1983, pp. 399-414.

80. D. A. Agievskii, V. A. Pronin and V. I. Kvashonkin, "Pore Size Distribution in a Disordered Particle Packing Model," *Industrial Laboratory (USSR)*, 1984, vol. 50, pp. 373-377.
81. I. Fatt, "The Network Model of Porous Media I. Capillary Pressure Characteristics," *Transactions of AIME, Petroleum Transactions*, 1956, vol. 207, pp. 144-159.
82. H. A. M. Van Eekelen, "The Random Spheres Model for Porous Materials," *Journal of Catalysis*, 1973, vol. 29, pp. 75-82.

Chapter Twelve

Capillarity and Liquid Interactions with Packed Particles

Chapter Outline

- **Overview and Definitions**
  - **Contact Angle**
  - **Spreading**
  - **Hysteresis**
- **Capillary Force**
- **Condensation**
- **Imbibition and Wicking**
- **Drainage**
- **Evaporation and Drying**
- **Fingering in Liquid Displacement**
- **Summary**
- **References**

Overview and Definitions

One area of great concern in dealing with packed particles is their interaction with wetting fluids. This interaction has importance for many phenomena, such as filtration, drying, soil drainage, ceramic agglomeration, and oil recovery. Indeed, the simple behavior of foams on carbonated beverages can be understood in terms of the liquid interaction with pores: the thinning of the foam on a beer after being poured into a glass is the result of liquid drainage from between the bubbles. A common feature of such problems is the capillary action associated with a wetting liquid in a highly curved pore structure. In the next chapter, the flow of fluids in pore structures will be covered, with special attention given to gas permeability.

This chapter discusses both the thermodynamics and kinetics of liquid interactions with packed particles. The treatment starts with a review of wetting and the capillary forces associated with a wetting liquid, and includes some of the kinetic aspects of liquid flow. The basic interaction in packed particles between fluids and pores is outlined in Figure 12.1. Two major variations, liquid flow into and out of the pores, are noted. In Chapter Eleven, one portion of this topic was introduced, that of mercury penetration porosimetry for the characterization of pore structures. This represents the case of pressurized pore filling with a nonwetting liquid. In this chapter, we will treat the other situations. The first is vapor penetration of the pores leading to condensation inside the structure. The second is direct liquid penetration via wicking or

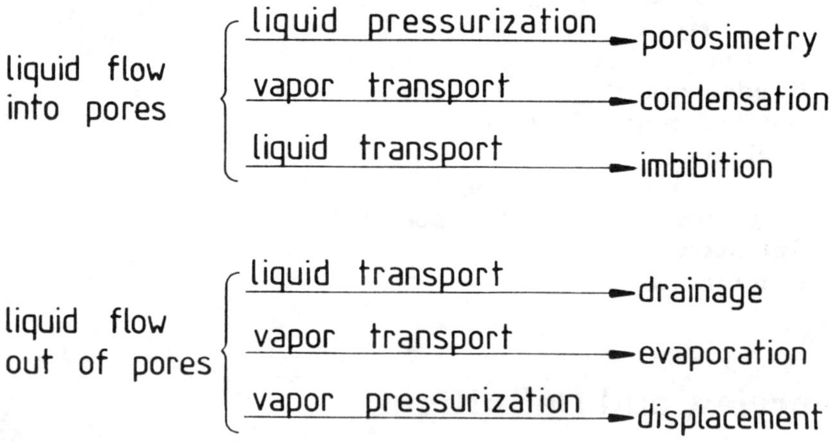

Figure 12.1 A definitional chart showing the classification of interactions between fluids and the pores associated with packed particles.

imbibition, an action analogous to a sponge soaking up water. Once a pore structure is saturated with a liquid, one must turn attention to the reverse situation. Initial liquid removal is by drainage, the reverse of imbibition. However, not all liquid can be removed by drainage; thus, the fourth process, evaporation, needs to be examined. The drying of a porous structure is actually a mixture of drainage and evaporation processes. Finally, special mention will be made of the case where a low viscosity fluid is used to displace a high viscosity fluid, and the conditions that lead to fingering. Largely, this presentation will emphasize the structure of randomly packed beds of spherical particles, yet mention will be made of the differences encountered with loose irregular and consolidated powders.

It will be evident that there are interrelations between the processes of evaporation, drainage, condensation, and imbibition that depend on the pore structure, fluid characteristics, and processing conditions. In most cases, the thermodynamic aspects of the problem dominate, although the kinetics are important to understanding many practical situations.

Besides the processes defined in Figure 12.1, a few other

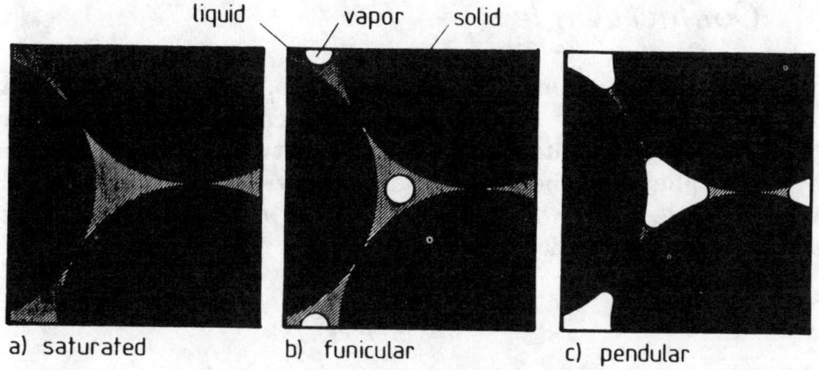

Figure 12.2 The three forms of liquid distribution in the pores between packed particles: a) saturated pores, b) the funicular state, and c) the pendular state.

terms should be specified. The use of the term capillarity will be restricted to descriptions of the equilibrium force associated with the liquid on the curved surface of the particles. Saturation will be used to describe the relative filling of the pores between packed particles by a fluid, typically a liquid. At 100% saturation, all of the interparticle pores will contain liquid. With partial saturation, the liquid will exhibit two other structures termed funicular and pendular. Figure 12.2 diagrams the differences in these structures for three spheres in point contact. At saturation all of the pore is filled. The limiting case for low liquid saturations is the pendular case, where a ring of liquid exists at the point contacts between neighboring particles. These rings do not touch one another. At intermediate saturations is the funicular state which corresponds to a connected liquid structure with capillaries of vapor in the liquid. The difference between the funicular and pendular states is in the amount of liquid and in the connection versus isolation of the liquid. In the funicular state the liquid rings wetting the particle contacts are connected to one another. The transition from pendular to funicular states occurs at saturations ranging from 6% for close-packed monosized spheres to approximately 12% for random loose packed monosized spheres [1-4]. However, in some cases of polydisperse powders, the transition has been reported at saturations as high as 24%.

Contact Angle

A main determinant of the interaction between packed particles and liquids is the contact angle. The contact angle is a measure of the relative interfacial energies between the solid, liquid, and vapor phases. The interfacial energy γ is defined as the change in free energy F with area A at constant temperature, pressure, and composition (T, P, X):

$$\gamma = \frac{\partial F}{\partial A} \Big|_{T,P,X} . \qquad (12.1)$$

For this treatment, subscripts on the interfacial energy will designate the type of interface; for example, SL will be the solid-liquid interface. The Young equation defines the contact angle in terms of the thermodynamic equilibrium of the solid-liquid-vapor system. Consider the equilibrium partially wetting situation diagramed in Figure 12.3,

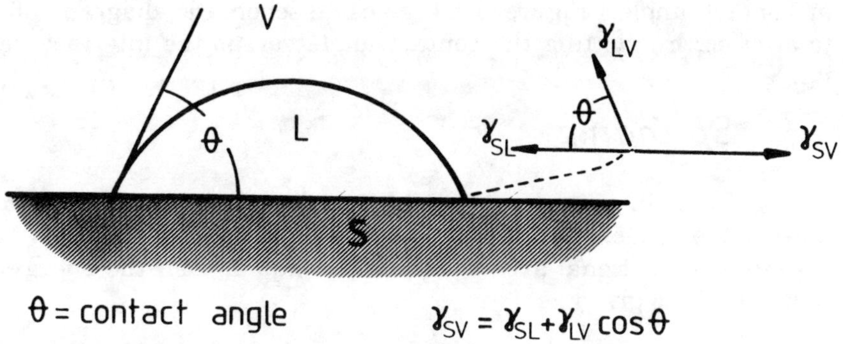

Figure 12.3 A partially wetting liquid on the solid-vapor surface with a contact angle θ that represents the force balance shown at the triple point shown on the right.

$$\gamma_{SV} = \gamma_{SL} + \gamma_{LV} \cos(\theta) \tag{12.2}$$

where θ is the contact angle. Note that the geometry away from the contact point of the three phases has no bearing on the contact angle. A liquid will cover a solid when the contact angle is 0°; this usually occurs when there is a favorable change in atomic bonding. At contact angles between 0 and 90°, the liquid will partially cover the solid. Larger contact angles are termed nonwetting and are not treated in this chapter.

The contact angle is usually presented as a system constant. Actually, the surface energies change with temperature, so the contact angle is temperature dependent. Because of surface roughness, fluid viscosity, and contamination there is a hysteresis in the contact angle [5]. An advancing liquid will have a higher contact angle than a receding liquid [6]. In part, this is because an advancing liquid has to perform work to coat the solid and because the solid surfaces are far from flat on an atomic scale. Alternatively, a receding liquid leaves a thin layer of contamination behind on the solid, which reduces its contact angle. Surface roughness (including protrusions and debris) can alter the equilibrium contact angle by as much as 20°. Furthermore, the rate of advance or recession in a fluid over a surface can contribute to the hysteresis in contact angle. Figure 12.4 gives a schematic diagram of the total effect by plotting the contact angle versus the interface velocity.

Spreading

Spreading is related to the surface energies, and is a measure of the tendency of a liquid to cover a solid surface. If X_A represents the fractional area covered by a liquid, then the energy per unit area E is given as

$$E = \gamma_{SV}(1 - X_A) + X_A(\gamma_{SL} + \gamma_{LV}) . \qquad (12.3)$$

A favorable change in the fractional coverage due to liquid spreading would be associated with a decrease in energy

$$dE/dX_A = - \gamma_{SV} + \gamma_{SL} + \gamma_{LV} \leq 0 . \qquad (12.4)$$

If the spreading energy change dE/dX_A is less than zero, then a liquid will spontaneously spread over the solid surface and will

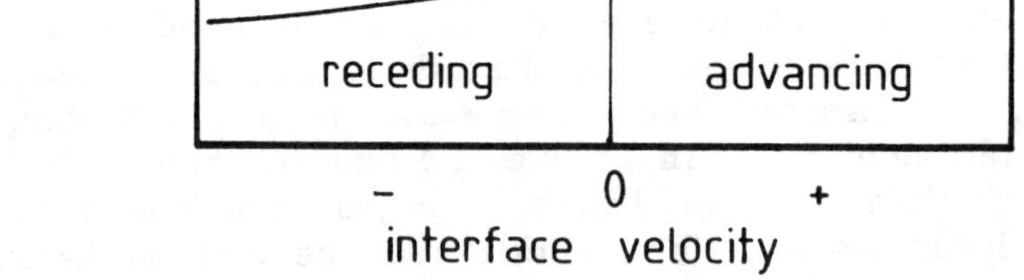

Figure 12.4 The contact angle hysteresis associated with the velocity of interface motion; there is a discontinuity between advancing and receding interfaces.

eliminate the solid-vapor interface. The rate of spreading is proportional to the covered area. Qualitatively, the magnitude of the spreading energy change correlates with the rate of liquid expansion over a solid surface. Actually, spreading involves diffusive transport of the liquid ahead of the actual bulk liquid. The diffusive transport can be through the vapor phase or on the solid surface. Thus, the structure consists of a thin, flat liquid-like layer ahead of the bulk liquid that forms the actual contact angle with the solid. This layer is probably less than 100 nm in thickness. The advancing thin layer of liquid does not have the properties of the bulk liquid. On a rough surface, the advancing liquid will spread in a discontinuous manner, and may even follow preferential capillary grooves [5,7]. Protrusions, debris, and surface roughness on the scale of 50 nm in size have a large effect on nonuniform spreading. Hence, surface roughness increases the uncertainty in contact angle.

Hysteresis

Contact angle hysteresis, coupled with the distribution in curvatures in randomly packed particles, leads to several problems in liquid motion. For one thing, the drainage of a liquid can never be complete. There will always be an irreducible saturation because of liquid trapped at particle contact points (pendular bonds) and in small pores [8]. Isolation of pores filled with liquid is common during drainage of randomly packed particles. Thus, the imbibition and drainage curves of saturation versus capillary pressure will never overlap. Likewise, evaporation and condensation will exhibit a hysteresis. In many instances an understanding of this hysteresis can aid in optimizing industrial processes. Figure 12.5 contrasts the difference in imbibition and drainage behavior for a small pore cluster [9]. As long as the liquid wets the solid, the liquid will have lower pressure than the pore space. This pressure deficit due to the surface energy and interfacial curvature leads to capillary attraction.

Other characteristics of concern with respect to the liquid wetting of pores are the surface energies, liquid viscosity, pore geometry, pore size distribution, particle shape, porosity, and various gradients. The gradients of importance include gravity, composition, permeability, and temperature [10]. When the pore sizes are large (over approximately 10 μm) the gravitational forces are important. Alternatively, when the pore size is below

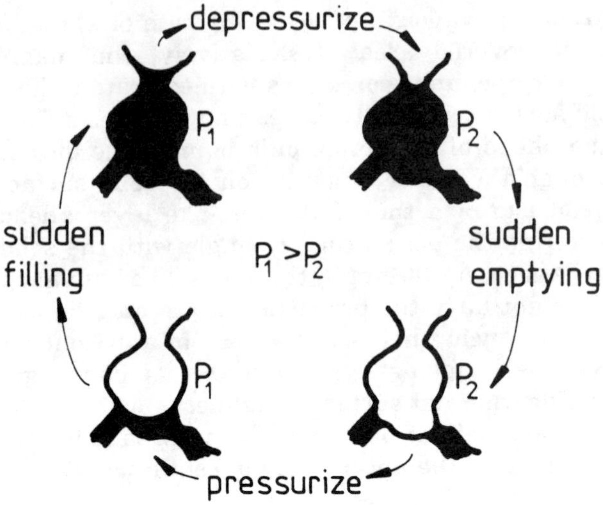

Figure 12.5 A schematic diagram showing the hysteresis of a pore with emptying and filling after pressure changes due to the contact hysteresis [9].

approximately 1 μm the low permeability of the pore structure has a dominant effect with respect to fluid flow. Based on the pore size and shape variations in particle packing, we can understand the hysteresis effects associated with drainage and imbibition.

Capillary Force

An attractive force exists between solid objects that have a intervening liquid bridge of wetting liquid. The pendular bond shown in Figure 12.2c is an example of a wetting liquid bridge that induces capillary attraction between the spheres. As the contact angle becomes smaller, there is more wetting of the spheres by the liquid. Accordingly, a minimization of energy is possible if the particles touch. The attraction arises from the need to minimize the liquid-vapor surface area while maximizing the liquid-solid surface area. To satisfy both needs while maintaining constant pressure in the liquid, the pendular bond places an attractive force on the particles. For small contact angles and relatively small liquid contents, the minimum energy occurs when the spheres touch [11].

The source of the capillary force can be understood in terms of the interfacial energy. As defined in Equation 12.1, the interfacial energy is the change in free energy per unit area at constant composition, pressure, and temperature. This capillary force is easy to observe with a thin diameter glass tube inserted into a wetting liquid. In the classic capillary rise experiment, the capillary tube is positioned vertically in the fluid as shown in Figure 12.6. As long as the fluid wets the tube bore, it will rise in the tube above the external height. The height of rise h depends on the contact angle θ, capillary tube diameter d, liquid density ρ, and surface energy γ_{LV} as follows:

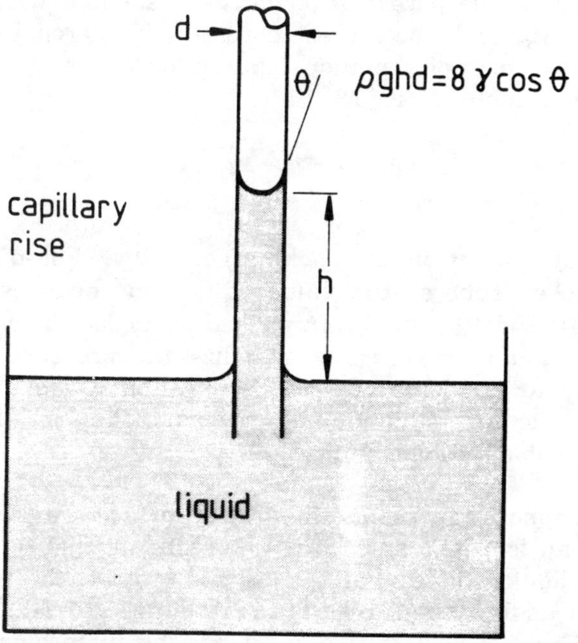

Figure 12.6 Capillary rise in a narrow tube is based on the balanced surface energy and gravitational forces as illustrated by this diagram.

$$h = 4\ \gamma_{LV} \cos(\theta)/(\rho\, g\, D) \tag{12.5}$$

with g equal to the gravitational acceleration. The reason that the fluid rises is that it has a lower pressure under the concave surface in the tube. The height of rise represents the balance between the pressure deficiency from curvature and the pressure head due to elevation of the column of liquid.

This same behavior can be attained with packed particles where the pores between the particles act in an analogous manner to the capillary tube. A wetting liquid is pulled into the pores. The distance of penetration increases as the pore size becomes smaller, or equivalently as the particle size becomes smaller. In a randomly packed structure, a wetting liquid will preferentially flow into the smallest accessible pores. Alternatively, a liquid with a large contact angle will seek the largest pores (as in mercury porosimetry). The modified form of Equation 12.5 applicable to packed monosized spheres of diameter D is [12]

$$\rho\, g\, h\, D = C_p\ \gamma_{LV} \cos(\theta) \tag{12.6}$$

with C_p dependent on the packing geometry. For monosized, randomly packed spheres the value of C_p has been estimated to be between 10 and 12 [12,13]. In packed particles, the contact angle is preserved and flow occurs to equalize the pressure in the liquid. This leads to problems of liquid separation at partial saturations because of the distribution in pore sizes and shapes. This behavior is most noticeable during draining.

A sphere represents the lowest surface-area-to-volume ratio. Therefor an isotropic surface a minimum in energy occurs with a spherical liquid shape. For a spherical droplet, the work of expansion from adding liquid results in a change in radius and surface area which must perform work against the surface energy. Accordingly, the increase in internal pressure of the droplet due to the curved surface is given as

$$\Delta P = 2\ \gamma/R \tag{12.7}$$

where ΔP is the pressure increase over ambient, R is the radius of the sphere, and γ is the surface energy. Thus, there is an increase in pressure associated with the curved surface of a sphere at equilibrium. This pressure varies with the inverse of the radius of curvature. It is a characteristic of fluids that the pressure is equal everywhere, otherwise flow occurs to relieve the pressure gradient. The internal increase in pressure associated with a curved surface is termed the capillary pressure, and the forces are the capillary forces.

A flat surface has no curvature and is defined as the equilibrium surface. Accordingly, a liquid with a convex surface will have a pressure higher than equilibrium and will act to smooth the raised surface. Alternatively, concave surfaces have pressures lower than equilibrium and will flow to fill the concavity if they are unconstrained. Thus, as shown in Figure 12.7, the presence of a concavity or convexity will lead to flow that smoothes the surface. According to the Laplace equation for a general surface of rotation the pressure change depends on the two principle radii of curvature [14]

$$\Delta P = \gamma_{LV} (1/R_1 + 1/R_2) \tag{12.8}$$

where R_1 and R_2 are the orthogonal (perpendicular) radii of curvature at a point on the surface. If both R_1 and R_2 are centered inside the liquid the pressure is positive, whereas, if they are both centered outside the liquid the pressure is negative.

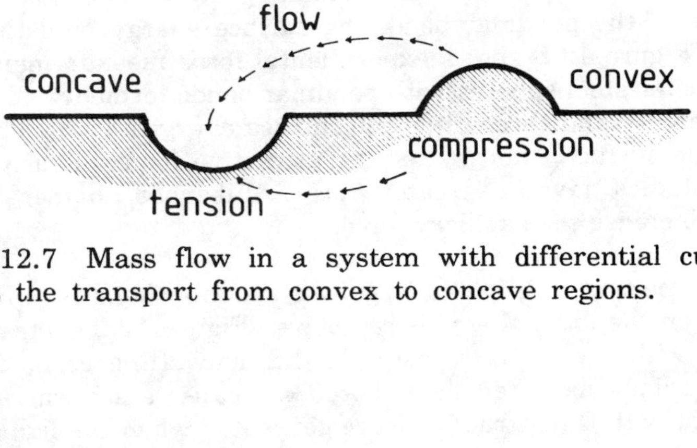

Figure 12.7 Mass flow in a system with differential curvature involves the transport from convex to concave regions.

With a saddle surface like the pendular bond, where one radius is centered inside the liquid and the other is centered outside the liquid, the sign of the pressure is dependent upon the smaller radius. A liquid bond will form such a saddle surface between particles. This bond exerts an attractive force on the spheres when the liquid wets the particles. One cause for the attractive force is the pressure deficiency in the liquid due to the general concave shape of the surface. A second cause is the surface tension along the contact perimeter of the liquid with the spheres. These two effects combine to force the two spheres together. In studies in which one of these forces has been missing or in which various simplifying geometric assumptions have been made, an improper capillary force has resulted [14-16]. For a surface of revolution described by an equation $y(x)$, the equilibrium configuration must satisfy the following condition:

$$\frac{P}{\gamma_{LV}} = \frac{d^2y/dx^2}{(1 + (dy/dx)^2)^{3/2}} - \frac{1}{y(1 + (dy/dx)^2)^{1/2}} \,. \qquad (12.9)$$

The pendular bond formed between two spheres has a complex saddle shape in three dimensions; the equation that describes this bond is termed a nodoid [15]. In some treatments, a circular arc profile has been incorrectly assumed for the bond shape, often with a substantial error [14]. The nodoid is the solution to Equation 12.9 where there is a constant curvature at all points on the surface. This ensures that the pressure in the liquid is the same everywhere.

The magnitude of the capillary force depends on the curvature of the pendular bond, the surface energy, and the liquid volume. Figure 12.8 shows experimental force measurements on 4 mm diameter steel spheres with pendular bonds formed with various liquid contents and three different surface energies [17]. The more curved the surface, the greater the attractive force. Thus, smaller liquid volumes give a greater force. Likewise, higher surface energies increase the capillary force.

If the contact angle is large, the pendular bond is convex and the force on the two spheres is repulsive. Figure 12.9 contrasts this situation with the wetting case. In the nonwetting case, the presence of a liquid between the spheres will cause a separation force, in contrast with the attractive force generated when the liquid has a

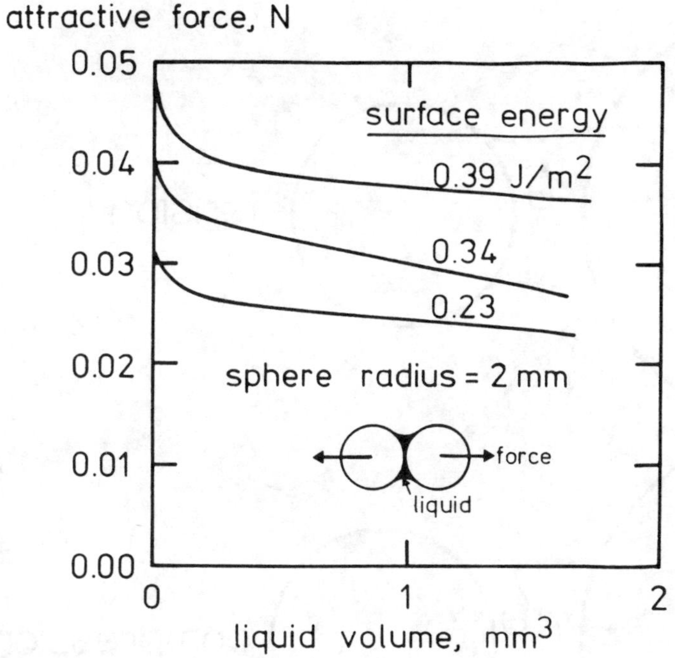

Figure 12.8 The variation in attractive force with liquid volume between 2 mm radius spheres with a pendular bond formed from three different liquid surface energies [17].

low contact angle. The transition between attraction and repulsion is dependent on the contact angle [15]. Figure 12.10 plots the normalized capillary force versus the distance of separation between two spheres for contact angles of 0 and 85°, both with the same liquid contents. The force is attractive at all separations for low contact angles, but repulsive in some situations for high contact angles. Accordingly, for the wetting case, because the attractive force increases as the separation decreases, there is a gradient at all separations. This means that with a wetting liquid bridge the two spheres will form a compressive contact when they are free to move. In contrast, the 85° contact angle exerts a repulsive force on the spheres at small separations and an attractive force at large separations. Hence, as expressed in Figure 12.11, there is an equilibrium separation at which there is neither an attractive nor repulsive force. This point of equilibrium depends on the liquid

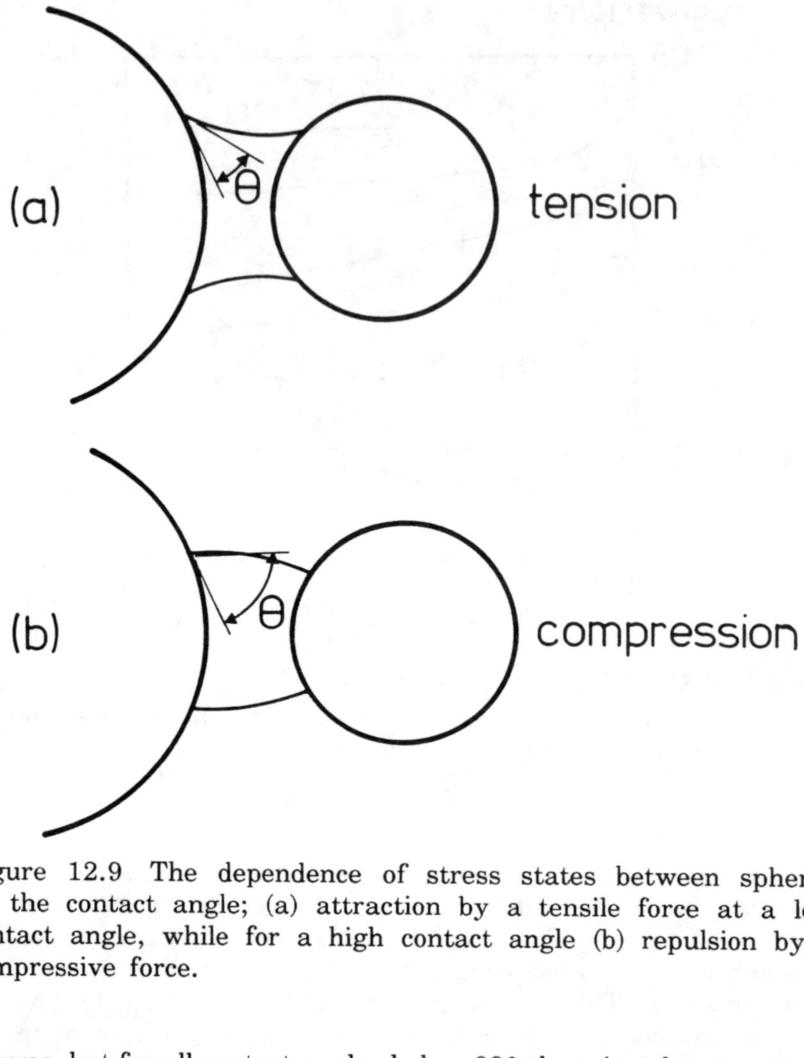

Figure 12.9 The dependence of stress states between spheres on the contact angle; (a) attraction by a tensile force at a low contact angle, while for a high contact angle (b) repulsion by a compressive force.

volume, but for all contact angles below 60° there is only attraction.

Experimental measurements show that the attractive force passes through a peak as the separation is increased [18]. Should two particles be forced apart, the liquid meniscus will not be stable at large separations. Instead, the pendular bond ruptures, thereby lowering the energy of the configuration by forming two spherical caps rather than a highly elongated bond [14,15]. A sketch of this structure is given in Figure 12.12. The initial force stretches the

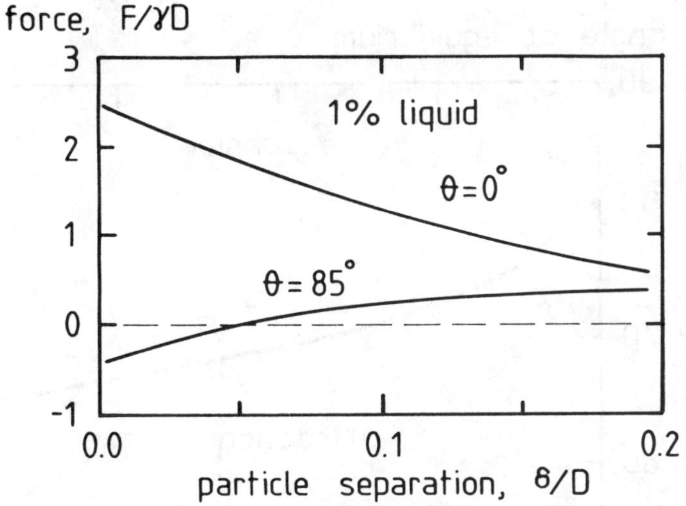

Figure 12.10 A plot of the interparticle force versus the separation distance (both parameters are normalized), for contact angles of 0 and 85° and 1% liquid.

liquid bond, but eventually the liquid ruptures and separates into the two caps, with no restoring force. Because the liquid surface curvature has reversed, evaporation will be faster after the bond ruptures.

One result of capillary forces is that particles agglomerate in the presence of a wetting fluid. The capillary action not only pulls particles together into a dense packing, but the liquid lubricates the surfaces and aids particle rearrangement. This is used beneficially in agglomerating small particles, where a slight deficiency of liquid improves packing density. Figure 12.13 shows two scanning electron micrographs of agglomerated molybdenum powder. In this case polyvinyl alcohol was used to agglomerate the individual particles into large clusters. The capillary pressure increases as the packing density increases. Another consequence of the capillary force dependence on curvature is that small pores will preferentially retain a liquid with a small contact angle. For irregular shapes, a wetting liquid will cause a volume expansion at low saturation levels. Nonsymmetrical particles will experience a torque from a wetting

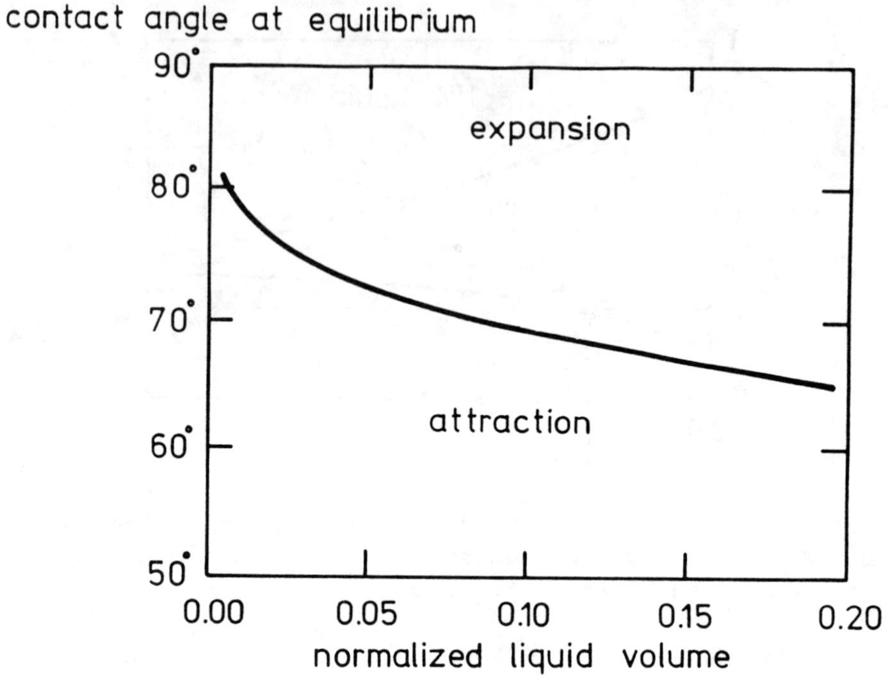

Figure 12.11 The relation between the contact angle and the volume of liquid (as normalized to the particle volume), showing the conditions for expansion and attraction between contacting particles [14].

liquid. As a consequence, the particles are pulled into lower density packing configurations. Under stress, a packing of saturated irregularly shaped particles will expand. When the stress is removed, the capillary suction associated with the relaxed structure will create a lower saturation surface. This is why after walking on wet sand, the sand will be initially lighter in color due to the decreased saturation at the top.

In this chapter several consequences of curved surfaces are discussed. One effect is obviously the agglomeration of particles in the presence of a wetting liquid. For particles over 100 μm in size, capillarity dominates agglomeration in comparison with van der Waals forces [19]. Another effect is that the vapor pressure over a

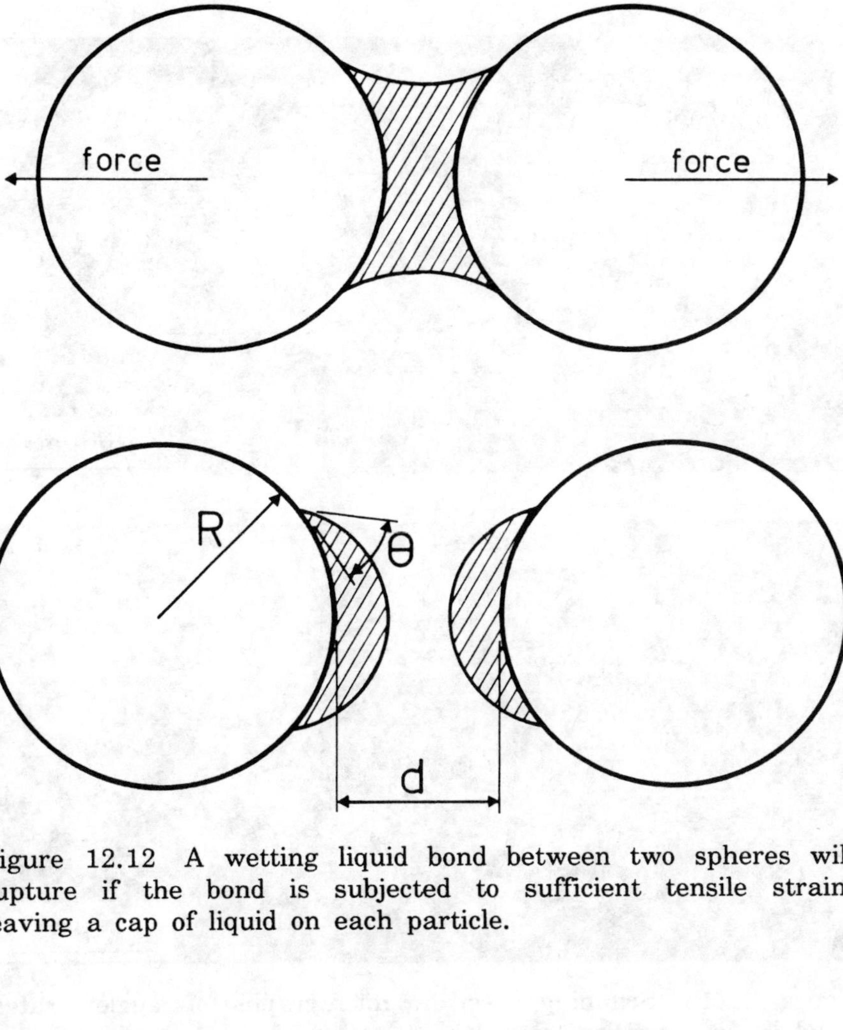

Figure 12.12 A wetting liquid bond between two spheres will rupture if the bond is subjected to sufficient tensile strain, leaving a cap of liquid on each particle.

curved surface will be different from the equilibrium value. For the typical case of interest, the liquid is curved with a concave surface and the vapor pressure over that surface is less than the equilibrium value. This has importance for the removal of liquids from packed particles by drying, since small, highly curved surfaces will resist drying [20].

An extrapolation of the capillary effect to very small particle sizes is not valid. When the capillary dimensions approach the

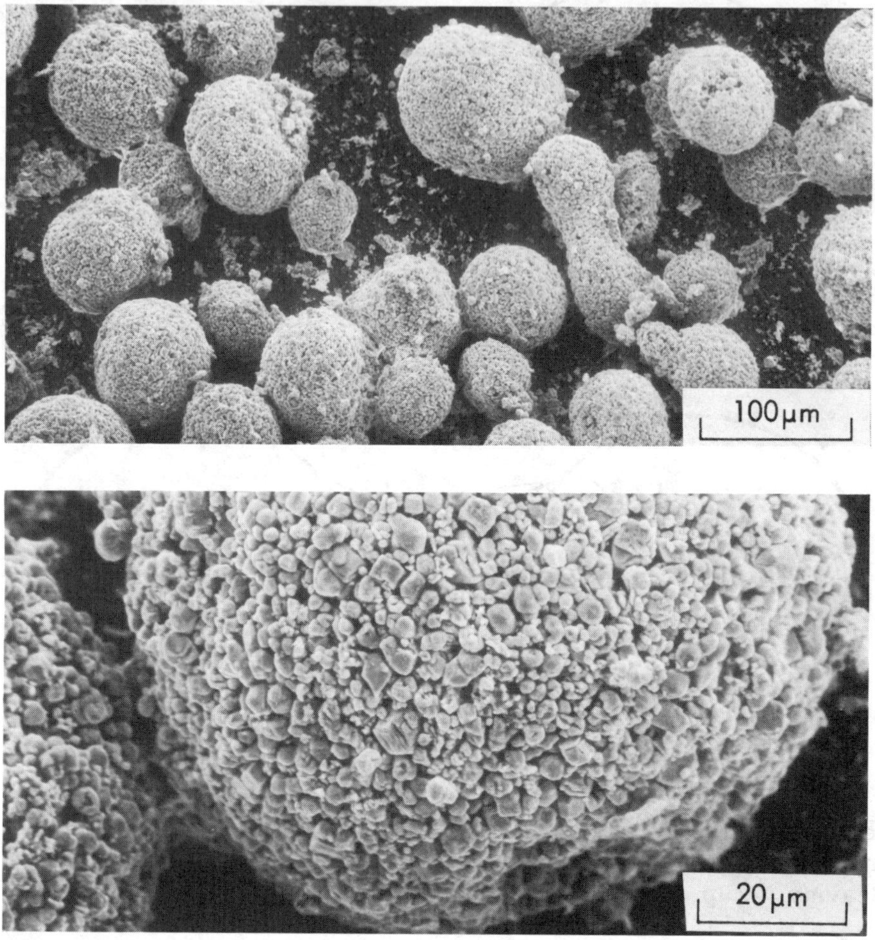

Figure 12.13 Scanning electron micrographs of agglomerated molybdenum particles formed by spray drying a mixture of particles and binder.

atomic scale the interfacial properties no longer match those of the bulk liquid [21]. Otherwise, very minimal quantities of wetting liquid would give rise to enormous pressures. However, at larger sizes the capillary force does vary systematically with the particle size, surface energy, volume fraction of liquid, porosity, and interparticle separation. The capillary force is high for low contact angles, small particle sizes, and small liquid quantities. This force increases as the particles approach one another, and leads to a minimum energy when wetted particles are in contact with one another. For small particles, the magnitude of the capillary force is very high. As an example, for a contact angle of zero, particle size of 20 μm, liquid density of 10 g/cm^3, and liquid volume of 10%, the capillary force would be a million times the liquid weight. Such a strong force will have a profound effect on the behavior of packed particles, especially at partial saturations. In contrast, packings of irregularly shaped particles swell when saturated with a wetting liquid [15]. The angularity of the particles leads to an increase in porosity as the pores saturate with liquid [22]. The degree of swelling decreases as the volume of liquid increases. As demonstrated in the balance of this chapter, the action of liquid in pores between packed particles is highly influenced by the capillary force.

Condensation

The rate of intrusion of a liquid into packed particles is often controlled by vapor condensation ahead of the advancing liquid. The liquid evaporates and then condenses at particle contacts (or other concavities on the solid surface). This is especially true when the liquid has a low heat of vaporization. Hence, vapor condensation is important to both vapor and liquid flow into a bed of packed particles. Preferential condensation at interparticle contacts is a precursor to the formation of pendular bonds.

In practical situations, the condensation of humidity within a powder bed can be troublesome. The water condenses at the particle contacts, causing agglomeration while increasing the interparticle friction. The condensation initiates at the point contacts, forming pendular bonds which grow with increasing condensate volume [23]. One monolayer usually forms on the particle when the relative vapor pressure is between 0.1 and 0.3 of the saturation vapor pressure. As the liquid saturation increases, the vapor pressure of the liquid also increases as illustrated in Figure 12.14 [24]. For water it is estimated that condensation and the formation of pendular bonds starts at a relative humidity of between 65

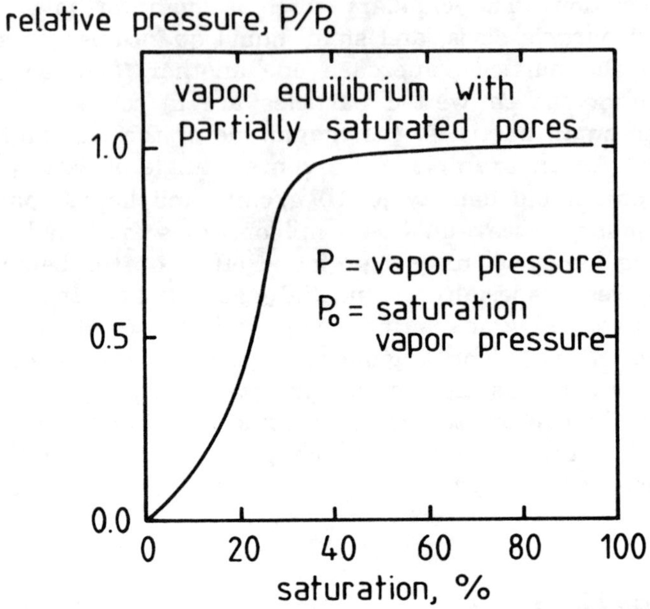

Figure 12.14 The relative partial pressure of vapor in equilibrium with a partially saturated pore structure showing the change in vapor pressure with saturation [24].

and 80% [25]. Likewise, depending on the surface roughness and particle size, there is a lower limit of humidity below which no condensation will occur.

The Kelvin equation provides an important description of the surface curvature effect on the vapor pressure

$$k\,T\,ln(P/P_o) = \gamma_{LV}\,\omega\,[1/R_1 + 1/R_2]\,. \qquad (12.10)$$

In this equation P_o is the equilibrium vapor pressure, P is the vapor pressure over the curved surface described by the curvatures R_1 and R_2, k is Boltzmann's constant, T is the absolute temperature, γ_{LV} is the liquid-vapor surface energy, and ω is the atomic volume. Preferential condensation at the particle contacts is explained by the Kelvin equation. After one monolayer of liquid is adsorbed on the particle surface, the liquid forming the pendular bond (a sharply concave surface) will have a much lower vapor pressure than the bulk liquid. The resulting vapor pressure deficiency at the contact points leads to a pressure gradient and preferential condensation at the pendular bonds, with concomitant exaggerated growth [26,27]. The pressure gradient associated with the liquid curvature further aids growth of the pendular bonds, since the liquid will also flow over the surface to equalize pressure.

In actual pores, bridging will occur as liquid films condense on the pore walls. This leads to rapid filling of liquid into the pores at points ahead of an advancing liquid front. The preferential growth of pendular bonds by condensation continues until neighboring bonds merge and form the funicular state [28]. Thus, saturation provides a guide to the vapor pressure inside a particle packing: low saturations indicate low vapor pressures.

During evaporation, such as occurs in drying, the liquid meniscus will exhibit a different curvature than during condensation. Consequently, a hysteresis effect is evident when evaporation and condensation events are compared. Indeed, packed particles with liquids trapped in the pores will exhibit liquid redistribution over long time periods as evaporation and condensation act to establish a constant vapor pressure everywhere.

As the size of the surface concavity becomes very small, the Kelvin equation is no longer valid. The measured vapor pressure in equilibrium with small pores (below 10 μm) exceeds the expected pressure by factors of up to 80 [20,29]. The degree of error in the Kelvin equation depends on the liquid polarity and the pore size. However, it is characteristic of liquid interactions with packed particles that vapor transport plays a role in distributing the liquid at low levels of saturation.

Imbibition and Wicking

A wetting liquid will flow into the pores between packed particles because of a reduction in the free energy as the liquid covers the solid. The precise reduction in energy depends on the liquid profile and the curvature of the solid surface. In this section the concern is with prediction of the flow rate for wetting liquids penetrating into the interstices between packed particles. Most treatments of liquid imbibition into pores generally start with the Poiseuille equation for viscous flow in a capillary tube [30]

$$\frac{dh}{dt} = \frac{d\ \gamma_{LV}}{8\ \mu\ h} - \frac{d\ g\ \rho}{32\ \mu} \tag{12.11}$$

where h is the distance, t is the time, g is the gravitational acceleration, ρ is the fluid density, μ is the fluid viscosity, γ_{LV} is the surface energy, d is the pore size, and θ is the contact angle. Although this equation is not valid for the pores between packed particles, it does provide a basis for estimating the wicking or imbibition behavior [31].

The maximum rise height h_m for a wetting liquid in a cylindrical pore is given by the Washburn equation [32],

$$h_m = 4\ \gamma_{LV} \cos(\theta)/(d\ g\ \rho)\ . \tag{12.12}$$

A combination of Equations 12.11 and 12.12 gives the time dependent capillary liquid imbibition [33]

$$h^2/(h_m - h) = d^2\ g\ t\ \rho/(32\ \mu)\ . \tag{12.13}$$

For large particle sizes, where gravitational effects are significant, Equation 12.13 must be corrected for the orientation of the direction of flow with respect to gravity. In all cases, this equation reduces to a general form where the depth of liquid penetration varies with the square root of time. From Equation 12.13, the volumetric rate of liquid ingression into the pores can be estimated from the compact surface area and surface porosity. The mass or volume gain will depend on the square root of time. Several variations on this basic equation involve corrections for the pore geometry, gravitational contribution, and flow viscosity [30,31,33-35]. Actual porous systems give a volume gain that depends on time to an exponent of between 0.4 and 0.5. Figure 12.15 illustrates this behavior with results of penetration depth versus time from ethanol wicking into porous yarn [31]. The actual behavior is slower than predicted by

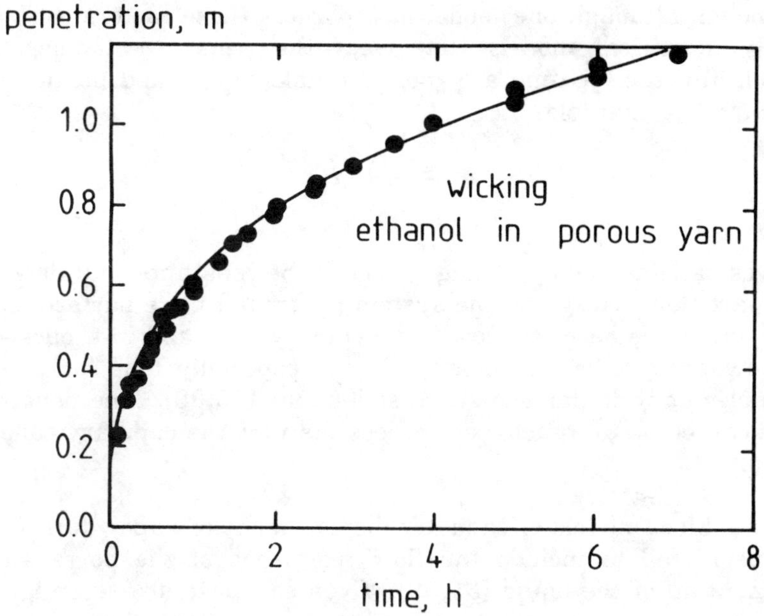

Figure 12.15 Wicking penetration of porous yarn by ethanol, showing the penetration depth versus time [31].

the capillary induced flow model for capillary tubes.

As illustrated above, the movement of a wetting fluid into a capillary tube has a time dependence that relates to the fluid viscosity μ, surface energy γ_{LV}, pore size d, and contact angle θ. These factors appear in all of the models and are dominant with respect to imbibition of liquids. In wicking, the liquid distribution is not uniform and the fluid does not fully saturate the pores near the leading edge of the liquid. The initial rate of liquid penetration is faster than predicted while the latter rate is slower than predicted. The inclusion of appropriate corrections for the nonuniform pore shape, liquid viscosity, flow streamlines, pore tortuosity, trapped vapor, and fluid density would improve the models. The role of vapor condensation ahead of an advancing liquid meniscus is also recognized as being important, but again it is not included in the current models. The effects of pore size distribution are relatively unknown, although one model incorporates these factors [34]. Generally, empirical models that treat the penetration depth as a simple function of time are most suitable for packed particles [32, 36,37]. For example,

$$h^2 = A\,t - B \tag{12.14}$$

proves satisfactory by letting A and B be constants that depend on the particle packing and the system parameters like surface energy. The difference between flow in a capillary tube and in a packed bed of powders can be significant. This is especially true for irregular particles or wide particle size distributions [13,38]. The penetration rate can be up to a factor of ten less than for the capillary tube situation.

Most models of capillary displacement of vapor during liquid intrusion fail to include the flow resistance of the pores and the momentum of the liquid [35,39]. Even so, these are secondary corrections. The general behavior of the penetration distance varying with the square root of time is characteristic of most systems. It is noteworthy that the capillary force is enhanced by particle motion during fluid intrusion, such as by vibration. The increase in the capillary force and penetration rate can be very significant in such situations [40].

Drainage

After a structure is saturated with liquid, the next concern is to remove the liquid. The two general techniques for removing a liquid are to either evaporate or drain the liquid. Evaporative drying processes are initially controlled by liquid flow to the compact surface. Thus, in both processes the initial focus is on liquid removal from the interstices of a particle packing by liquid flow. This process is termed drainage, reflecting its historical importance in the study of soils.

Drainage occurs because a suction (or other pressure) is placed on the wetting liquid, causing it to move from the pores. The pressure needed to remove a liquid from a pore is higher than the pressure associated with flow into the pore. Figure 12.16 plots such hysteresis, showing saturation as a function of capillary pressure [41]. In part, this hysteresis reflects the smaller contact angle of a retracting interface compared to an advancing interface. The smaller the pore size, the greater the difference between the pressures of pore filling and draining, similar to that seen earlier with mercury porosimetry.

No liquid is removed until the drainage pressure equals the capillary breakthrough pressure for the largest pore. At drainage pressures (suction) above this level, the pore structure usually drains over a narrow pressure range. First the funicular state is formed, and as liquid drains from the packed particles, pendular bonds are left behind at the particle contacts, as diagramed in Figure 12.17. Besides forming pendular bonds, it is also possible for

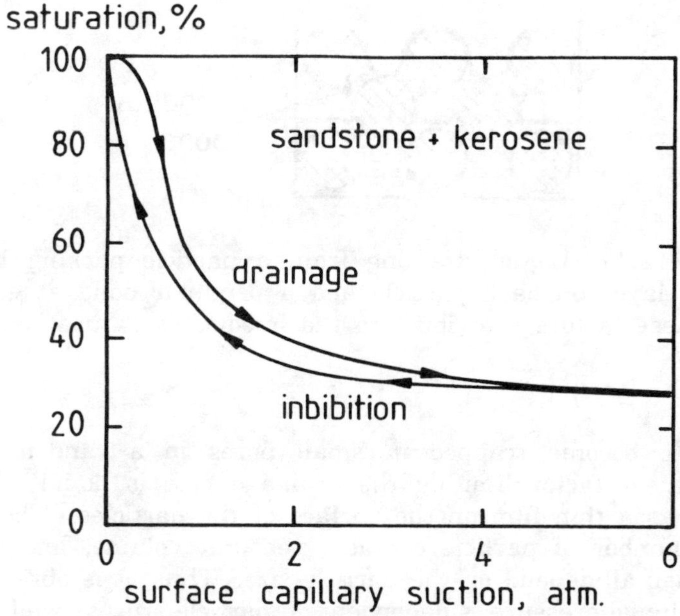

Figure 12.16 The saturation versus capillary suction pressure for kerosene drainage and imbibition in sandstone [24].

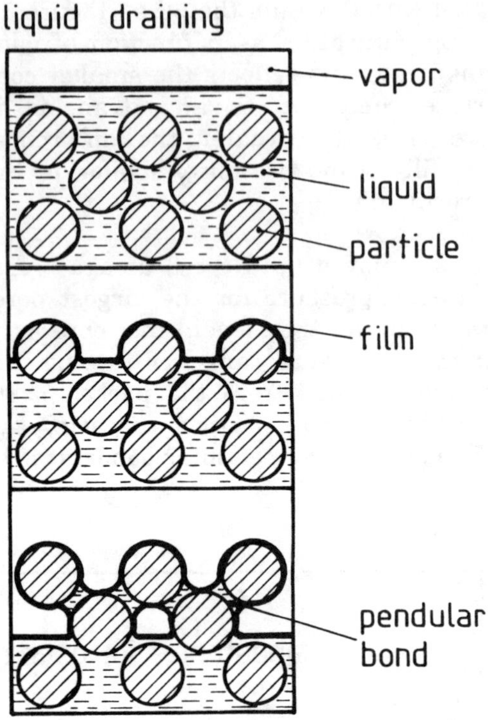

Figure 12.17 Liquid draining from a particle packing leaves a surface layer on each particle and a pendular bond at each contact; these factors contribute to the irreducible saturation.

liquid to become trapped in small pores in a random particle packing. A factor limiting the drainage rate is liquid remaining behind as a thin film on the surface of the particles. The surface area, number of particle contacts per unit volume, and capillary attraction all depend on the particle size. Thus, it is observed that the drainage pressure is dependent on particle size, as well as particle shape and fluid viscosity. Up to a point, the greater the force causing drainage, the larger the volume drained [42,43]. This is shown by the equation

$$S = C\,[\gamma_{LV} \cos(\theta)/(a\,P)]^{0.264} \tag{12.15}$$

where S is the saturation, C is a constant ranging between 0.025 and 0.012, γ_{LV} is the surface energy, θ is the contact angle, P is the applied pressure, and a is the permeability of the pore structure. There will be more residual saturation with a low permeability pore structure, high surface energy liquid, and small contact angle.

Unfortunately, in structures composed of packed particles there is an irreducible or connate saturation representing the immobile liquid which remains as disconnected segments in small pores or at pendular bonds [2-4,24,28,44,45]. Such behavior is in direct contrast with capillary tubes, which drain completely. Figure 12.18 illustrates the drainage behavior for a random packing, showing saturation as a function of the capillary pressure [24]. The liquid remaining between packed particles must be removed by evaporation since its discontinuous character makes it insensitive to the external capillary pressure. The level of irreducible saturation depends on the permeability of the pore

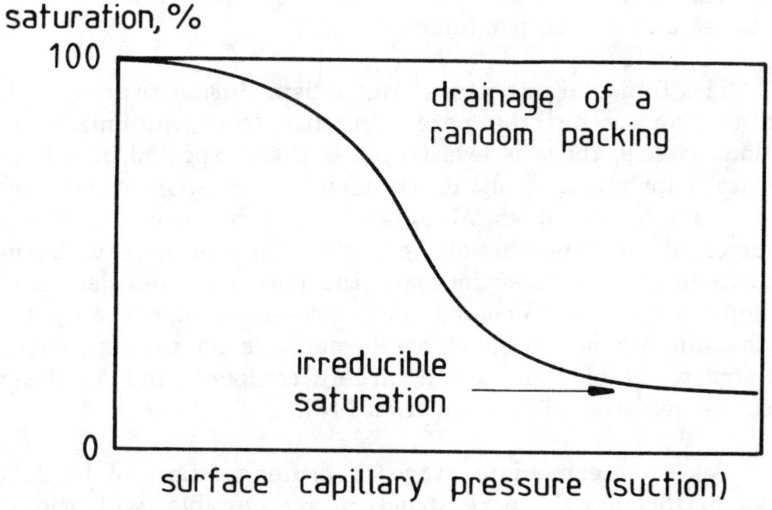

Figure 12.18 Drainage from a random packing, showing the saturation decreasing to the irreducible level as the surface capillary pressure increases.

structure, decreasing as the permeability increases. In his extensive experimentation, Morrow [2] included variations in the interfacial surface energy, fluid viscosity, fluid density, contact angle, particle shape, particle size, permeability, and bed porosity. From his measurements, he concluded that the irreversible saturation was generally 6 to 10% of the pore volume. This fluid was trapped in pores as well as at pendular bonds. When the powder was consolidated, the porosity was decreased and the irreversible saturation was increased. Inhomogeneities in the particle packing increase the irreducible saturation, so that multimodal particle mixtures increase the irreducible saturation.

In contrast with Morrow, other investigators of irreducible saturation claim that there are effects of surface energy and particle size [28,43,46,47]. The amount of liquid trapped in the pores increases with the surface energy and the inverse of the particle size. Accordingly, a wide particle size distribution or porosity in the particles will impede drainage and increase the irreducible saturation. Examinations of the irreducible saturation have shown levels ranging from 4 to 15% [3,4,45,46], distributed as 1/4 in filled pores and 3/4 in pendular bonds.

The funicular structure that exists during drainage allows liquid to flow within the pore structure, thus minimizing liquid isolation. Hence, there is less trapping than expected based on the pore size distribution, and the irreducible saturation is independent of most packing variables, although it does increase as a structure is subjected to compaction or sintering [48]. Furthermore, increases in porosity also increase the pore size and decrease the coordination number. Since capillary pressures depend on pore size and the number of pendular bonds depends on packing coordination, structural differences are largely compensating factors with respect to the irreducible saturation [45].

At an intermediate stage of drainage, flow of both liquid and vapor through the pore structure are possible, with the vapor pressure being essentially constant to maintain equilibrium in the structure. To maintain a constant vapor pressure, the liquid-vapor curvature must satisfy the Kelvin equation (Equation 12.10), while the relative permeability of the liquid and vapor phases depend on the saturation [9,49]. Until the funicular state forms, vapor permeability is not possible since a continuous vapor phase will not exist in the pores.

One way that drainage can be induced in packed particles is to put the liquid in contact with a packing of smaller particles. The difference in capillarity between the two pore structures will induce wicking of the liquid from the pores in the packing of larger particles into the pores between the smaller particles. Another technique is to apply a centrifugal force to a packing. In this case the pressure depends on the acceleration due to spinning. The application of gas pressure is another possible approach to forcing a liquid out of the pores between the particles. This technique is important to oil recovery. However, as shown later in this chapter, gas pressure induced liquid displacement can lead to unstable and nonuniform penetration by fingering.

One application of irreducible saturation is powder agglomeration, where the liquid remaining as pendular bonds provides strength to particle clusters after draining. But often the disadvantage of incomplete liquid removal is of greater concern.

Evaporation and Drying

During the drying of a packed powder, two mechanisms contribute to liquid egression from the pores - liquid flow and evaporation. Initially, the liquid will evaporate from the external surface of the packed particles. As drying continues pores open in the structure and the evaporation continues largely from the compact surface while liquid flows in the interparticle capillaries. The small pores at the surface region will provide a capillary attraction on the internal liquid in larger pores, leading to liquid flow and evaporation at the surface [1,8,50]. This assumes the pore size is small enough that gravitational effects can be ignored. This is the case if the particles are less than 100 μm in diameter. Otherwise, the liquid will show a variation with depth in the compact due to gravity. As shown in Figure 12.19, the surface liquid concentration will remain above a minimum saturation as long as capillary flow is faster than evaporation [1]. This figure shows the nonuniform distribution of water in packed sands undergoing surface drying. The water distribution versus depth is indicated for several saturation levels. For most of the drying process, the liquid flows to the external surface prior to evaporation.

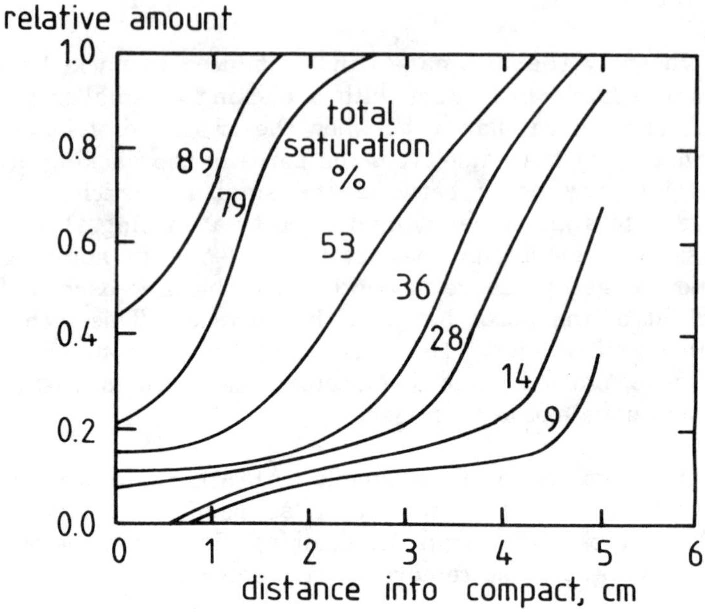

Figure 12.19 The relative placement of water in sand that is undergoing surface drying [1]. The total saturation decreases, but considerable flow to the surface occurs by capillary migration.

Because the specific volume of the liquid is greater than that of the vapor, the mass of liquid moving to the surface is much greater than would be possible as a vapor, even if the viscosity of the liquid is very great. Concurrent with this liquid flow, larger pores are opening and the vapor phase is penetrating into the compact. Figure 12.20 sketches the drying process in two dimensions. Liquid wicking and evaporation at the surface is favored over internal vaporization as long as pendular bonds are not forming in the surface region. This means that the smallest surface pores dominate drying. As progressively smaller pores are emptied, the surface concentration of liquid can no longer sustain the funicular state, and the liquid becomes stranded so that evaporation must occur from within the compact. The surface then dries by evaporation of the near-surface pendular bonds, the vapor being removed by evaporation through the open pore channels. The

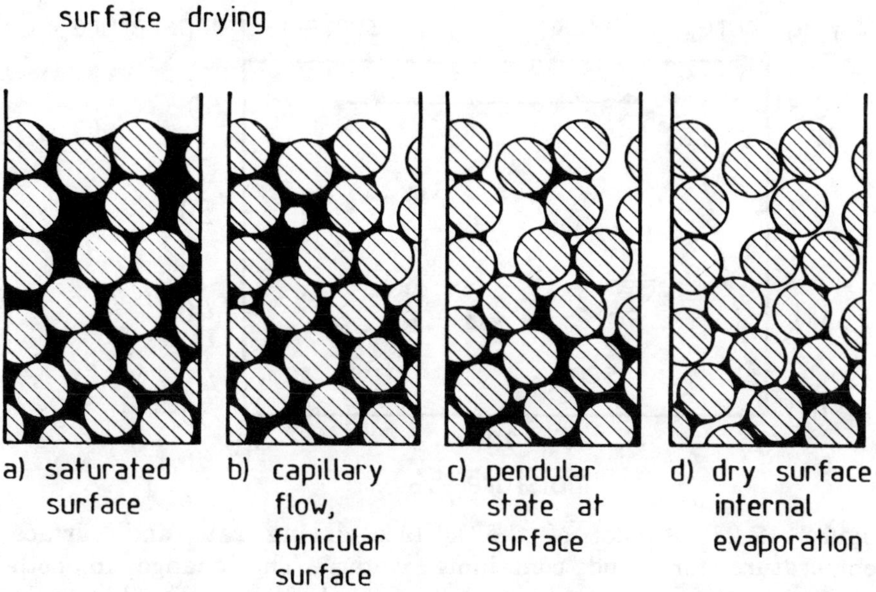

Figure 12.20 A sketch of the surface drying process where an initially saturated particle packing undergoes a major change in the surface liquid distribution as the saturation decreases.

vapor pressure depends on both the temperature and saturation (meniscus curvature). Thus, the evaporation interface moves deeper into the compact in later stages of drying, and gas permeability becomes a concern, especially with large compacts. A thin liquid film is left behind on the pore surfaces after the bulk liquid has been removed. Depending on the partial pressure, the film thickness will range from one to five atomic layers [51]. Evaporation of this final film requires that the vapor pressure in the pores be very low.

During the period of drying when liquid flow feeds the compact surface, the drying rate will remain constant. Later, as shown in Figure 12.21, the rate of drying decreases [1]. Once the external surface has been depleted of liquid, the rate of drying declines, due to an increase in the depth from which evaporation occurs. In some cases, there is a transition zone of intermediate

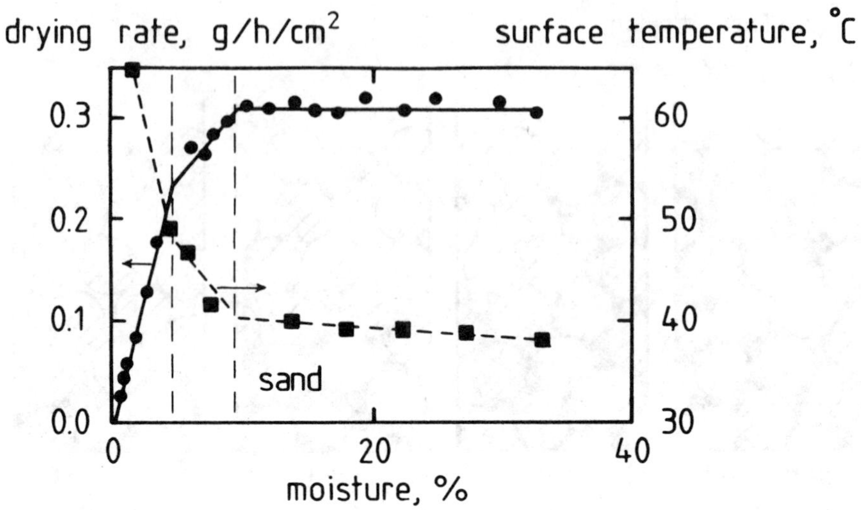

Figure 12.21 A plot of the surface drying rate and surface temperature for sand containing water. The change in both parameters corresponds to a change in the liquid distribution in the pores [1].

but declining drying rate that represents a combination of evaporation inside the compact and liquid flow to the surface. Extension of the period of constant drying rate is dependent on maintaining the funicular state at the compact surface. It has been estimated that this will persist until the surface saturation reaches approximately 9%. The actual transition saturation depends on the pore size, contact angle, and surface tension. Figure 12.22 shows how the drying rate, effective capillary suction pressure, and surface temperature vary with saturation for 24 μm silica particles [10,50]. As long as liquid is able to flow to the surface the temperature remains constant. Both the capillary pressure and drying rate establish plateau behavior until the pendular state forms at the compact surface. This indicates that saturation is a primary determinant of the capillary pressure. In the pendular state, the temperature and suction pressure increase, while the drying rate decreases. As the liquid front moves into the compact, evaporation in the interior and permeation to the exterior are necessary. The vapor pressure in the compact is dictated by the curvature of the liquid-vapor interface, which is essentially constant, while the

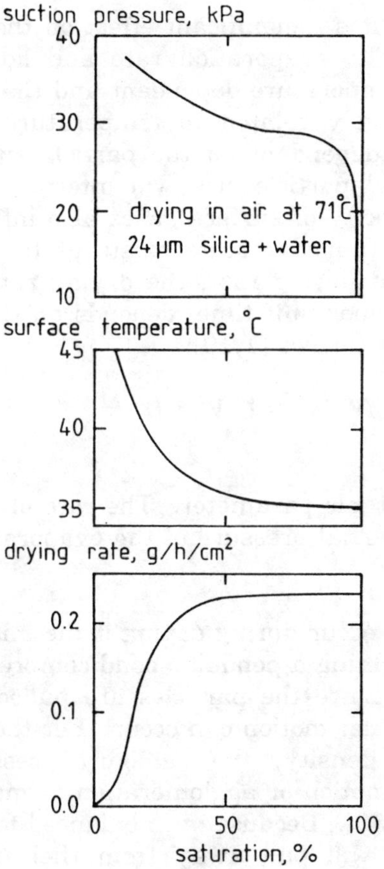

Figure 12.22 The behavior of silica during surface drying in air, showing the changes in surface suction pressure, surface tension, and drying rate versus the mean saturation [10,50].

depth of permeation is continually increasing. Thus, a decreased rate of drying occurs as the interface moves into the compact. Because of the dependence on permeability, drying is faster given more porosity and larger pores. A low surface energy, higher temperature, more surface motion of atmosphere (agitation), and a lower surface vapor pressure for the evaporating species also aid drying.

Temperature has a significant effect on the drying rate and liquid distribution. The evaporation rate and liquid viscosity are both exponentially temperature dependent and the liquid-vapor surface energy is inversely related to temperature. Since capillary forces are inversely dependent on the particle size, it is expected that temperature and particle size will interact to determine the drying rate. The velocity of surface gases also influence the drying rate as long as the funicular state exists at the surface. During the period of constant drying rate, the drying rate dW/dt (weight change per unit area per unit time) depends on the gas velocity V and particle size D as follows [1,10]:

$$dW/dt = C\, V^{0.8}\, D^{0.11} \tag{12.16}$$

where C is an adjustable parameter. The rate of drying is directly proportional to the partial pressure of the evaporate at the compact surface.

Cracking can occur during drying if the initial solids content is low. The liquid forming a pendular bond compresses the particles and forces contact. Since the particles are pulled together by the capillary pressure, a net motion can occur. For the cases where the packing is low in density, the capillary pressure will induce nonuniform particle motion or agglomeration to maintain contact at the pendular bonds [52]. Because of imbalanced forces in a random mix, some particles will pull away from their neighbors, giving cracks.

The hysteresis between advancing and receding contact angles leads to conditions for pore filling by condensation that are different than those for pore emptying by evaporation. There are also effects from trapped liquid (liquid distribution in a random packing), contact area, and entropy factors that contribute to the hysteresis. Figure 12.23 shows an example of the hysteresis between condensation and evaporation in bundled capillary tubes [26]. This figure plots the relative volume in the pores versus the vapor pressure, showing that substantial hysteresis occurs over the entire pressure range. Another contribution to the evaporation/condensation hysteresis is the delay in emptying of small pores by evaporation. With a wide pore size distribution, the liquid meniscus will attempt to maintain a constant curvature to

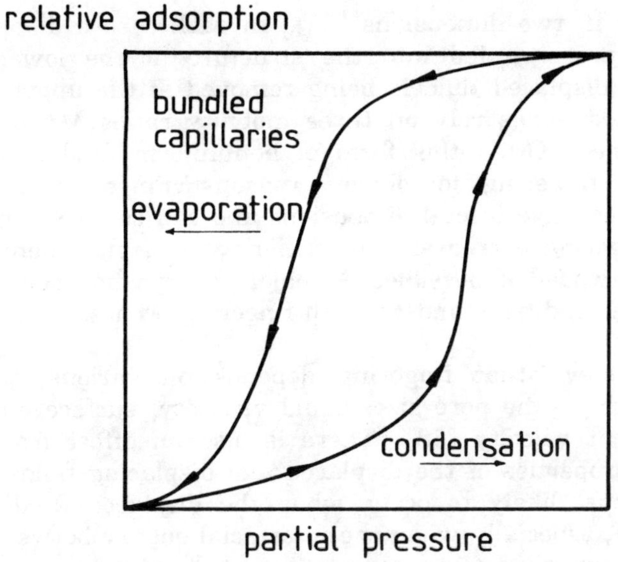

Figure 12.23 The relative adsorption versus vapor partial pressure for bundled capillaries, illustrating the typical hysteresis between evaporation and condensation [26].

maintain pressure equilibration. This will delay the emptying of smaller pores. Finally, if the liquid is composed of two components of differing vapor pressures, then the liquid wetting and viscosity will continually change during drying. This greatly complicates the analysis of drying behavior.

Fingering in Liquid Displacement

Because a wetting liquid preferentially penetrates into pores, a pressure is needed to remove the liquid. This pressure depends on the surface energy, contact angle, and inverse of the pore size. In some cases, drainage can be affected by centrifugal forces, gravity, or capillary attraction of the liquid to smaller pores. This section considers the events associated with pressurized displacement of the liquid using a second fluid, typically a gas. During displacement of a wetting liquid by a low viscosity fluid, instabilities occur [27,53-57]. Figure 12.24 sketches such instability in two dimensions [55]. In each case a displacing fluid is being pressure fed into the structure at the lower left corner and the displaced fluid is being removed at the upper right corner. These drawings rely on three mobility ratios M to illustrate the changes. Often this form of nonuniform fluid displacement is termed fingering, for obvious reasons. Fingers formed in porous structures have fractal dimensions and can be very complex in reality. Fingering decreases the efficiency of displacement processes and is avoided if possible. A major concern has been with determining the conditions under which fingering occurs.

It is known that fingering depends on various system parameters such as the pore size, liquid viscosity, surface energy, and displacement velocity [58]. There is also an effect from the differences in properties of the displaced and displacing fluids [59]. Fingering is most likely to occur when the displaced fluid is of higher viscosity, especially when the interfacial energy between the two fluids is low; thus, fingering is expected when the viscous forces exceed the capillary forces. Other factors leading to fingering include density differences that induce a high density fluid to penetrate into a low density fluid and significant differences in fluid permeabilities. A generalized condition for fingering is given

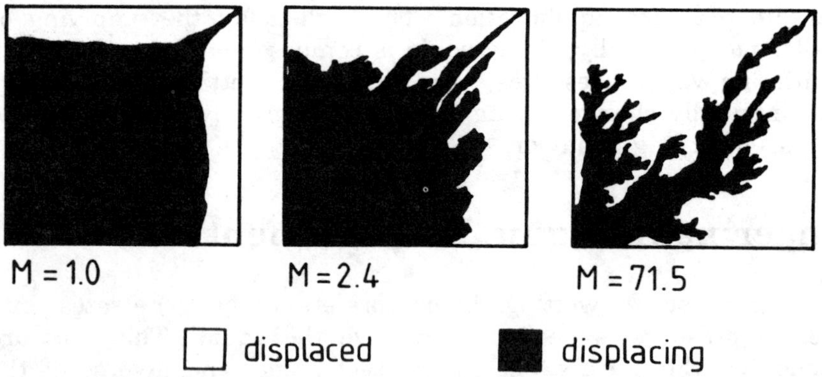

Figure 12.24 Sketches of fingering by a displacing fluid in a pore structure containing another fluid (three mobility ratios). Filling occurs from the lower left corner to the sink at the upper right corner.

by the following inequality [53,57]:

$$g(\rho_1 - \rho_2) + (A\ V/d^2)\,[\mu_1/a_1 - \mu_2/a_2] \leq 0 \qquad (12.17)$$

where the subscript *1* indicates the displaced fluid and the subscript *2* indicates the displacing fluid. The parameters in this equation are the fluid density ρ, permeability a, and viscosity μ, as well as the velocity V, gravitational constant g, pore size d, and a constant A estimated to be near 12. For situations where the viscosity difference is dominant, the prediction of fingering is based on the mobility ratio M given as [55]

$$M = \mu_1\ a_2\ /\ (\mu_2\ a_1)\ . \qquad (12.18)$$

When the mobility ratio is below unity fingering is not expected. Alternatively, as the mobility ratio increases over one, the interface becomes more unstable; hence, the degree of fingering should increase with the mobility ratio.

The most common occurence of fingering is the situation when gas is used to force a wetting liquid from pores. The large viscosity difference creates an instability where the gas penetrates preferentially along a finger. Fingering is also expected in the presence of localized permeability variations; this corresponds with uneven packing. Regions of high permeability (low packing density) will initiate fingers. A temperature variation can also induce fingering, because of viscosity changes. If mixing occurs between the two fluids, then the viscosity will change at the leading edge of a finger. The consequential gradation in viscosity will lead to side branching fingers.

A certain roughness is impressed on a displacement front by the very character of a pore structure. However, the actual spacing of fingers changes with time. Initially, the spacing between fingers depends on the pore size, surface energy, viscosities, and flow velocity [57]. The fastest growing fingers are predicted to have a spacing of slightly less than twice the spacing when fingering initiates. After the initiation, the number of fingers decreases with time and the longest finger grows in proportion to time to the 1.6 power. Initially, there is no impoverishment of fingers in the neighborhood of the fast growing fingers. Except during slow displacement conditions, the width of a finger is approximately one-half the pore channel size, though at low velocities the finger width is larger. In all cases, there is a residual film of liquid left on the pore surface by the displacing fluid. This film is unstable and will rearrange with time.

A common problem involving fingering is in the recovery of oil from packed soils. Gas (or another fluid) is used to pressurize the oil at one site and the output is collected at another site. The occurrence of fingering reduces the recovery possible by pressurized displacement. After a certain degree of displacement, the oil is penetrated by the finger and allows escape of the displacing gas. Initially, the gas permeability is zero in the oil saturated soil, but due to fingering the gas can eventually bypass the oil, limiting the effective recovery.

One problem similar to fingering is the motion of a gas bubble through a porous structure that is saturated with a liquid. In this case the bubble shape depends on the pore dimensions and the velocity. Generally, the bubble velocity times the width is a minimum for a given bubble surface area [54].

Summary

A wetting fluid provides an attractive force on particles that explains the various aspects of liquid interactions with packed particles. At low saturations, the wetting liquid will preferentially flow into the smallest pores (those with the largest surface area to volume ratio). This behavior is part of the general capillary effect. In a capillary tube, the capillary pressure is measured by the height of rise of liquid against gravity. When the liquid is in the form of a pendular bond, the capillary pressure results in a compressive stress which forces the particles into contact. This compressive force explains agglomeration of particles in the presence of humidity. The vapor pressure of a fluid in a porous structure depends on the curvature of the liquid surface, which in turn depends on the particle size, contact angle, and saturation.

In this chapter several aspects of liquid motion into and out of pores between packed particles were treated. It is characteristic of the liquid-pore interaction that a wetting liquid flows to reduce the energy of the system. Initial flow is by condensation or wicking of liquid into the pores. During filling there is preferential flow into small pores and formation of pendular bonds. During withdrawal of the fluid these same small pores and pendular bonds are most resistant to drainage and evaporation. Efforts to enhance drainage by use of a displacing fluid are often frustrated by fingering. Once a liquid penetrates into packed particles, evaporation is the only way to attain complete removal. Only partial removal is obtainable by wicking or displacement. In all aspects of liquid-pore interactions there is a hysteresis between flow into and out of the pores.

References

1. N. H. Ceaglske and O. A. Hougen, "The Drying of Granular Solids," *Transactions of the Institution of Chemical Engineers*, 1937, vol. 33, pp. 283-314.
2. N. R. Morrow, "Irreducible Wetting Phase Saturations in Porous Media," *Chemical Engineering Science*, 1970, vol. 25, pp. 1799-1815.
3. R. D. Wyckoff and H. G. Botset, "The Flow of Gas-Liquid Mixtures Through Unconsolidated Sands," *Physics*, 1936, vol. 7, pp. 325-345.
4. W. B. Haines, "Studies in the Physical Properties of Soil. V. The Hysteresis Effect in Capillary Properties, and the Modes of Moisture Distribution Associated Therewith," *Journal of Agricultural Science*, 1930, vol. 20, pp. 97-116.
5. C. Huh and S. G. Mason, "Effects of Surface Roughness on Wetting (Theoretical)," *Journal of Colloid and Interface Science*, 1977, vol. 60, pp. 11-38.
6. E. B. Dussan and S. H. Davis, "Stability in Systems with Moving Contact Lines," *Journal of Fluid Mechanics*, 1986, vol. 173, pp. 115-130.
7. J. F. Oliver and S. G. Mason, "Microspreading Studies on Rough Surfaces by Scanning Electron Microscopy," *Journal of Colloid and Interface Science*, 1977, vol. 60, pp. 480-487.
8. F. A. L. Dullien and V. K. Batra, "Determination of the Structure of Porous Media," *Flow Through Porous Media*, American Chemical Society, Washington, DC, 1969, pp. 2-30.
9. E. E. Miller and R. D. Miller, "Physical Theory for Capillary Flow Phenomena," *Journal of Applied Physics*, 1956, vol. 27,

pp. 324-332.
10. J. F. Pearse, T. R. Oliver and D. M. Newitt, "The Mechanism of Drying of Solids Part I -The Forces Giving Rise to Movement of Water in Granular Beds, During Drying," *Transactions of the Institution of Chemical Engineers*, 1949, vol. 27, pp. 1-8.
11. K. S. Hwang, R. M. German and F. V. Lenel, "Capillary Forces Between Spheres During Agglomeration and Liquid Phase Sintering," *Metallurgical Transactions A*, 1987, vol. 18A, pp. 11-17.
12. C. Y. Liu, K. Murakami and T. Okamoto, "Effect of Capillary Pressure on Interdendritic Liquid Flow," *Acta Metallurgica*, 1986, vol. 34, pp. 159-166.
13. D. M. Newitt and J. M. Conway-Jones, "A Contribution to the Theory and Practice of Granulation," *Transactions of the Institution of Chemical Engineers*, 1958, vol. 36, pp. 422-442.
14. K. S. Hwang, R. M. German and F. V. Lenel, "Capillary Forces in the Early Stage of Liquid Phase Sintering," *Reviews in Powder Metallurgy and Physical Ceramics*, 1986, vol. 3, pp. 113-164.
15. R. M. German, *Liquid Phase Sintering*, Plenum Press, New York, NY, 1985, pp 52-57.
16. J. C. Melrose, "Chemical Potential Changes in Capillary Condensation," *Journal of Colloid and Interface Science*, 1972, vol. 38, pp. 312-322.
17. V. N. Eremenko, Y. V. Naidich and I. A. Lavrinenko, *Liquid Phase Sintering*, Consultants Bureau, New York, NY, 1970.
18. G. Mason and W. C. Clark, "Liquid Bridges Between Spheres," *Chemical Engineering Science*, 1965, vol. 20, pp. 859-866.
19. M. C. Coelho and N. Harnby, "Moisture Bonding in Powders," *Powder Technology*, 1978, vol. 20, pp. 201-205.
20. J. L. Shereshefsky and C. P. Carter, "Liquid-Vapor Equilibrium in Microscopic Capillaries I. Aqueous System," *Journal of Chemical Physics*, 1950, vol. 72, pp. 3682-3686.
21. A. M. Schwartz, "Capillarity Theory and Practice," *Industrial and Engineering Chemistry*, 1969, vol. 61, pp. 10-21.
22. H. J. Fraser, "Experimental Study of the Porosity and Permeability of Clastic Sediments," *Journal of Geology*, 1935, vol. 43, pp. 910-1010.
23. R. Castells, J. Meda, J. Caprari and M. Damia, "Particle Packing Analysis of Coatings Above the Critical Pigment Volume Concentration," *Journal of Coatings Technology*, 1983, vol. 55, no. 505, pp. 53-59.

24. R. E. Collins, *Flow of Fluids Through Porous Media*, Reinhold Publishing Corp. , New York, NY, 1961.
25. M. C. Coelho and N. Harnby, "The Effect of Humidity on the Form of Water Retention in a Powder," *Powder Technology*, 1978, vol. 20, pp. 197-200.
26. D. H. Everett and J. M. Haynes, "Model Studies of Capillary Condensation I. Cylindrical Pore Model with Zero Contact Angle," *Journal of Colloid and Interface Science*, 1972, vol. 38, pp. 125-137.
27. J. Nittmann, G. Daccord and H. E. Stanley, "Fractal Growth of Viscous Fingers: Quantitative Characterization of a Fluid Instability Phenomenon," *Nature*, 1985, vol. 314, pp. 141-144.
28. W. O. Smith, "The Final Distribution of Retained Liquid in an Ideal Uniform Soil," *Physics*, 1933, vol. 4, pp. 425-438.
29. M. Folman and J. L. Sherehfesky, "Liquid-Vapor Equilibrium in Microscopic Capillaries. II. Non-Aqueous Systems," *Journal of Physical Chemistry*, 1955, vol. 59, pp. 607-610.
30. G. P. Martins, D. L. Olson and G. R. Edwards, "Modeling of Infiltration Kinetics for Liquid Metal Processing of Composites," *Metallurgical Transactions B*, 1988, vol. 19B, pp. 95-101.
31. W. B. Palmer, "The Advance of a Liquid Front Along a Glass Yarn," *Journal of the Textile Institute*, 1953, vol. 44, pp. T391-T400.
32. E. W. Washburn, "The Dynamics of Capillary Flow," *Physical Review*, 1921, vol. 17, pp. 273-283.
33. R. D. Laughlin and J. E. Davies, "Some Aspects of Capillary Absorption in Fibrous Textile Wicking," *Textile Research Journal*, 1961, vol. 31, pp. 904-910.
34. R. L. Peek and D. A. McLean, "Capillary Penetration of Fibrous Materials," *Industrial and Engineering Chemistry*, 1934, vol. 6, no. 2, pp. 85-90.
35. D. H. Bosanquet, "On the Flow of Liquids into Capillary Tubes," *Philosophical Magazine*, 1923, vol. 45, pp. 525-531.
36. R. Muller and D. L. Clegg, "Paper Chromatography," *Analytical Chemistry*, 1951, vol. 23, pp. 396-411.
37. L. W. De Backer, "Characterization of Some Retention and Transfer Properties of Granular Materials in Relation with Geometrical and Liquid Properties," *Advances in the Mechanics and the Flow of Granular Materials*, vol. 1, M. Shahipoor (ed.), Trans Tech Publ., Clausthal, Federal Republic Germany, 1983, pp. 399-414.
38. J. Van Brakel, "Pore Space Models for Transport Phenomena in

Porous Media Review and Evaluation with Special Emphasis on Capillary Liquid Transport," *Powder Technology*, 1975, vol. 11, pp. 205-236.
39. R. G. Larson and N. R. Morrow, "Effects of Sample Size on Capillary Pressures in Porous Media," *Powder Technology*, 1981, vol. 30, pp. 123-138.
40. D. N. Mazzone, G. I. Tardos and R. Pfeffer, "The Behavior of Liquid Bridges Between Two Relatively Moving Particles," *Powder Technology*, 1987, vol. 51, pp. 71-83.
41. A. E. Scheidegger, *The Physics of Flow Through Porous Media*, revised edition, University of Toronto Press, Toronto, Ontario, Canada, 1960.
42. F. M. Tiller and C. D. Tsai, "Theory of Filtration of Ceramics: I, Slip Casting," *Journal of the American Ceramic Society*, 1986, vol. 69, pp. 882-887.
43. J. Marecek, M. Krpata and F. Moudry, "The Effect of Particle Size Distribution on the Removal of the Liquid Phase from a Centrifuged Filter Cake," *International Chemical Engineering*, 1972, vol. 12, pp. 384-388.
44. J. S. Goodling, R. I. Vachon, W. S. Stelpflug, S. J. Ying and M. S. Khader, "Radial Porosity Distribution in Cylindrical Beds Packed with Spheres," *Powder Technology*, 1983, vol. 35, pp. 23-39.
45. C. C. Harris and N. R. Morrow, "Pendular Moisture in Packings of Equal Spheres," *Nature*, 1964, vol. 203, pp. 706-708.
46. M. Tschapek, S. Falasca and C. Wasowski, "The Undrainable Water in Quartz Sand and Glass Beads," *Powder Technology*, 1985, vol. 42, pp. 175-180.
47. G. H. Fricke, D. Rosenthal and G. A. Welford, "Removal of Liquid from a Packed Column of Spherical Beads Under Centrifugal Force," *American Institute of Chemical Engineers Journal*, 1973, vol. 19, pp. 991-998.
48. G. Mason, "Desaturation of Porous Media I. Unconsolidated Materials," *Journal of Colloid and Interface Science*, 1972, vol. 41, pp. 208-227.
49. I. Mahgoub, C. Prost and J. A. Dodds, "Experimental Relative Permeability Curves Compared with a New Cutting and Joining Model of Porous Media with a Distribution of Pore Sizes," *Powder Technology*, 1982, vol. 31, pp. 143-151.
50. T. R. Oliver and D. M. Newitt, "The Mechanism of the Drying of Solids Part II - The Measurement of Suction Potentials and Moisture Distribution in Drying Granular Solids," *Transactions*

of the Institution of Chemical Engineers, 1949, vol. 27, pp. 9-18.
51. C. S. O'Neil, "Computation of Nitrogen Sorption Pore Volume Distributions," *American Laboratory*, 1985, June, pp. 41-48.
52. G. Mason, "Formation of Films from Latices, A Theoretical Treatment," *British Polymer Journal*, 1973, vol. 5, pp. 101-108.
53. P. G. Saffman and G. Taylor, "The Penetration of a Fluid into a Porous Medium or Hele-Shaw Cell Containing a More Viscous Liquid," *Proceedings of the Royal Society of London*, 1958, vol. A245, pp. 312-329.
54. G. Taylor and P. G. Saffman, "A Note on the Motion of Bubbles in a Hele-Shaw Cell and Porous Medium," *Quarterly Journal of Mechanics and Applied Mathematics*, 1959, vol. 12, pp. 265-279.
55. B. Habermann, "The Efficiency of Miscible Displacement as a Function of Mobility Ratio," *Transactions of AIME, Petroleum Transactions*, 1960, vol. 219, pp. 264-272.
56. J. P. Heller, "Onset of Instability Patterns Between Miscible Fluids in Porous Media," *Journal of Applied Physics*, 1966, vol. 37, pp. 1566-1579.
57. J. V. Maher, "Development of Viscous Fingering Patterns," *Physical Review Letters*, 1985, vol. 54, pp. 1498-1501.
58. L. Paterson, "Fingering with Miscible Fluids in a Hele Shaw Cell," *Physics of Fluids*, 1985, vol. 28, pp. 26-30.
59. T. Maxworthy, "The Nonlinear Growth of a Gravitationally Unstable Interface in a Hele-Shaw Cell," *Journal of Fluid Mechanics*, 1987, vol. 177, pp. 207-232.

Chapter Thirteen

Flow of Fluids Through Packed Particles

Chapter Outline

Introduction

Fluid flow through packed particles is important in several situations, including water seepage through soils, lubrication of bearings, slip casting of ceramics, debinding of injection molded components, product recovery in filtration, carbonation of fluids, flow in heat pipes, and oil extraction from geological structures. Because of this widespread interest, considerable literature dealing with flow through porous structures has accumulated. Besides packed particles, various natural and fabricated materials exhibit open pore structures that are permeable; bricks, concrete, leather, sintered metals, and paper are a few examples. Thus, there is extensive scope to the problem of fluid flow in pores. The desire is to predict flow behavior in terms of simple structural parameters such as the pore size. In turn, the flow rate can be estimated if the pressure drop is known, or the pressure drop can be estimated for a needed flow rate.

The two fundamental types of fluids, liquids and gases, differ in compressibility. The flow rate in the pore structure, as a further subdivision, determines the interaction between the fluid and pore, thereby controlling the flow resistance [1]. At very low flow rates, especially for a gas at low pressures, the mean free path between molecular collisions exceeds the dimensions of the pores. This condition is termed diffusional flow. A characteristic aspect of diffusional flow is its independence from the fluid viscosity. It is most applicable to very small pores, low flow rates, and low absolute pressures. At higher flow rates, pressures, and pore sizes, the fluid viscosity effect becomes dominant. This is because a fluid passing through a pore will momentarily stick to the pore surface, leading to drag. In this case the interaction between the pore wall and the fluid combines with the fluid viscosity to determine the flow resistance. Viscous flow occurs with both liquids and gases.

The transition between diffusion and viscous flow is termed slip flow, and involves both effects simultaneously. At very high flow velocities there is substantial turbulence and energy loss for fluids passing through tortuous pore structures. Thus, a condition is reached where the output is independent of the applied pressure. Between turbulent flow and viscous flow is a region termed inertial flow, characterized by an increasing energy loss as the flow velocity increases.

Figure 13.1 provides a graphical view of the subdivision of the flow conditions described above. Nominal regions are indicated on this plot of pressure and particle size. Although this is a schematic diagram, it does emphasize the partitioning of flow regions as a function of two important variables. Because of the subdivision, the mathematical treatments of fluid flow involve

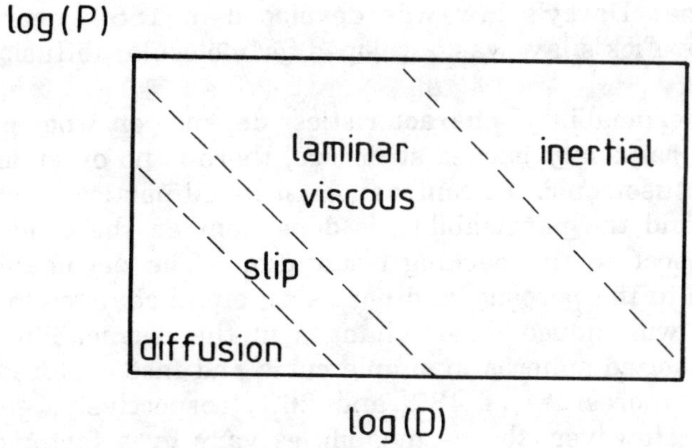

Figure 13.1 A schematic plot of the fluid pressure versus the particle size illustrating the changes in flow mechanisms associated with these two variables.

several variations dependent on the particle size, packing density, pore surface roughness, surface area, particle shape, fluid viscosity, fluid compressibility, fluid pressure, fluid velocity, temperature, and pore saturation.

The case of laminar flow of a viscous fluid through pores is the most widely encountered case in practice. Darcy's (D'Arcy's) law proves applicable where the velocity of flow is directly proportional to the gradient in applied pressure. For an incompressible fluid the appropriate form of Darcy's law is

$$Q = - \alpha A \; \Delta P/(\mu L) \tag{13.1}$$

where Q is the volumetric flow rate (m^3/s), A is the cross-sectional area for flow, L is the thickness through which flow is occurring, ΔP is the pressure difference ($P_1 - P_2$) over the distance L, μ is the fluid viscosity, and α is the permeability. The negative sign indicates that flow is down the pressure gradient, from the high pressure P_1 to the low pressure P_2. The important material factor in Darcy's law is the permeability coefficient, which is analogous to the conductivity for electrical current flow in a metal. It is noteworthy that Darcy's law was developed in 1856, nearly the same time that Fick's law was developed for molecular diffusion [2].

The permeability characteristics depend on the particle packing. In a randomly packed structure, there is no order and the permeability is isotropic. In contrast, an ordered particle packing is not isotropic and the permeability is dependent on the direction of flow with respect to the packing orientation. The permeability is very sensitive to the porosity and pore size; small changes in either characteristic will induce major changes in the permeability. For example, monosized spheres in simple cubic and face-centered cubic packings have porosities of 48% and 26%, respectively, giving a ratio of 1.84. However, the permeabilities vary by a factor of 6.8, with the lower porosity face-centered cubic structure having the lower permeability [3].

Another factor of concern is the flow of two immiscible fluids in a pore structure. For example, what is the flow rate of a gas in packed particles which are partially saturated with a liquid? The liquid saturation will determine the connection between vapor-filled pores. At high levels of liquid saturation the vapor-filled pores are discontinuous and there is no permeability for the gas. At intermediate levels of saturation the connectivity of each phase determines its relative permeability. A treatment of this problem is given in the section on mixed phase flow.

Several techniques are applicable to permeability measurements of packed particles [3]. Most typically the permeability is measured by determining the fluid flow volume, mass, or velocity versus the applied pressure differential. Various levels of sophistication are needed to perform measurements in the different flow regimes. Wall effects are not usually significant because of the low surface area of the container compared to the particles. However, in some studies, higher velocities were noted at the wall regions [4,5]. The appropriate measurement techniques are discussed in several of the textbooks and references cited at the end of this chapter.

Diffusion

At low gas pressures the mean free path between molecular collisions can be longer than the distance between collisions of the particles. Typically, when the mean free path is at least five times the pore diameter there is no effect of viscosity on flow and Darcy's law is inapplicable. Such situations occur with small pore sizes (smaller particle sizes), low gas pressures, and low molecular weight molecules. In these situations, gas motion through the pores of a packed powder will depend on the diffusion rate and the mean distance between collisions with the pore walls. The mean free path λ is dependent on the gas density M, measured as molecules per unit volume, and molecular diameter δ as follows:

$$\lambda = 1/(2^{1/2} \pi M \delta^2) . \tag{13.2}$$

For example, water vapor under equilibrium conditions at 50°C in a vacuum has a partial pressure of 12 kPa (0.012 atmospheres) and a mean free path of 0.2 μm. Necessarily, diffusion requires that collisions with the pore walls are more frequent than collisions with other molecules. This form of gas diffusion is sometimes referred to as Knudsen diffusion, in which the flux J depends on the pressure gradient $\Delta P/L$ and diffusivity K as follows [6]:

$$J = - K \Delta P /(k T L) \tag{13.3}$$

with k equal to Boltzmann's constant and T equal to the absolute temperature. The flux is in terms of moles of vapor per unit time and area. The negative sign in Equation 13.3 indicates that flow is from the high pressure to low pressure regions. Diffusion controlled transport is similar to electrical and heat conduction through a porous material or ionic migration in electrolytes [7,8].

For an unobstructed fluid there will be an inherent diffusivity that is dependent on the temperature and fluid. The presence of solid particles will degrade the inherent diffusivity in proportion to the amount of solid. The solid affects diffusivity through its tortuosity and pore shape [9,10]. In a packing of monosized spheres, the mean free path λ depends on the fractional density f as follows:

$$\lambda = 4\,(1 - f)/S_v \qquad (13.4)$$

where S_v is the surface area per unit volume. Accordingly, the diffusivity will vary with the product of the mean free path and the mean molecular velocity. Maxwell provided an estimate of the diffusivity in a packing of spheres. The ratio of diffusivity in a particle packing to the inherent diffusivity for the fluid is denoted as Γ and depends on the fractional packing density f as follows [8,11]:

$$\Gamma = 2\,(1 - f)/(2 + f)\,. \qquad (13.5)$$

This form is applicable to diffusion through a packing of uniform porosity with an arbitrary particle size distribution. Figure 13.2 plots Equation 13.5, showing the decreasing effective diffusivity as the fractional density increases. The requirement for applying Equation 13.5 is that the pore size be substantially smaller than the mean free path of the fluid. Alternative derivations relate the diffusivity ratio to the particle size, packing density, and tortuosity. Thus, the rate of mass transport depends on the gradient, tortuosity, diffusivity, and on molecular weight of the gas [11,12]. The flux varies with the square root of the molecular weight and the diffusivity depends on the pore structure [13]. The wider the pore size distribution, the lower the pressure at which diffusion control is lost. This is because the large pores exhibit viscous flow at lower pressures than the small pores.

Various models relating the diffusivity K to the fractional packing density f, tortuosity τ, particle size D, and pore size d have been proposed [6,8-16]. Table 13.1 provides a summary of some of these relations, which generally rely on a mean capillary pore size. For random packings of loose spheres, the tortuosity is near 3 and the value increases to near 7 for dense random packings [16].

Finally, in cases involving very high surface area structures (small particle sizes), significant migration over the surface of the solid by surface diffusion can occur [2]. The flux in this case is the sum of the contributions from permeation, diffusion, and

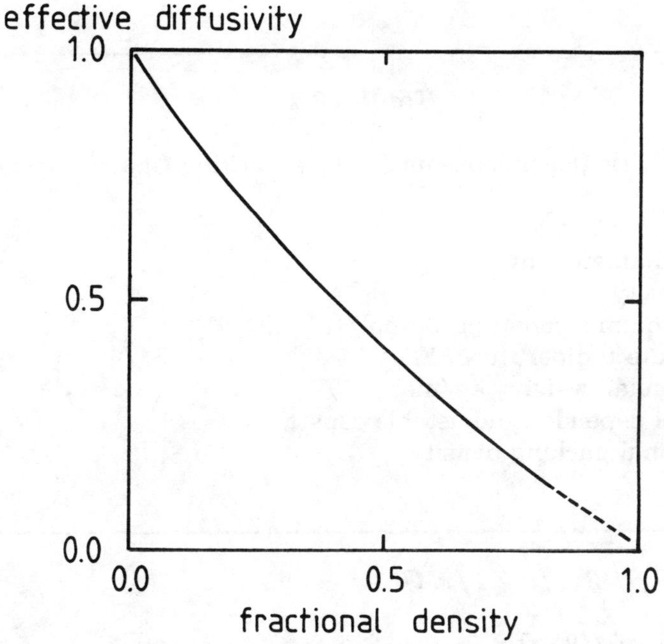

Figure 13.2 A plot of the effective diffusivity versus the fractional packing density according to the model of Equation 13.5.

surface diffusion. There is a further effect from adsorption in the pores as the surface area increases and the pore size decreases. Here it is possible that a contribution from liquid flow may add significantly to the overall mass flow through the pore structure.

Slip Flow

When the pore size is approximately the same as the mean free path of the fluid, then slip flow occurs. This process is usually encountered in gas passage through small pore sizes at moderate (atmospheric) pressures. Slip flow is actually a hybrid that combines the aspects of diffusion and viscous flow into one permeability parameter. As the gas pressure increases, the mean free path decreases and there is more of a dependence of flow rate on

Table 13.1

Diffusivity Dependence on Particle Packing Characteristics

K = diffusivity, m^2/s
d = pore diameter, m
τ = tortuosity
k = Boltzmann's constant, J/mol/K
T = absolute temperature, K
M = molecular weight, kg/mol
A = model dependent adjustable constant
f = fractional packing density

$K = d\,(1 - f)/(3\tau)\,[8\,k\,T/\pi M]^{1/2}$	[15]
$K = A\,2\,(1 - f)/(2 - f)$	[8,13]
$K = d\,(k\,T/M)^{1/2}$	[6]
$K = d\,(1 - f)/\tau^2$	[6]
$K = d\,(A - 0.1\,f)$	[16]

the viscosity. Slip flow is encountered in this region of transition between pure diffusion and pure viscous flow.

Under conditions in which slip flow occurs, the effect is variable with pressure. The mean free path of the gas molecules decreases as the pressure increases. At high pressures the gas velocity at the pore surface is zero, in accord with viscous drag theory. As the pressure is decreased, the velocity of the gas at the pore surface is not necessarily zero, yet there is still a viscosity component to flow. At very low pressures the gas flow is independent of viscosity since all flow resistance is from collisions between the molecules and the pore walls [17].

The usual treatment of slip flow is to modify the permeability a for a pressure dependence according to the equation

$$a_s = a(1 + b/\bar{P}) \qquad (13.6)$$

where a_s is the modified slip permeability, $\bar{P}$ is the mean gas pressure, and b is a factor dependent on the pressure, temperature, and gas. With this modification, Darcy's law, as applicable to gases, can be applied with a corrected permeability, giving

$$Q = -\,a_s A\,(P_1^2 - P_2^2)/(2\,\mu\,L\,P_2) \qquad (13.7)$$

where the volumetric flow rate Q is measured at the downstream pressure P_2. In Equation 13.7, A is the cross-sectional area, L is the compact thickness, and μ is the gas viscosity. Experimentally, several factors influence the mass flow rate under conditions of slip flow: porosity, pressure, pressure gradient, temperature, particle size, gas viscosity, compact thickness, molecular weight, and pore tortuosity [12,17,18].

Viscous Flow

Incompressible Fluids

Darcy's law is generally assumed to be valid for pressure driven flow of fluids through packed particles [19-24]. The fluid viscosity contributes to a sticking of individual molecules at the particle surfaces, leading to a drag effect on the fluid. The applicability of Darcy's law at high velocities of flow must be judged by the dimensionless Reynolds number associated with the fluid flow. The assumption is that only laminar flow is occurring when the Reynolds number is below approximately 2 to 4. The Reynolds number R_n for this situation can be estimated as either [18,25]

$$R_n = \rho\,V/(\mu\,S_v) \qquad (13.8)$$

or

$$R_n = \rho D V/(\mu (1 - f)) \qquad (13.9)$$

where V is the flow rate in m/s (the volumetric flow rate divided by the cross-sectional area), ρ is the fluid density, μ is the liquid viscosity, D is the particle diameter, and S_v is the surface area per unit volume. These forms are only approximations, since turbulence in the pores is the important parameter and Equations 13.8 and 13.9 use the external velocity, not the velocity in the pores. Actually, because of the tortuous and divergent-convergent character of pores, an accurate estimate of the Reynolds number is difficult.

Since permeability of packed particles depends on the pore shape and size, then particle size, particle shape, packing density, particle roughness, and compact surface area all affect the flow rate [19,21-29]. Various treatments have incorporated corrections for such factors into Darcy's law. These are not sufficiently general to handle all cases and do not greatly improve the model. The two dominant features of packed particles are the porosity and the particle size. The resistance to fluid flow increases rapidly as the porosity decreases. One estimate shows the permeability as being proportional to the fourth power of the porosity [23]. Generally, the permeability varies with the inverse square of the particle size. A normalization is presented in Figure 13.3 showing the permeability and particle size relation to the packing density [24]. The ordinate is the logarithm of the particle diameter divided by the square root of the permeability. The generality of this curve is very striking. On the basis of this relation, one can predict incompressible flow behavior through many powder packing geometries. Note that for a given particle size Figure 13.3 shows that a major change in permeability will occur with relatively small changes in packing density.

In a mixture of particles the higher porosity around the large particles results in high permeability zones. Consequently, the addition of a low concentration of large particles in a small particle bed will give an increase in permeability. This occurs in spite of an increase in density due to the mixed particle sizes. Alternatively, as shown in Figure 13.4, the addition of small particles to a large particle bed will fill pores and result in a rapid decrease in permeability. The packing density slightly increases with the addition of smaller particles, but the permeability undergoes a large reduction.

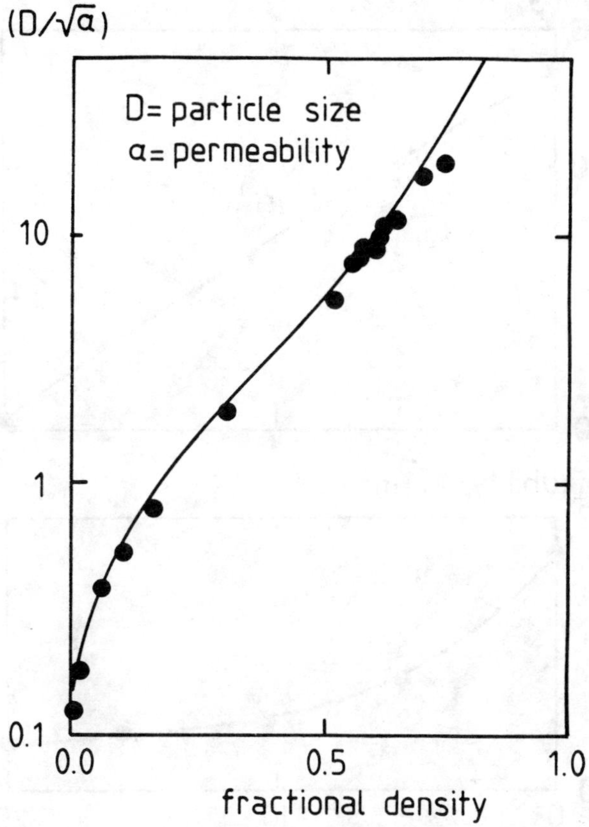

Figure 13.3 The permeability in monosized spheres is linked to the particle size and fractional density as indicated by this curve [24].

Compressible Fluids

Darcy's law must be modified to account for the change in fluid volume with pressure when the fluid is compressible. In this case, Equation 13.1 takes the form

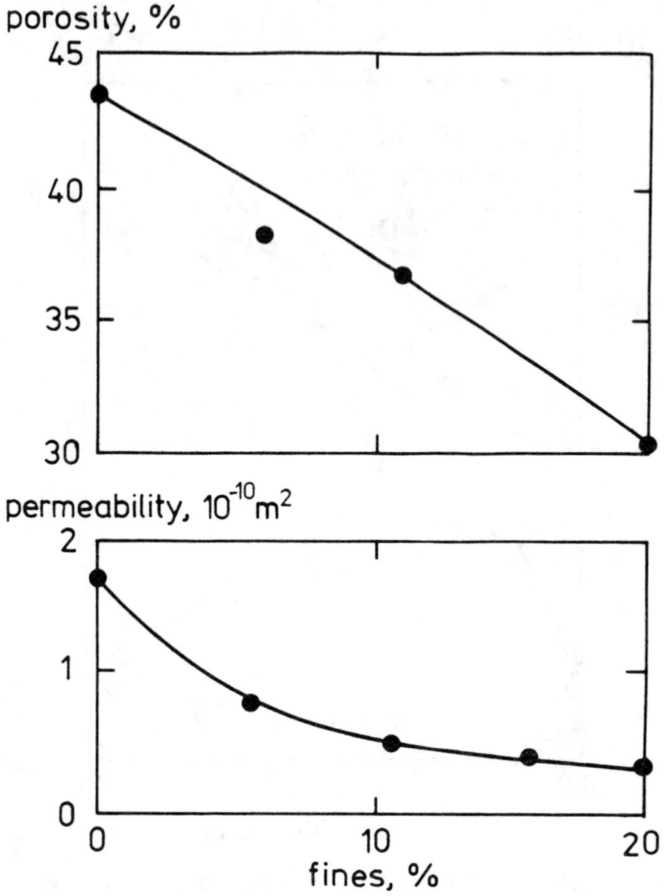

Figure 13.4 The effects of mixing small particles into a bed of large particles. The upper plot shows the associated decrease in porosity and the lower plot shows the resulting permeability decrease.

$$Q = - \alpha A (P_1^2 - P_2^2)/(2 \mu L P_2) \tag{13.10}$$

where the volumetric flow rate Q is measured at the pressure P_2. An alternative form recognizes that the mean pressure $\bar{P}$ equals one-half of the sum of the upper and lower pressures, thus,

$$Q = - \alpha A \bar{P} \Delta P/(\mu L) \qquad (13.11)$$

where ΔP is the pressure difference. Because of the compressibility of the fluid, it is important to specify the pressure of the flow measurement.

The permeabilities for compressible and incompressible flow are usually not the same. This represents differences in the pressure gradients, surface friction, and atomic motion. Generally, for conditions of modest flow velocities, corresponding to Reynolds numbers near unity or less, Equation 13.10 is very successful in fitting experimental data. At high flow velocities, inertial effects impede gas flow and a second term is needed to represent the nonlinear energy loss [30].

The effect of particle characteristics has been the subject of much experimentation and modeling. The main focus is on predicting the permeability from easily measured characteristics such as the porosity, particle size, and compact surface area [4,30-38]. An inherent assumption behind Darcy's law is that the pore size is small with respect to the sample size. In a manner similar to the case of incompressible flow, the permeability proves to be very sensitive to changes in packing density. Generally, the permeability varies with the porosity to a power ranging from 2.5 to 5.0. As the pore shape becomes more irregular (reflecting particle shape effects), the role of porosity becomes even larger. The permeability varies inversely with the surface area, and accordingly has a dependence on the particle size. In a packed bed with a range of particle sizes there may be preferential flow paths near the larger particles (larger pores). Figure 13.5 illustrates how permeability varies with composition for a ternary powder mixture [39]. This plot shows the velocity contours on the ternary composition plot for a constant pressure gradient. Underlying the permeability changes in such mixtures are the effects of particle size, surface area, and packing density. In the region of high packing density along the large-small axis, the permeability is low. This behavior illustrates the dominant effect of powder packing density on permeability. Irregular particles usually have a lower packing density but more surface area than spheres. From a practical view, these are essentially offsetting factors, making permeability fairly insensitive to particle shape differences for randomly packed structures.

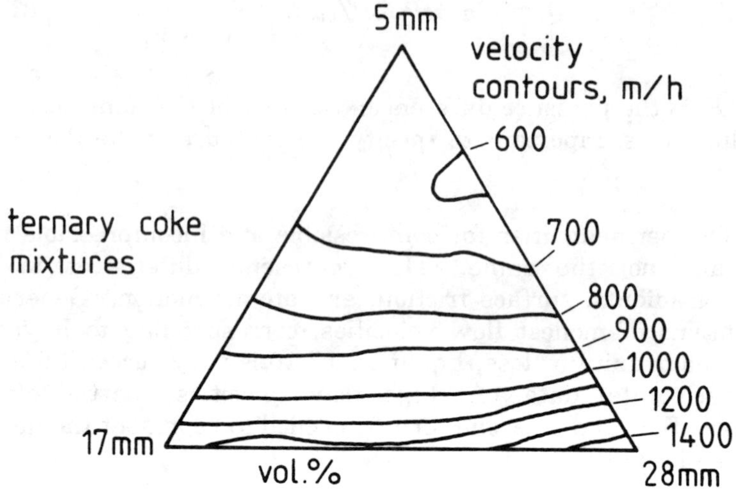

Figure 13.5 Velocity contours plotted against composition for ternary mixtures of coke particles of 5, 17, and 28 mm size. The maximum permeability occurs for the largest particles [39].

Mixed Phases

It is possible for both the vapor and the liquid phases to be interconnected within the pores for a partially saturated particle packing. Figure 13.6 sketches a two-dimensional model of two interpenetrating and connected phases occupying the pores between packed particles. Because of wetting, the liquid resists displacement by the vapor and each fluid establishes its own tortuous flow channel [40,41]. Mixed flow is possible under these conditions of partial saturation, where the liquid permeates through the liquid skeleton and the gas permeates through the vapor skeleton. Alternatively, under certain conditions involving simultaneous gas and liquid flow, foaming can occur. These problems are of great importance to the recovery of oil and natural gas, and in the behavior of fuel cells [22,28,42-45].

The variation in permeability with the degree of liquid saturation follows the behavior shown in Figure 13.7 [22]. At an intermediate level of saturation, under steady-state conditions, the

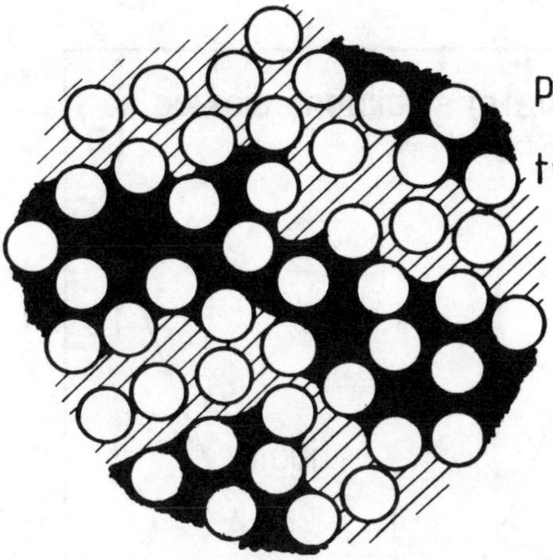

Figure 13.6 A sketch of a porous packing of spherical particles with two interpenetrating fluid phases.

relative permeabilities are unchanging with time. The sum of the relative permeabilities is less than unity for all partial saturations [41]. In the terminal regions, either the liquid or vapor is disconnected. Each phase reduces the permeability for the other phase, but is unable to form a connected pathway. As a consequence, the terminal permeability of the minor phase is zero and the sum of permeabilities is not unity. If the liquid structure is out of equilibrium, then the relative permeabilities will change with time, giving some secondary effects. The ratio of gas and liquid flow rates uniquely defines the saturation level as dictated by the partial pressure ratio, independent of the pore structure [22,38]. The behavior depicted in Figure 13.7 actually has a hysteresis character as well. The permeability of the liquid during draining is less than the permeability during filling at any given saturation level. As a consequence, up to a 30-fold difference in permeabilities is possible at a given saturation, depending on whether draining or imbibition is occurring [46].

The flow of two phases simultaneously through a porous structure is a problem best treated by using percolation theory. For a mixture of oil and gas, the liquid exhibits flow at a saturation of

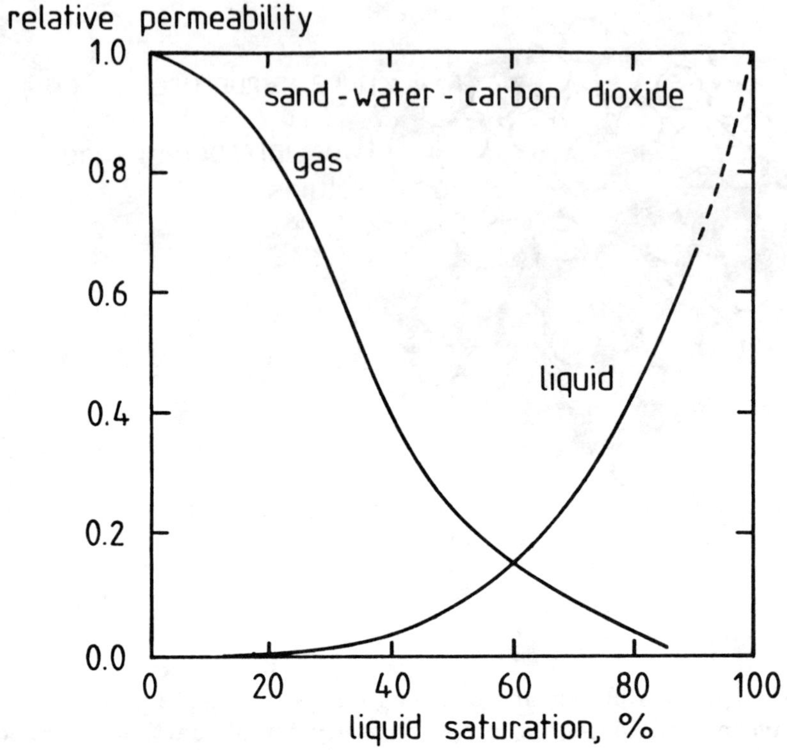

Figure 13.7 The relative permeability of water and carbon dioxide in sand, showing the relative values versus the liquid saturation of the pores [22].

20%, while the gas has no permeability beyond 90% saturation [44]. Archie [28] proposed that at the intermediate saturations the relative permeability would vary with the square of the saturation S,

$$a \sim a_o S^2 \tag{13.12}$$

where a/a_o is the relative permeability of the liquid. The wetting fluid in the pores affects the gas permeability by partially blocking pathways for vapor flow and by changing the pore shape. This is another problem that is best treated in terms of percolation theory (see Chapter Ten). Mason [47] has provided a statistical model that relates the meniscus curvature to the saturation level and shows that the relative permeability roughly follows the expected behavior.

An important situation occurs under conditions where the gas is pressurized with the intent of displacing the liquid from the pore. As remarked earlier in Chapter Twelve, liquid displacement by a low viscosity gas results in fingering, and possibly even foaming. The formation of preferential vapor fingers can occur anytime there are differing densities, viscosities, or flow rates [48]. Thus, even though both fluids are pressurized and flowing, it is possible to create an instability in mixed phase flow that leads to fingering. The onset of such instabilities in mixed phase flow depends on the pore structure, flow rate, and fluid properties [41,49]. This can obviously complicate the analysis of mixed phase flow.

Permeability Coefficient

Predicting flow through a random packing of particles is difficult. For both compressible and incompressible fluid flow the parameter in greatest doubt is the permeability of the pore structure. The treatment in this section has assumed that the flow conditions consist of laminar, viscous flow without significant turbulence or inertial energy losses. Under these conditions, Darcy's law is a valid description of the flow behavior. In both Equations 13.1 and 13.10 the volumetric flow rates can be normalized to the cross-sectional area, giving the superficial flow velocity V as

$$V = Q/A \,. \tag{13.13}$$

The permeability represents a material constant that sets the proportionality between the pressure gradient and the flow velocity. A prediction of the permeability based on particle packing characteristics has considerable importance for predicting flow behavior.

The early models for pores assumed a structure of parallel cylindrical capillary tubes [50]. For laminar flow of liquids in narrow cylindrical tubes, Poiseuille's equation,

$$Q = - \pi d^4 \Delta P/(128 \mu L) \qquad (13.14)$$

is applicable. In this equation d is the diameter of the capillary tube and L is the length. The total flow rate Q_t in a porous structure can be modeled by assuming N capillary tubes per unit area, with the total cross-sectional area equal to A, giving

$$Q_t = N A Q = - \pi N A d^4 \Delta P/(128 \mu L) . \qquad (13.15)$$

For a collection of capillary tubes, the fractional density f is given as

$$f = 1 - (\pi d^2 N/4) . \qquad (13.16)$$

A combination of Equations 13.15 and 13.16 gives a form similar to the basic Darcy law:

$$Q_t = - (1 - f) d^2 A \Delta P / (32 \mu L) . \qquad (13.17)$$

This provides the following capillary tube equivalence for the permeability coefficient:

$$\alpha = (1 - f) d^2/32 . \qquad (13.18)$$

Thus, for a first model, the permeability of a porous structure varies with the porosity and the square of the pore size.

This is the simplest model for a pore structure. However, for particle packings the capillary model is insufficient and many other models have been proposed [1,9,13,25,26,36,39,51-68]. The flow constrictions between particles provide a convergent-divergent contribution to the flow that increases the resistance to flow. The resulting velocity and pressure within the pore structure causes the flow behavior to vary considerably over small distances. The lack of an adequate pore model coupled with the complexities of flow within the pores has led several investigators to determine empirical models.

The shape of the actual pore structure is highly variable from point-to-point. Thus, since flow through a random particle packing involves passage through multiple interconnections (constrictions), the predictions of flow based on easily measured parameters is error prone [54]. Although there are numerous models for permeability based on flow data, the predictions can come only within 50% of the true permeability. The best known of the models linking permeability to powder characteristics are those of Kozeney, Carman, and Ergun [25,31,57]. Table 13.2 is a collection of several models for the permeability α that rely on parameters such as the particle size D, surface area per unit volume S_v, and packing density f. This table demonstrates considerable differences between the available models linking permeability to pore structure. Some of the differences are attributable to particle shape effects. Many of the early models dealt with flow through randomly packed beds of monosize spheres. These usually substantiate the claim that permeability is related to porosity and surface area as power functions. The most successful of these models is attributed to Ergun [31,57], and is given by the equation

$$\alpha = A\,(1 - f)^3\, D^2/(f^2) \tag{13.19}$$

where A is a constant and the particle diameter D is determined from the ratio of the volume to surface area for nonspherical particles. Since monosize spheres have a predictable surface area based on the particle size, models using either parameter have been presented [25]. For packings composed of a range of particle sizes the mean size is used [41,69,70]. Figure 13.8 illustrates the packing density and flow variations for a ternary mixture of particles. In this study, the drop in pressure was maintained as a constant, and the flow velocity was measured versus the composition. The results indicate that the mean particle size is an adequate first basis for predicting flow in mixed powder systems. However, a complication arises if there is size segregation in the mixture. Thus, better predictability of flow behavior is observed with narrow particle size distributions.

A wide particle size distribution gives a higher packing density and a smaller pore size; thus, the permeability decreases rapidly as the width of the particle size distribution increases. Correlations based on the pore size have been employed in such

Table 13.2

Permeability Models

a = permeability, m^2
f = fractional density
S_v = surface area, m^2
D = particle size, m
d = pore size, m
A = constant, varying according to the model
τ = tortuosity

$a = A\,(1 - f)^3/S_v^{\,2}$	[19,68]
$a = A\,(1 - f)^{4.7}/S_v^{\,2}$	[68]
$a = A\,(1 - f)^4\,D^2/f^2$	[6]
$a = A\,D^2$	[26,41,62,66]
$a = A\,(1 - f)^4/S_v^{\,3}$	[56]
$a = A\,((1 - f)^2/S_v)^{2.4}$	[56]
$a = A\,((1 - f)\,d)^{1.9}$	[56,57,68]
$a = A\,(1 - f)^3\,D^2/f^2$	[67]
$a = A\,(1 - f)^2\,D^2/f$	[26]
$a = A\,(1 - f)^3/(\tau\,S_v^{\,2})$	[18,20]
$a = A\,(1 - f)^3/(\tau\,f^2 S_v^{\,2})$	[13,18,61]
$a = A\,(1 - f)^3/(S_v^{\,2}\,f^2)$	[1,19,26,33,65]
$a = A\,(1 - f)^3/(\tau\,S_v^{\,2})$	[60]
$a = A\,(1 - f)^{4.8}/(S_v^{\,2.4})$	[18]

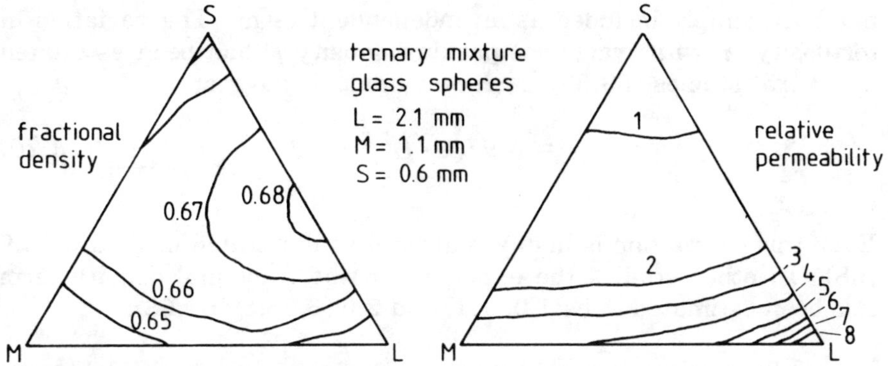

Figure 13.8 Ternary mixtures of glass spheres showing the fractional density and relative permeability versus composition [69,70].

cases [53]. One problem is that such correlations overemphasize the effect of small particles when the packing contains a low concentration of small particles. When only some of the pores are filled with small particles, then the flow is dominated by the larger pores. This is particularly evident at low packing densities where flow channeling has been observed [5].

Other models of permeability ignore the particle size and shape effects by using parameters such as the surface area. Usually, such an approach utilizes statistical regression models. These models incorporate the obvious factors of packing density and surface area, often to noninteger powers. An alternative approach is to measure the pore size. As noted already, the largest pores usually dominate the flow behavior, so models relying on an effective or mean pore size are also possible [39].

An important correction to the early models was to include the tortuosity [13,20,60]. With an irregular pore structure the distance a fluid travels in passing through the pores exceeds the compact thickness. The tortuosity is a measure of the ratio of the actual flow distance to the geometric compact thickness. In some models the tortuosity is included directly in the permeability through an estimate based on the packing density, while in other models it has been simply included as an independent term. The variation in tortuosity τ with fractional packing density f has been estimated in several studies [59,68]. A recent analysis suggests

$$\tau = 0.9/(1 - f)^{1.3} . \qquad (13.20)$$

Even this correlation is highly scattered as illustrated in Figure 13.9 [68]. In other studies the exponent on the fractional density term has been estimated to be 1.0, 1.1, and 2.0 [6,56,59,61,68].

Two other factors important to estimating permeability are the particle shape and the surface roughness. Rough particles have a greater surface area and contribute more resistance to flow

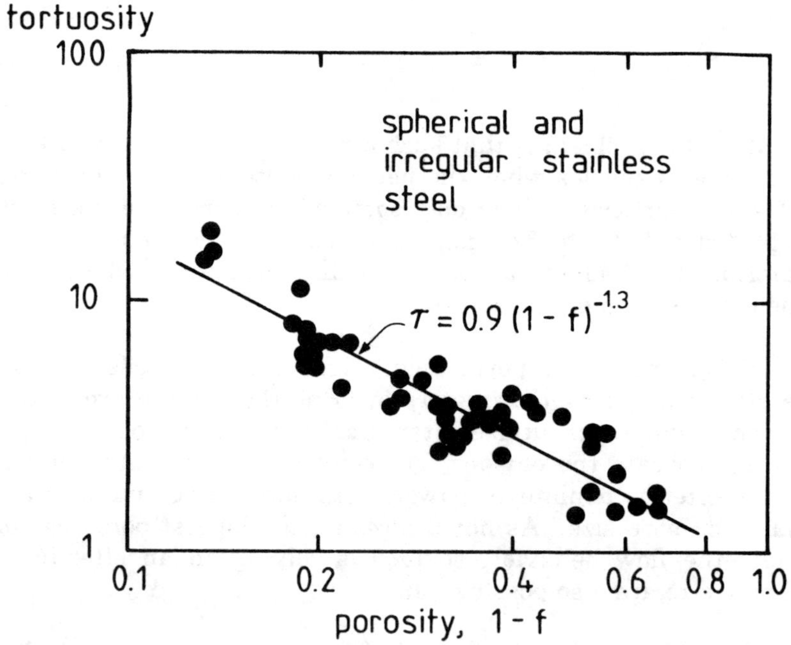

Figure 13.9 The flow path tortuosity as a function of the porosity for both spherical and irregular stainless steel particles [68].

[68,71]. Indeed, the effect of particle shape on permeability is dependent on the packing density, as illustrated in Figure 13.10 [1]. This figure plots the relative permeability correction with fractional density for some simple shapes. There is roughly twice as much flow resistance from an irregular or rough particle as from a sphere of the same size. Additionally, for elongated particles there is more opportunity for packing orientation and anisotropic permeability.

The connectivity of the pores is another factor that influences the permeability. Usually, it is assumed that all pores are participating in flow. However, as the fractional density of a powder compact increases (especially above 0.85), an increasing fraction of the pores do not contribute to flow. Since most models assume all pores are contributing to flow, a large error results when they are applied to high density compacted or sintered particle structures.

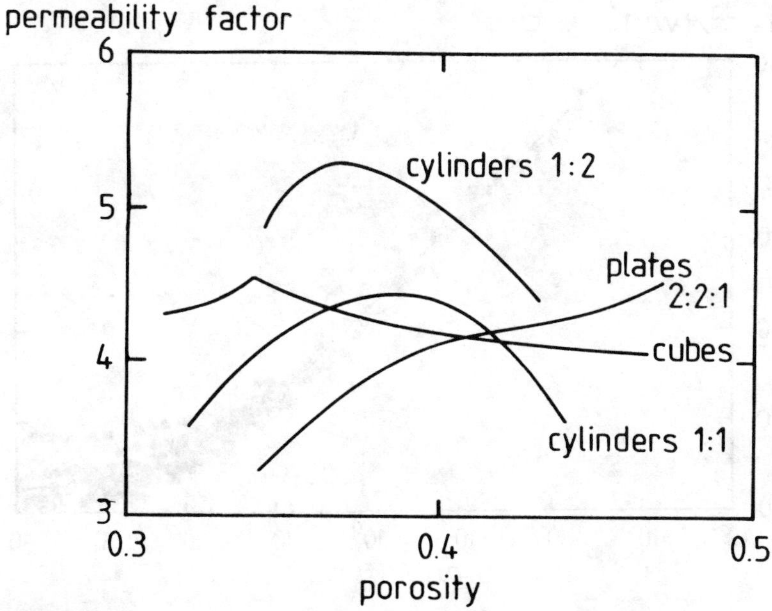

Figure 13.10 The permeability variation with porosity for various nonspherical shapes [1].

The results of the application of percolation theory to this problem show that as the interconnections between pores become more variable, the overall bulk permeability is increased [72]. A thorough examination of the permeability models shows that most provide a reasonable basis for predicting flow behavior. Considering the range of particle sizes, size distributions, particle shapes, packing densities, and fluids, the best models can only be accurate to within 50% of the measured permeability. Figure 13.11 illustrates the general variation in permeability with porosity for several materials [55]. This figure plots the permeability from many measurements on a range of materials. The scatter illustrates the basic problem in isolating a single model for permeability based solely on particle size and packing density.

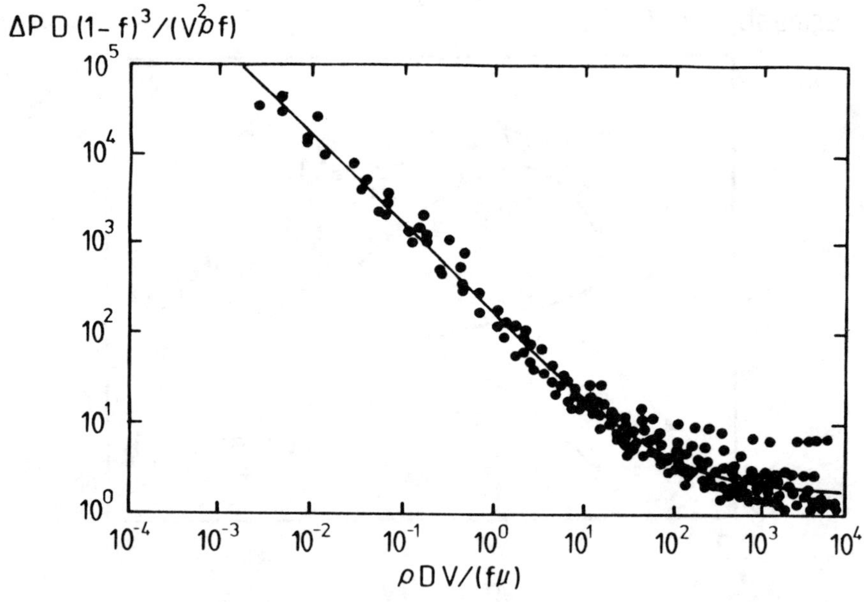

Figure 13.11 A master plot of data from several studies showing permeability behavior over a wide range of experimental conditions [55].

Inertial and Turbulent Flow

As the pressures and rates of fluid flow are increased, a condition where the laminar flow assumption becomes invalid is eventually reached. Just as there is a transition region between molecular diffusion and laminar viscous flow (slip flow), there is likewise a transition region between laminar viscous flow and turbulent flow. This transition region is termed inertial flow. The transition is smooth over a range of flow rates, reflecting the natural variations in pore characteristics and velocities in the pores. The initial instabilities can be seen at Reynolds numbers as low as 0.1 [73]. The flow is not truly turbulent in the inertial region; however, the expansion and contraction of the fluid through a tortuous pore structure couples with the inertial energy loss due to changes in flow direction to impede flow. As shown in Figure 13.12, the actual flow rate is below that expected from Darcy's law. This figure indicates the behavior for porous stainless steel in terms of the flow rate versus pressure drop [20]. The decrement associated with inertial flow is indicated by the shaded region.

Thus, at high flow rates Darcy's law is invalid because nonlinear energy losses reduce the flow rate. In the inertial range, the resistance to flow has two components, viscous drag which is proportional to the velocity, and a kinetic energy loss which is proportional to the velocity squared. The overall relation between the pressure drop ΔP and flow velocity V is as follows [73,74]:

$$\Delta P = K\, V^n \qquad (13.21)$$

where K is a constant and n is usually between 1.7 and 2.0, with a typical value of 1.85. The decrease in velocity arises from the repeated directional change and associated inertial energy loss as the fluid passes through the tortuous flow path. For a compressible fluid, the modified flow equation needed to compensate for inertial energy losses was first suggested by Forchheimer as follows [20,33,60]:

$$\Delta P\, \bar{P}/P_2 = L\, \mu\, V/\alpha + \rho\, V^2/\beta \qquad (13.22)$$

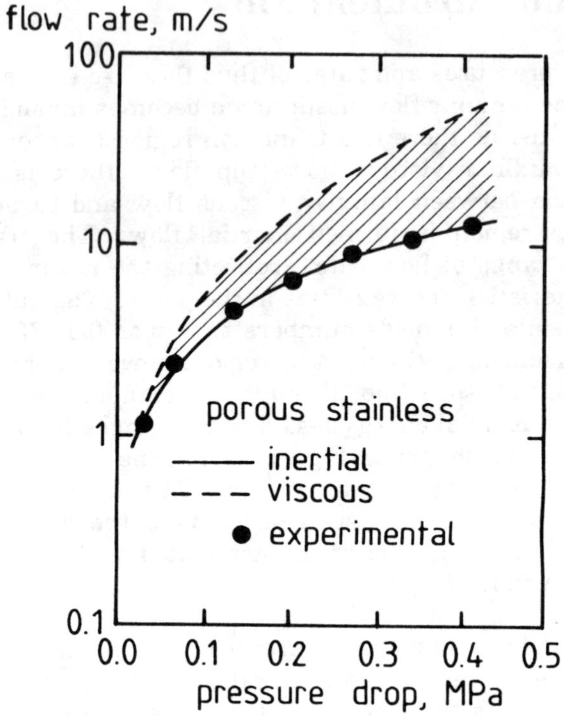

Figure 13.12 The flow rate versus the pressure drop for porous stainless steel, comparing the experimental data with the models for viscous and inertial flow [20].

with V equal to the superficial flow velocity (volume per unit time per unit area), L equal to the compact thickness, ΔP equal to the pressure drop $(P_1 - P_2)$, $\bar{P}$ equal to the mean pressure, P_2 equal to the lower pressure, ρ equal to the fluid density, and α equal to the permeability. The new coefficient β has units of length and is termed the inertial coefficient. It represents the shape characteristics of the pore structure and is sensitive to the tortuosity.

In general, Equation 13.22 provides a good model for flow through porous structures with nominal superficial flow velocities between 0.1 and 20 m/s, depending on the actual pore characteristics. It is the fluid velocity in the pores which determines the applicability of Equation 13.22, but for most treatments, it is assumed that the inertial energy losses become significant for Reynolds numbers over 4. Equation 13.22 is not valid for Reynolds numbers over 200.

Table 13.3 compiles some of the prior correlations between the inertial coefficient and the pore structure [18,57,75]. The

Table 13.3

Models for the Inertial Coefficient

β = inertial coefficient, m
f = fractional density
S_v = surface area per unit volume, m^2/m^3
D = particle size, m
A = constant, varying with the model
τ = tortuosity

$\beta = A(1-f)^3 D/f$	[57]
$\beta = A(1-f)^3/(f S_v)$	[18,75]
$\beta = A(1-f)^3 D/f$	[55,66,67]
$\beta = A(1-f)^5/S_v$	[68]
$\beta = A(1-f)^3/(f S_v)$	[18,75]
$\beta = A(1-f)^3/S_v$	[59,60,68]
$\beta = A(1-f)^{5.1}/S_v$	[68]
$\beta = A(1-f)^{6.4} D^{0.9}$	[60]

desire has been to link the inertial coefficient to easily measured properties of the packed particles. In all cases, estimates of the inertial coefficient are more error prone than those for the permeability coefficient. One such set of correlations is shown in Figures 13.13 and 13.14 for various irregular and spherical stainless steel particles [68]. The first figure plots the product of the permeability coefficient and the square of the surface area versus the fractional porosity, giving a reasonably good correlation that is independent of the particle shape. Figure 13.14 plots the product of the inertial coefficient and the surface area as a function of the fractional porosity. In this case there is a behavior difference associated with the particle shape. It is unclear which of the several models is the most accurate. The relative errors are similar, and no basis exists

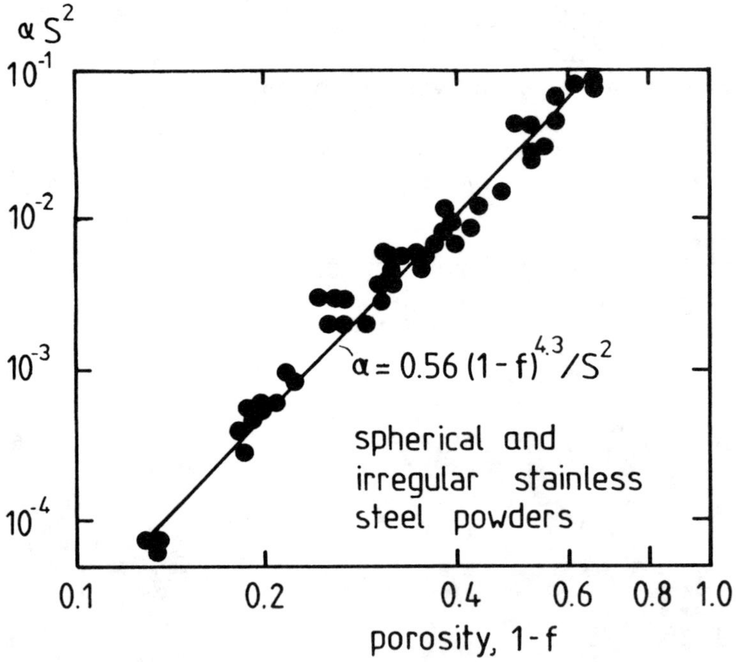

Figure 13.13 The product of the permeability and the square of the surface area versus the fractional porosity for stainless steel particles [68].

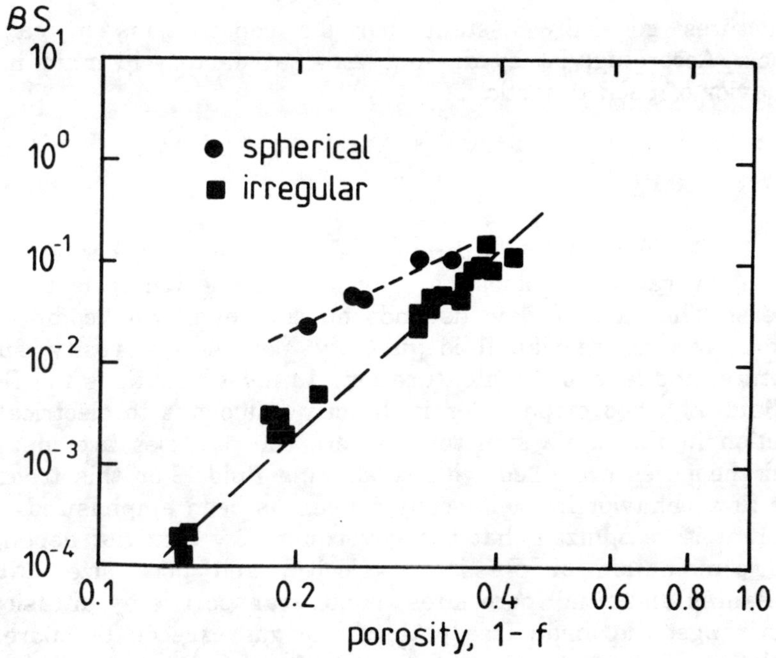

Figure 13.14 A plot of the inertial flow coefficient times the surface area versus the fractional porosity, showing a definite particle shape effect [68].

for selecting one model over another. Empirical observations of the permeability and inertial coefficients indicate some possible links. The ratio of β/a provides an index of the pore shape and tortuosity. In some cases the inertial coefficient has been predicted most accurately from estimates based on the measured permeability coefficient.

At high flow velocities (high Reynolds numbers), turbulent conditions exist. In the turbulent flow range, the drop in pressure is proportional to the velocity to the 1.85 power [76]. The permeability associated with turbulent conditions exhibits a dependence on the square root of the particle size and the porosity to the 1.5 power [77]. At very high pressures, choked flow occurs. In this range, turbulence is extensive and the flow rate is independent of the pressure gradient. The models of turbulent flow in pore structures generally assume periodic constrictions in capillary tubes. Accordingly, they are not very satisfactory in matching the behavior of packed particles.

Summary

The flow of fluids through porous structures has applicability to several technologies, ranging from the design of heat pipes to filters. The rate of flow depends on the fluid type (compressible versus incompressible), fluid viscosity, pore size, porosity, surface texture, and level of fluid saturation. In most situations the flow of a fluid in a bed of particles is directly analogous to electrical conduction in the same structure, where the particles are insulators and the pores are filled with conducting fluid. For this treatment the flow behavior in the viscous range has been emphasized. However, it is recognized that the governing flow behavior depends on the combination of pressure, velocity, and pore size. At low pressures and small pore sizes, vapor transport is by diffusion. In this range, the mean free path of the gas exceeds the pore size, and flow depends on the diffusivity. As the pressure or the pore size increases, then viscous effects become more significant. Initially, this arises as slip flow, but eventually as the mean free path becomes much smaller than the pore size true viscous flow occurs. For laminar viscous flow, Darcy's law is applicable. This is the most typical condition for fluid flow through a bed of particles.

As the pressure driving the flow increases, turbulence and inertial energy losses arise. Darcy's law must then be modified for the nonlinear flow rate dependence on velocity. Finally, at very high flow rates, the flow is turbulent and choked flow conditions prevail. The Reynolds number and mean free path provide guidance for deciding on the dominant region of flow.

One important problem has been the determination of the links between the permeability and inertial flow coefficients and the packing characteristics of the particles. Tables 13.2 and 13.3 summarize various treatments. Such models are desired to give a basis for estimating behavior, but in the best cases the errors are relatively large. The effects of packing density and particle size are clearly dominant. Relatively small changes in either parameter can result in significant variations in both α and β. One unanswered question is whether there is a link between the two coefficients. Should such a link exist, then the mathematical models for flow over a wide range of pressures could be predicted with relatively little effort.

Fluid flow in a porous structure depends on several variables, including the pressure drop, fluid properties, porosity, and particle size. Of the various factors, porosity and particle size (actually pore size) are dominant. Differences in particle shape are less influential on flow. In most models, particle shape has a small effect on the permeability coefficient compared to the effects of porosity and particle size.

This chapter has touched on the case of mixed phase flow in porous media. The case of most interest is simultaneous gas and liquid flow. The relative permeability of each phase is reduced by the presence of the other phase because of the disruption in the connectivity. The sum of the relative permeabilities is less than unity for all intermediate liquid saturations.

The viscosity of most fluids varies with temperature. For a liquid, the viscosity decreases with heating while for a gas the viscosity increases with heating. Because of this temperature effect, the temperature of the fluid and the pore structure must be considered to accurately model flow behavior. At high flow rates, a gas can cool the pore structure adiabatically and in turn affect the permeability. Experiments at varying temperatures indicate the permeability coefficient is more temperature sensitive than the inertial coefficient.

The treatment of fluid flow in porous materials usually begins with a capillary tube model. Unfortunately, the pores between packed powders have very irregular shapes, with a high level of connectivity. Consequently, real flow behavior is far from that of the capillary tube. In this chapter, the validity of these simple models has been reviewed and the necessary modifications for extension to real particle packings have been introduced.

References

1. J. M. Coulson, "The Flow of Fluids Through Granular Beds: Effect of Particle Shape and Voids in Streamline Flow," *Transactions of the Institution of Chemical Engineers*, 1949, vol. 27, pp. 237-157.
2. R. M. Barrer, "Surface and Volume Flow in Porous Media," *The Solid-Gas Interface*, vol. 2, E. A. Flood (ed.), Marcel Dekker, New York, NY, 1967, pp. 557-609.
3. L. C. Graton and H. J. Fraser, "Systematic Packing of Spheres With Particular Relation to Porosity and Permeability," *Journal of Geology*, 1935, vol. 43, pp. 785-909.
4. J. M. Campbell and R. L. Huntington, "Heat Transfer and Pressure Drop in Fixed Beds of Spherical and Cylindrical Solids," *Petroleum Refiner*, 1951, vol. 30, no. 12, pp. 127-133.
5. H. Happel and N. Epstein, "Cubical Assemblages of Uniform Spheres," *Industrial and Engineering Chemistry*, 1954, vol. 46, pp. 1187-1194.
6. R. M. German, "Theory of Thermal Debinding," *International Journal of Powder Metallurgy*, 1987 vol. 23, pp. 237-245.
7. L. Zagar, "Methods of Determination of Pore Structure Permeability and Diffusion," *Proceedings of the International Symposium on Pore Structure and Properties of Materials*, RILEM-IUPAC, Prague, Czechoslovakia, September 1973, pp. C249-C277.
8. G. H. Neale and W. K. Nader, "Prediction of Transport Processes Within Porous Media: Diffusive Flow Processes Within a Homogeneous Swarm of Spherical Particles," *American Institute of Chemical Engineers Journal*, 1973, vol. 19, pp. 112-119.
9. D. T. Wasan, "Analysis and Evaluation of Permeability Techniques for Characterizing Fine Particles Part I. Diffusion and Flow Through Porous Media," *Powder Technology*, 1976, vol. 14, pp. 209-228.
10. H. A. M. Van Eekelen, "The Random Spheres Model for Porous Materials," *Journal of Catalysis*, 1973, vol. 29, pp. 75-82.
11. H. L. Weissberg, "Effective Diffusion Coefficient in Porous Media," *Journal of Applied Physics*, 1963, vol. 34, pp. 2636-2639.
12. R. B. Evans, G. M. Watson and E. A. Mason, "Gaseous Diffusion in Porous Media at Uniform Pressure," *Journal of*

Chemical Physics, 1961, vol. 35, pp. 2076-2083.
13. J. van Brakel, "Pore Space Models for Transport Phenomena in Porous Media Review and Evaluation with Special Emphasis on Capillary Liquid Transport," *Powder Technology*, 1975, vol. 11, pp. 205-236.
14. N. S. Lidorecko, A. S. Kagan, S. P. Chizhik, A. E. Kovalshii, S. M. Prokoshina and O. V. Sokolov, "Packing Texture of Brittle Materials," *Soviet Powder Metallurgy and Metal Ceramics*, 1969, vol. 8, pp. 840-841.
15. Z. M. Allawi and D. J. Gunn, "Flow and Diffusion of Gases Through Porous Substrates," *American Institute of Chemical Engineers Journal*, 1987, vol. 33, pp. 766-775.
16. J. W. Evans, M. H. Abbasi and A. Sarin, "A Monte Carlo Simulation of the Diffusion of Gases in Porous Solids," *Journal of Chemical Physics*, 1980, vol. 72, pp. 2967-2973.
17. P. J. Dees, T. G. Tjan and J. Polderman, "Determination of Pore Diameter by Permeability Measurements," *Powder Technology*, 1980, vol. 27, pp. 29-36.
18. R. M. German, "Porous Materials," *Advances in Powder Technology*, G. Y. Chin (ed.), American Society for Metals, Metals Park, OH, 1982, pp. 225-251.
19. A. E. Scheidegger, *The Physics of Flow Through Porous Media*, revised edition, University of Toronto Press, Toronto, Ontario, Canada, 1960.
20. R. M. German, "Gas Flow Physics in Porous Media," *International Journal of Powder Metallurgy and Powder Technology*, 1979, vol. 15, pp. 23-30.
21. M. Leva and M. Grummer, "Pressure Drop Through Packed Tubes Part III Prediction of Voids in Packed Tubes," *Chemical Engineering Progress*, 1947, vol. 43, pp. 713-718.
22. H. J. Fraser, "Experimental Study of the Porosity and Permeability of Clastic Sediments," *Journal of Geology*, 1935, vol. 43, pp. 910-1010.
23. H. E. Rose, "An Investigation into the Laws of Flow of Fluids Through Beds of Granular Materials," *Proceedings of the Institution of Mechanical Engineers*, 1945, vol. 153, pp. 141-148.
24. G. H. Neale and W. K. Nader, "Prediction of Transport Processes Within Porous Media: Creeping Flow Relative to a Fixed Swarm of Spherical Particles," *American Institute of Chemical Engineers Journal*, 1974, vol. 20, pp. 530-538.
25. P. C. Carman, "Fundamental Principles of Industrial

Filtration," *Transactions of the Institution of Chemical Engineers*, 1938, vol. 16, pp. 168-188.

26. T. Allen, *Particle Size Measurement*, second edition, Chapman and Hall, London, UK, 1974.
27. J. J. Martin, W. L. McCabe and C. C. Monrad, "Pressure Drop Through Stacked Spheres, Effect of Orientation," *Chemical Engineering Progress*, 1951, vol. 47, no. 2, pp. 91-94.
28. G. E. Archie, "The Electrical Resistivity Log as an Aid in Determining Some Reservoir Characteristics," *Transactions of the AIME*, 1942, vol. 146, pp. 54-62.
29. R. R. Rumer, "Resistance to Flow Through Porous Media," *Flow Through Porous Media*, R. J. M De Wiest (ed.), Academic Press, New York, NY, 1969, pp. 91-108.
30. A. R. Morcom, "Fluid Flow Through Granular Materials," *Transactions of the Institution of Chemical Engineers*, 1946, vol. 24, pp. 30-36.
31. S. Ergun and A. A. Orning, "Fluid Flow Through Randomly Packed Columns and Fluidized Beds," *Industrial and Engineering Chemistry*, 1949, vol. 41, pp. 1179-1184.
32. A. G. Kostornov and M. S. Shevchuk, "Effect of Porosity, Thickness and Differential Pressure on the Principal Properties of Permeable Fiber Materials," *Soviet Powder Metallurgy and Metal Ceramics*, 1973, vol. 12, pp. 752-756.
33. E. M. Cliffel, W. E. Smith and A. D. Schwope, "Theory and Applications of Controlled Permeability," *Modern Developments in Powder Metallurgy*, vol. 3, H. H. Hausner (ed.), Plenum Press, New York, NY, 1966, pp. 114-128.
34. H. E. Rose, "The Isothermal Flow of Gases Through Beds of Granular Materials," *Proceedings of the Institution of Mechanical Engineers*, 1945, vol. 153, pp. 148-153.
35. H. E. Rose, "On the Resistance Coefficient-Reynolds Number Relationship for Fluid Flow Through a Bed of Granular Material," *Proceedings of the Institution of Mechanical Engineers*, 1945, vol. 153, pp. 154-161.
36. R. M. German, *Powder Metallurgy Science*, Metal Powder Industries Federation, Princeton, NJ, 1984, pp. 228-235.
37. R. D. Wyckoff and H. G. Botset, "The Flow of Gas-Liquid Mixtures Through Unconsolidated Sands," *Physics*, 1936, vol. 7, pp. 325-345.
38. E. C. Childs, *An Introduction to the Physical Basis of Soil Water Phenomena*, Wiley Interscience, New York, NY, 1970.
39. P. R. Marshall, "Pore Size Distribution and Permeability of

Porous Metal Materials," *Proceedings of a Symposium on Powder Metallurgy*, Special Report number 58, Iron and Steel Institute, London, UK, 1956, pp. 71-80.

40. C. Y. Liu, K. Murakami and T. Okamoto, "Effect of Capillary Pressure on Interdendritic Liquid Flow," *Acta Metallurgica*, 1986, vol. 34, pp. 159-166.
41. R. E. Collins, *Flow of Fluids Through Porous Media*, Reinhold Publishing Corp., New York, NY, 1961.
42. J. C. Charpentier, "Hydrodynamics of Two-Phase Flow Through Porous Media," *Journal of Powder and Bulk Solids Technology*, 1978, vol. 2, pp. 53-60.
43. T. Katan and E. A. Grens, "The Effect of Liquid Configuration in Porous Gas Electrodes," *Journal of the Electrochemical Society*, 1971, vol. 118, pp. 1881-1886.
44. G. G. Brown, *Unit Operations*, John Wiley & Sons, New York, NY, 1950.
45. I. Mahgoub, C. Prost and J. A. Dodds, "Experimental Relative Permeability Curves Compared with a New Cutting and Joining Model of Porous Media with a Distribution of Pore Sizes," *Powder Technology*, 1982, vol. 31, pp. 143-151.
46. E. E. Miller and R. D. Miller, "Physical Theory for Capillary Flow Phenomena," *Journal of Applied Physics*, 1956, vol. 27, pp. 324-332.
47. G. Mason, "Desaturation of Porous Media I. Unconsolidated Materials," *Journal of Colloid and Interface Science*, 1972, vol. 41, pp. 208-227.
48. T. Maxworthy, "The Nonlinear Growth of a Gravitationally Unstable Interface in a Hele-Shaw Cell," *Journal of Fluid Mechanics*, 1987, vol. 177, pp. 207-232.
49. L. Paterson, "Fingering with Miscible Fluids in a Hele-Shaw Cell," *Physics of Fluids*, 1985, vol. 28, pp. 26-30.
50. A. M. Schwartz, "Capillarity Theory and Practice," *Industrial and Engineering Chemistry*, 1969, vol. 61, pp. 10-21.
51. A. C. Payatakes, C. Tien and R. M. Turian, "A New Model for Granular Porous Media: Part 1. Model Formulation," *American Institute of Chemical Engineers Journal*, 1973, vol. 19, pp. 58-67.
52. A. C. Payatakes, C. Tien and R. M. Turian, "A New Model for Granular Porous Media: Part 2. Numerical Solution of Steady State Incompressible Newtonian Flow Through Periodically Constricted Tube," *American Institute of Chemical Engineers Journal*, 1973, vol. 19, pp. 67-76.

53. M. K. Bo, D. C. Freshwater and B. Scarlett, "The Effect of Particle Size Distribution on the Permeability of Filter Cakes," *Transactions of the Institution of Chemical Engineers*, 1965, vol. 43, pp. T228-232.
54. G. R. Youngquist, "Diffusion and Flow of Gases in Porous Solids," *Flow Through Porous Media*, American Chemical Society, Washington, DC, 1970, pp. 58-69.
55. I. F. Macdonald, M. S. El-Sayed, K. Mow and F. A. L. Dullien, "Flow Through Porous Media - The Ergun Equation Revisited," *Industrial and Engineering Chemistry*, 1979, vol. 18, pp. 199-208.
56. D. W. Smith and T. Marth, "An Examination of the Effects of Pore Morphology on Gas Flow Through Sintered Compacts," *Modern Developments in Powder Metallurgy*, vol. 12, H. H. Hausner, H. W. Antes and G. D. Smith (eds.), Metal Powder Industries Federation, Princeton, NJ, 1981, pp. 835-854.
57. S. Ergun, "Fluid Flow Through Packed Columns," *Chemical Engineering Progress*, 1952, vol. 48, pp. 89-94.
58. L. Green and P. Duwez, "Fluid Flow Through Porous Metals," *Journal of Applied Mechanics*, 1951, vol. 73, pp. 39-45.
59. B. A. Meyer and D. W. Smith, "Flow Through Porous Media: Comparison of Consolidated and Unconsolidated Materials," *Industrial Engineering Chemistry Fundamentals*, 1985, vol. 24, pp. 360-368.
60. R. M. German, "Porosity and Particle Size Effects on the Gas Flow Characteristics of Porous Metals," *Powder Technology*, 1981, vol. 30, pp. 81-86.
61. C. S. Brooks and W. R. Purcell, "Surface Area Measurements on Sedimentary Rocks," *Petroleum Transactions of AIME*, 1952, vol. 195, pp. 289-296.
62. J. Marecek, M. Krpata and F. Moudry, "The Effect of Particle Size Distribution on the Removal of the Liquid Phase from a Centrifuged Filter Cake," *International Chemical Engineering*, 1972, vol. 12, pp. 384-388.
63. E. Klar and W. M. Shafer, "On Green Strength and Compressibility in Metal Powder Compaction," *Modern Developments in Powder Metallurgy*, vol. 9, H. H. Hausner and P. W. Taubenblat (eds.), Metal Powder Industries Federation, Princeton, NJ, 1977, pp. 91-113.
64. J. Happel, "Viscous Flow in Multiparticle Systems: Slow Motion of Fluids Relative to Beds of Spherical Particles," *American Institute of Chemical Engineers Journal*, 1958, vol. 4,

pp. 197-201.
65. J. M. Culson, "The Flow of Fluids Through Granular Beds: Effect of Particle Shape and Voids in Streamline Flow," *Transactions of the Institution of Chemical Engineers*, 1949, vol. 27, pp. 237-257.
66. A. T. Robinson, "Permeability of Tungsten Matrices as a Function of Density, Particle Size and Shape," *Transactions of the American Society for Metals*, 1964, vol. 57, pp. 650-657.
67. N. Standish and P. J. Leyshon, "The Permeability of Quaternary Particulate Mixtures," *Powder Technology*, 1981, vol. 30, pp. 119-121.
68. D. W. Smith, J. E. Smugeresky and B. A. Meyer, *The Dependence of Permeability and Filtration on Pore Morphology in Consolidated Particulate Media*, Technical Report SAND87-8227, Sandia National Laboratories, Livermore, CA, October 1987.
69. N. Standish and D. N. Collins, "The Permeability of Ternary Particulate Mixtures for Laminar Flow," *Powder Technology*, 1983, vol. 36, pp. 55-60.
70. M. Leitzelement, C. S. Lo and J. Dodds, "Porosity and Permeability of Ternary Mixtures of Particles," *Powder Technology*, 1985, vol. 41, pp. 159-164.
71. W. A. Gray, *The Packing of Solid Particles*, Chapman and Hall, London, UK, 1968.
72. C. H. Seager and G. E. Pike, "Percolation and Conductivity: A Computer Study. II," *Physical Review B*, 1974, vol. 10, pp. 1435-1446.
73. D. Cornell and D. L. Katz, "Flow of Gases Through Consolidated Porous Media," *Industrial and Engineering Chemistry*, 1953, vol. 45, pp. 2145-2152.
74. T. H. Chilton and A. P. Colburn, "Heat Transfer and Pressure Drop in Empty, Baffled and Packed Tubes II - Pressure Drop in Packed Tubes," *Industrial and Engineering Chemistry*, 1931, vol. 23, pp. 913-919.
75. D. B. Greenberg and E. Weger, "An Investigation of the Viscous and Inertial Coefficients for the Flow of Gases Through Porous Sintered Metals with High Pressure Gradients," *Chemical Engineering Science*, 1960, vol. 12, pp. 8-19.
76. H. Deresiewicz, "Mechanics of Granular Matter," *Advances in Applied Mechanics*, vol. 5, H. L. Dryden and T. von Karman (eds.), Academic Press, New York, NY, 1958, pp. 233-306.
77. N. Standish and D. G. Mellor, "The Permeability of Ternary

Coke Mixtures," *Powder Technology*, 1980, vol. 27, pp. 61-68.

Chapter Fourteen

Interparticle Strength

Chapter Outline

- **Overview**
- **Loose Particles**
    - **Adhesion and Cohesion**
    - **Shear Induced Flow**
- **Compacted Particles**
- **Role of Particle Characteristics**
- **Summary**
- **References**

Overview

Particulate materials have both solid and fluid characteristics. The individual particles are solid, yet the entire mass can flow under relatively low stresses to fill a container similar to a viscous liquid. Like a liquid, loose particles do not have a high resistance to shear. However, they exhibit plastic deformation and strength behavior similar to bulk solids. This combination of characteristics is due to the bonding forces that exist between loose versus compacted particles. The weak bonding between loose particles provides a resistance to deformation and sliding that can be overcome at relatively low stresses. However, after deformation (such as in compaction), the particle contacts are enlarged, and both chemical and mechanical bonds that greatly increase the interparticle strength are created. Thus, compacted particles hold shape and resist shear deformation like a solid.

A wetting liquid can be a source of interparticle bonding. Such a fluid will form a bridge between the particles and provide a capillary attractive force. Likewise, when particles are compacted the deformation and shear at the contact regions greatly increase the strength. As a consequence, the interparticle strength rapidly increases with the packing density.

For a loose powder, a first measure of the interparticle bond strength is given by the packing characteristics. Particles with strong bonds exhibit a resistance to flow and do not pack to very high densities. Alternatively, particles with weak bonds (or in some cases repulsive bonds) will pack to higher densities and exhibit easier flow through orifices. Additionally, the uniformity in particle packing relates to the degree of interparticle bonding. Powders with a low overall packing density also have a low packing homogeneity and wide variation in interparticle strength.

In dealing with the strength of particle aggregates, it is necessary to consider the effective stress. Because the contact points between particles are small, an external stress is concentrated at the particle contacts, giving a high effective stress. At the points of concentrated stress, the interparticle bonding resists sliding and deformation. Particles under a compressive stress resist shear because a dilation is needed to separate the particles. Thus, the bond strength depends on the total state of stress for the powder mass.

For particles, the response to an applied stress includes elastic strains, flow and rearrangement, and permanent plastic strains (particle deformation). The elastic strains are small and proportional to the applied stress. Yet, even the elastic behavior of a powder has significance for industrial processing. In powder compaction, elastic strains account for springback and possible compact delamination on ejection from the die. Rearrangement and flow of a powder is important to powder mixing, and particle shape changes due to plastic deformation are inherent to powder compaction. Thus, the interparticle strength has components of chemical bonding, cohesion (wetting liquids), mechanical bonding (shear and deformation), and other forces such as magnetic attraction.

These various contributors to the interparticle strength of loose and compacted particles are the subject of this chapter. The treatment will be separated into three main divisions. Initially, loose (uncompacted) particles will be treated with a focus on adhesive and cohesive forces. Subsequently, the treatment shifts to compacted powders with attention given to the tensile strength versus density. Finally, the treatment shifts to the particle characteristics and their effects on the interparticle bond strength.

Loose Particles

Adhesion and Cohesion

The strength of a particle system can be improved by several techniques. In the loose, dry state the weak van der Waals and electrostatic forces are the dominant contributors to the interparticle strength. Unfortunately, these weak forces are only effective with very small particle sizes. Alternatively, in the presence of a wetting liquid, the adhesive forces are much larger and are determined by the capillary attraction. In turn the capillary attraction depends on the contact angle, liquid quantity, and particle radius [1]. These capillary forces from a wetting liquid tend to be much larger than the weak forces. The adsorption of humidity from the atmosphere is a common source of surface liquid. Routinely, particles agglomerate on humid days with a decrease in the rate of particle flow. For large particles, in which these forces are too weak to be significant, only mechanical interlocking due to surface irregularities is significant [2]. For small particles, the interparticle cohesive forces are large and have a significant effect on the system characteristics including packing density, flow rate, and angle of repose.

A common measure of the interparticle bond strength for loose particles is the angle of repose [3]. This is the natural angle of inclination a powder forms with the horizontal when it is poured through a funnel, as shown in Figure 5.1. The angle of repose increases as the cohesion between particles increases. Thus, smaller particles and irregular particles have larger repose angles than large, spherical particles. Figure 14.1 shows the variation in the angle of repose for magnesia as a function of the particle size [4]. The angle of repose a decreases as the particle size increases. A general model for monodisperse particles is given as follows:

$$a = A + B/D \tag{14.1}$$

where D is the particle size, and A and B are constants. One problem with this equation is that the angle of repose does not approach zero as the particle size becomes very large. In reality, the angle of repose is typically below 30° for large spherical particles, and ranges up to 38° for free flowing powders. When the angle of repose exceeds approximately 45° the powder is termed cohesive, and it is not free flowing [5]. For a highly cohesive powder the angle of repose can be measured by mixing with it noncohesive spherical particles in various concentrations. Then the angle of repose is estimated by extrapolation to a zero added sphere concentration. More complex models for predicting the angle of repose, that incorporate

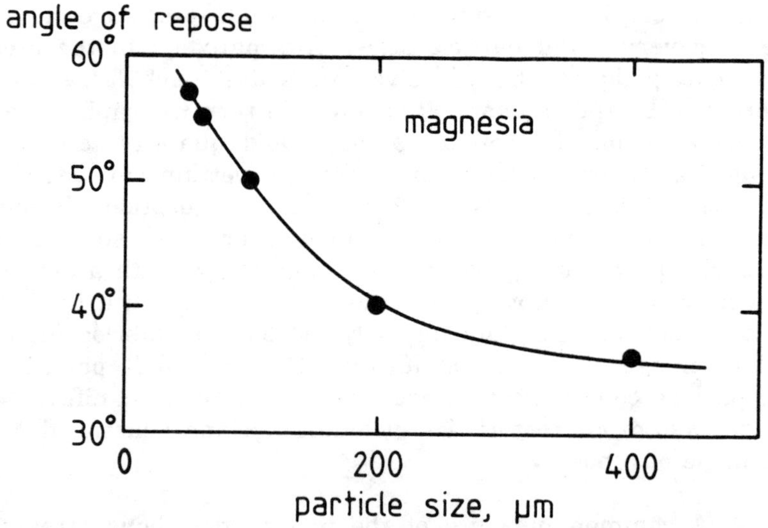

Figure 14.1 The angle of repose decreases with an increasing particle size, as illustrated with these data for magnesia particles [4].

particle size, cohesive forces, interparticle friction, and various material characteristics have been suggested [4].

Dry particles exhibit high angles of repose and substantial bonding by van der Waals forces when the particle size is below approximately 10 μm. The strength of the bond depends inversely on the interparticle separation δ, according to

$$\sigma = C/(\delta\, D^2) \tag{14.2}$$

where C is a system constant, σ is the interparticle strength, and D is the particle diameter.

In the presence of a wetting film, interparticle bonding is dependent on the surface energy of the liquid and the contact angle with the particle [6-9]. The surface films create pendular bonds which pull the particles together. The angle of repose can be quite high because of the large interparticle attractive forces from the wetting liquid. Here the strength varies with the fractional density f, particle size D, packing coordination number N_c, and cohesive force from the wetting film F_c, according to

$$\sigma = A\, f N_c\, F_c / D^2 \tag{14.3}$$

where A is a constant. In turn, the cohesive force depends on the particle size, surface energy, and amount of wetting fluid [6]

$$F_c = K\, D\ \gamma\, S^m \tag{14.4}$$

with K equal to a system dependent constant, γ equal to the liquid surface energy, and S equal to the saturation. The exponent m in Equation 14.4 depends on the particle size and particle shape; typical values for this exponent range from 0.25 to 1.00. If the void space is partially filled with liquid, then the strength is dependent on the curvature at the internal liquid-vapor interfaces. Figure 14.2 illustrates the variation in strength with density and saturation for glass beads containing two concentrations of water in the pores [6]. Although there is some scatter, the strength systematically increases with fractional density and water

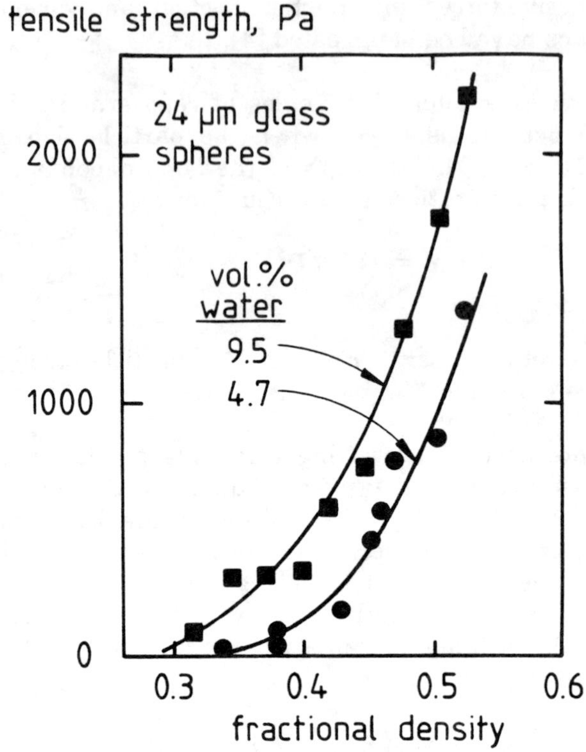

Figure 14.2 The variation in tensile strength with fractional density for spherical glass particles containing two different water contents [6].

saturation. Alternatively, if the pores are fully saturated with liquid, then the exterior liquid-vapor surface curvature determines the capillary force and agglomerate strength. Another estimate for powders containing pendular liquid bonds suggests that the strength varies with the inverse cube of the particle size, rather than the inverse square [7]. In either case, strength and agglomeration are increased by a wetting film on the particles. This increase in strength is important to particle agglomeration.

Beyond weak forces and capillary forces from a wetting liquid, there are other sources of interparticle strength. Strong bonds can be created by sintering, chemical reactions, crystallized salts, cementing agents, and magnetic forces. Each of these forces decreases in strength as the particle size increases [10].

Because of the natural variations in particle shapes and surface conditions, there is a distribution of interparticle adhesive forces within a mass of particles. Even in situations involving a wetting liquid, the homogeneity of the liquid has an effect on the overall strength. Inhomogeneities are the main influence on the strength of bulk powder. Thus, the measured strength will be below that calculated using Equations 14.1 or 14.3 because of the limitation from packing inhomogeneities.

Shear Induced Flow

The shear strength of loose particles is important to understanding the mixing, flow, packing, and discharge behavior of particles. Loose particles will flow under relatively low shear stresses by first dilating and then slipping over a shear plane [3,11-13]. The shear deformation concentrates on slip bands and is not uniform within a mass of particles. Figure 14.3 is a photograph of two different colored sands after initial mixing by rotating the transparent container. The shear bands are evident, showing that mixing is not homogeneous but rather involves sliding of particle groups over slip bands. At the beginning of slip the initial particle dilation is small and extends over a few particles. This is equivalent to forming a crack on a failure plane which propagates across the powder mass. The shear strength determines the conditions under which this initial crack forms and the stress needed for particles to slide past one another. Typically, shear occurs on the planes that have a locally high porosity. Thus, any conditions that produce local expansion in the powder packing lead to a weakened region and preferential dilation and shear.

Various tests have been proposed to measure the shear strength of loose particles [3]. The most common techniques involve the use of a core of particles held in a container or between fixed plates. Shear is induced in the powder, and the strength is measured based on the interfacial area and the load needed to cause failure. The presence of a small compressive force increases the shear strength, while a tensile force decreases the shear strength. Care is needed to account for the effects of the container wall and cohesive agents.

Figure 14.3 This photograph shows the banding due to localized slip on planes in a mixture of light and dark particles.

The strength of packed particles is dominated by the fractional density. Wetting fluids improve the interparticle cohesion and increase the resistance to shear. Similarly to Equation 14.3, the variation in tensile strength with fractional density and particle size can be approximated as [3,6,14-18]

$$\sigma_t = K f^m N_c F_c / D^2 \tag{14.5}$$

where σ_t is the tensile strength, K is a factor dependent on the particle shape, f is the fractional density, N_c is the packing coordination, m is an exponent typically near unity, F_c is the cohesive bond strength between particles, and D is the particle diameter. The packing density is important because it controls the number of contacts per unit volume. Beyond these factors, the strength is also sensitive to the particle shape and particle size distribution. The more irregular the particle shape, the higher the strength at a given fractional density. Figure 14.4 shows an example of the variation in strength with packing density for loose packed zinc powder [17]. In contrast with the behavior of moist glass spheres shown in Figure 14.2, the dry zinc powder exhibits a linear increase in strength as the packing density increases.

The presence of a compressive force on the powder increases the shear strength [19]. Various relations have been proposed between the shear strength τ and the normal stress σ, such as [14,17,20-22]

$$\tau^n = F_c (1 + \sigma/\sigma_t) \qquad (14.6)$$

where n is the shear index, F_c is the cohesive strength of the powder, and σ_t is the tensile strength. The cohesive strength is approximately twice the interparticle tensile strength. The shear

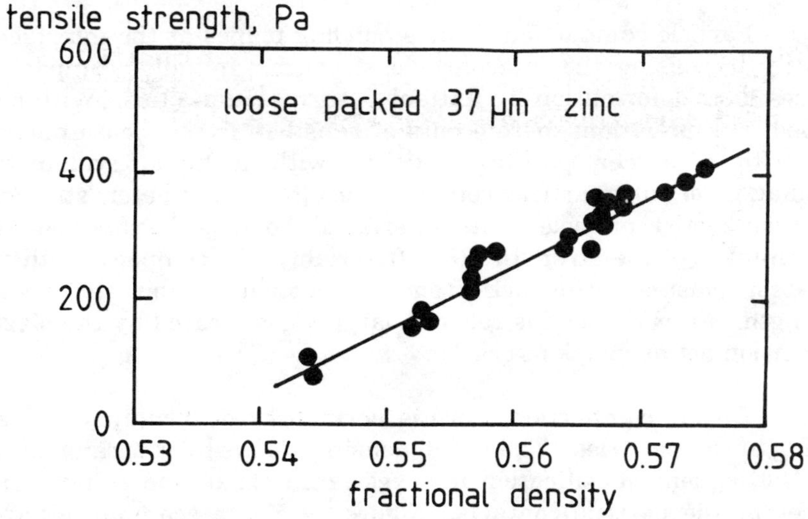

Figure 14.4 The tensile strength versus fractional density for loose packed zinc particles. The relation is nearly linear over the narrow density range [17].

index n ranges between one and two; for large spherical particles it is near unity, while for small particles it approaches two. It can be approximated as follows:

$$n = 1 + 0.5/D^{2/3} \tag{14.7}$$

where D is the particle diameter in micrometers. Thus, small particles with a high packing density and containing a cohesive fluid will have the highest resistance to failure by shear. The presence of a compressive load further increases the resistance to shear. Alternatively, should there be a tensile stress on the powder, then the resistance to shear is reduced. Equation 14.7 can be further modified to include the level of fluid saturation as well as particle shape and interparticle friction factors. Another factor not considered in this equation is the friction between the particles and the container.

Compacted Particles

Particle compaction creates bonding planes at the interfaces, greatly increasing the interparticle strength. Initially, compaction causes local deformation by flattening surface asperities. With continued pressurization, there occurs a transition from a loose particle mass to a higher packing density, with a higher volumetric population of interparticle contacts, resulting in greater strength. For compacted particles, the interfacial bonding becomes a key determinant of the strength [23]. Invariably, the compacts with the greatest elastic springback from compaction exhibit the lowest strength. After the load is relaxed, large voids created by the elastic expansion act as incipient cracks.

During compaction there is permanent deformation or fracturing of the particles. The initial densification is by rearrangement, but subsequent densification involves large shear and compressive forces at the particle contacts. Figure 14.5 is a scanning electron micrograph of the fracture surface from spherical copper particles after compaction showing the deformation at the contact points. Fracture in this case has occurred along the particle contacts since they remain the weak links in the compact. During compaction shear and deformation are localized at these particle contacts. Deformation increases the bonding area and the number of bonds,

Figure 14.5 A scanning electron micrograph of the fracture surface of compacted spherical copper particles. Deformation has caused the spheres to form polygonal shapes with little interlocking between particles.

and disrupts surface films that otherwise inhibit bonding. Thus, interparticle bonding improves as the compaction forces increase.

The strength of compacted powders depends largely on the size and quality of the contact zones (contamination) and the fractional density [24]. Because of surface contamination, shear stresses at the interparticle contacts are important to attaining high strengths. The initial deformation improves strength in proportion to the cube of the contact size. However, once a polygonal shape is established, little further strengthening occurs. Several investigators have examined the strength of compacted particles in hope of generating a universal relation [26-34]. Table 14.1 is a collection of some relations. Figure 14.6 gives an example of the strength and

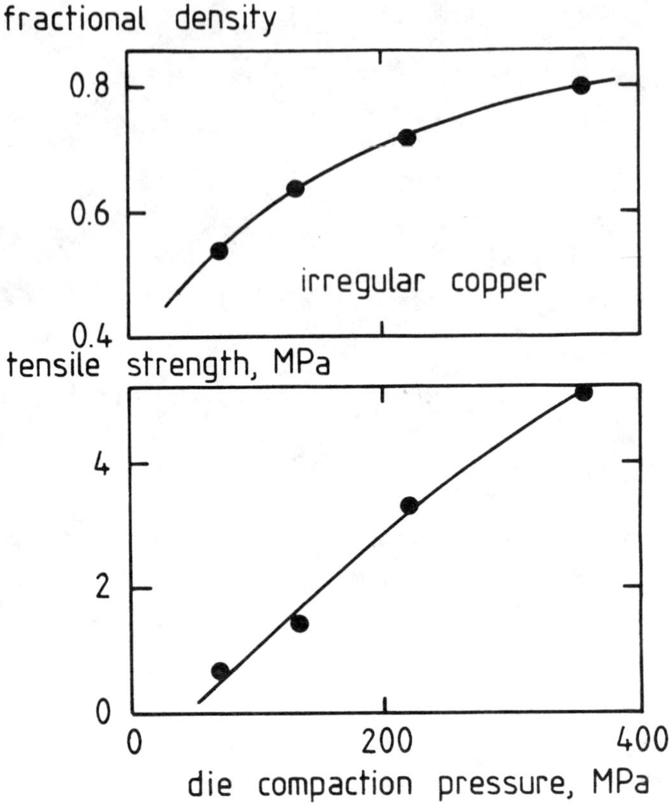

Figure 14.6 Fractional density and tensile strength versus the compaction pressure for irregular copper particles compacted in a uniaxial die [35].

fractional density variations with compaction pressure for an irregular, polydisperse copper powder [35]. The tensile strength and density both increase with compaction pressure as expected. One problem with comparing such results to the models is the large scatter in experimental measurements. A 20% variation between repeated measurements is common; thus, selection of a strength model is hindered by the low precision of the experimental results.

Table 14.1

Strength Relations for Compacted Particles

A, B, C, K, and m are constants that vary between the models
σ = the strength of the powder compact
f = the fractional density
f_o = the random loose fractional density

$\sigma = A\,(f - f_o)/(1 - f_o)$	[25]
$\sigma = (A\,f + B)/(C - K\,f)$	[26,29]
$\sigma = A\,[(f - f_o)/(1 - f_o)]^2$	[26,29]
$\sigma = A\,exp(K(f - 1))$	[27,28]
$\sigma = A\,(1 - f)^{2/3}$	[28]
$\sigma = A - B\,(1 - f)^{2/3}$	[32]
$\sigma = A + B\,f^{2/3}$	[28]
$\sigma = (A + B\,f)f^{2/3}$	[26,28,29]
$\sigma = A\,(1 - f)^a\,exp(K(f - 1))$	[28]
$\sigma = A - B\,exp(C\,f^m + K)$	[26,29,31]

The strength of the compacted particles is dictated by the strength of the interface between particles. Since deformation is more concentrated in small particle systems, there is a higher strength at a given density as the particle size decreases. However, there is greater difficulty in compacting smaller particles to a given density. Additionally, the strength at 100% density does not correspond to the handbook material strength because surface films between particles contaminate the bonding. As indicated in Figure 14.5, fracture occurs preferentially along the particle boundaries, termed the prior particle boundaries, especially those associated with high porosity levels [29,36]. Indeed, the strength is a measure of the success in bonding these surfaces.

For particulates, failure depends on the exact state of stress in the powder [37-39]. The interparticle strength is most dependent on the shear stresses during compaction. But failure occurs when the stress at the interparticle bond reaches a critical value that varies with the combination of normal and shear stresses. The pores between particles act to concentrate the stress; thus, the details of a strength model should include more than just the porosity. Most mechanical properties vary with the porosity, and similar models exist for ductility, elastic modulus, and hardness [29, 40-43]. However, the models for these other properties fail to include the details of pore shape, particle shape, particle size, and interparticle bond quality.

Role of Particle Characteristics

It is clear that the material and particle characteristics influence strength characteristics. Anyone who has attempted to compact a hard powder like alumina or boron knows the frustration associated with a low green strength. In contrast, a ductile metal like aluminum is much easier to form into a strong compact. To overcome the difficulties with low strength compacts of hard particles it is common to add cohesive fluids. As a good example of this, sand castles at the beach are much stronger when constructed from wet sand. Obviously, factors like hardness, plasticity, surface films and the basic mechanism of compaction influence the strength. Unfortunately, much of this area is poorly treated by current theory, although considerable empirical evidence providing a good qualitative guide is available.

The strength of a particulate mass is dictated by the number and size of the interparticle contacts. Accordingly, smaller particles or wide particle size distributions give a greater number of contacts and are associated with higher strengths at a given density [18,34, 44]. However, smaller particles also prove more difficult to compact. This is illustrated in Figures 14.7 and 14.8. The first figure compares the density versus compaction pressure for two particle sizes of aluminum [26]. The smaller particle size gives a lower fractional density at all compaction pressures, while in Figure 14.8 the strength versus density plot shows that the smaller particle size gives a higher strength at all densities. Various treatments have suggested a dependence of strength σ on the particle size D according to the relation

$$\sigma \sim D^{-N} \tag{14.8}$$

where N is a exponent estimated to range between 1 and 3 [10,15, 16,30]. Because of this effect, it is common to use wide particle size distributions to mix the benefits of large particles (good

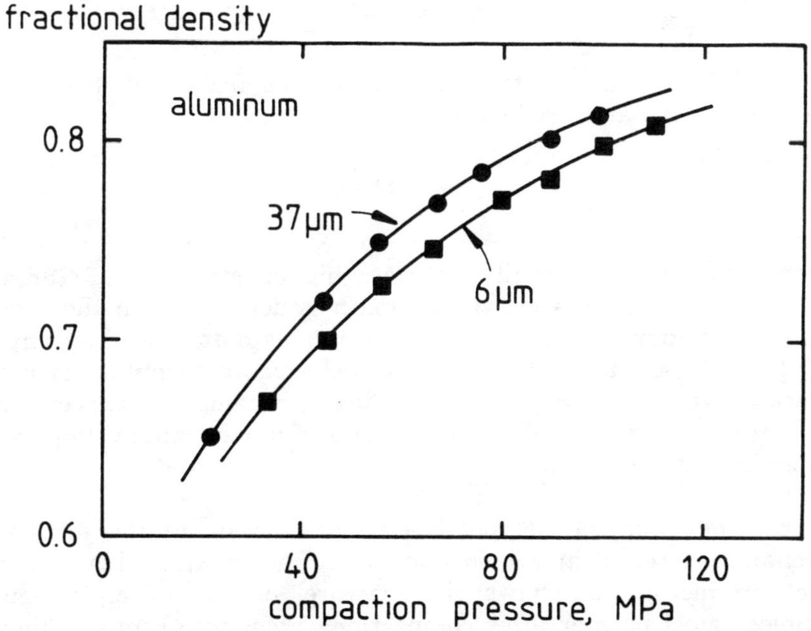

Figure 14.7 The variation of fractional density with compaction pressure for two particle sizes of aluminum. The smaller particle size is more resistant to compaction [26].

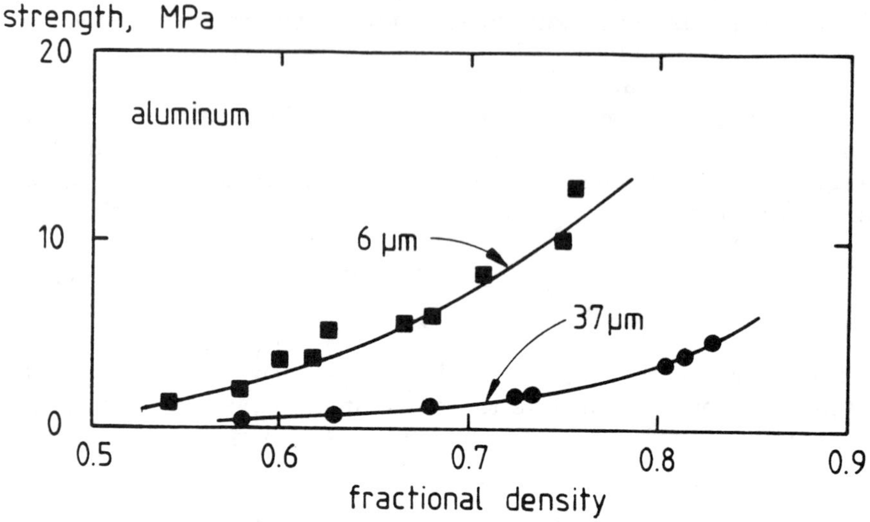

Figure 14.8 A plot of the strength versus fractional density for the aluminum particles in Figure 14.7 [26].

compactability) and small particles (higher strength) [34,38,45]. With mixed particle sizes, the strength is dependent on the size of the smallest particle that forms a continuous matrix in the compact [10,46]. Thus, the width of the particle size distribution is not a direct gauge of the strength. Accordingly, connectivity concepts are important to understanding the strength of particulate systems with wide size distributions.

The effect of the particle shape is seen in the degree of mechanical interlocking, as is evident in Figure 14.9. This scanning electron micrograph shows the fracture surface of an irregular stainless steel powder after compaction. Mechanical interlocking of the particles is evident, especially when this figure is contrasted with Figure 14.5. Additionally, irregular particles give more localized contact deformation and smaller pores. Both factors are significant with respect to compact strength [47]. As an example of the shape effect, Figure 14.10 compares the strengths of two copper powders with different sphericities [24]. Also indicated on this

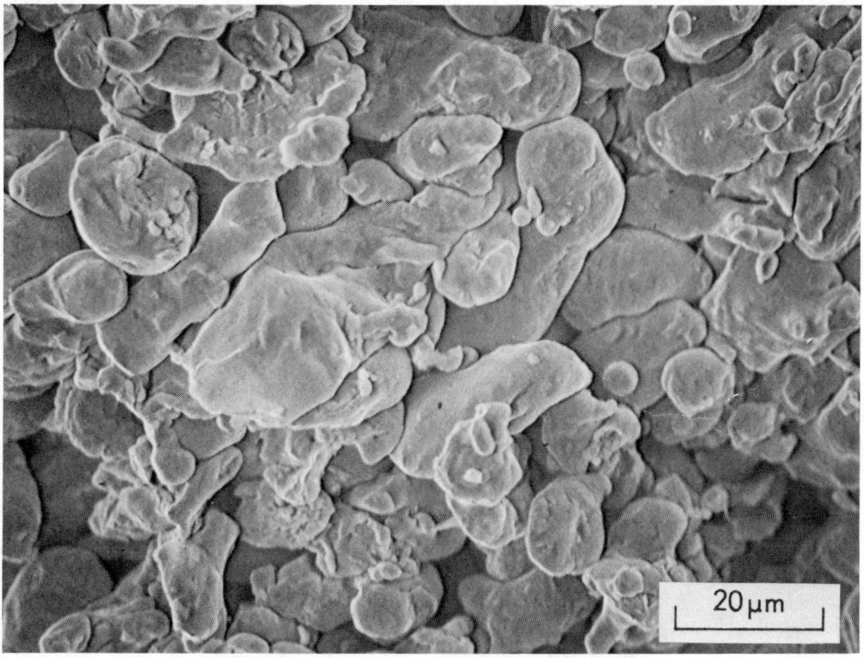

Figure 14.9 A scanning electron micrograph of the fracture surface of irregular stainless steel powder compacted to 280 MPa. Mechanical interlocking between particles is apparent in contrast with the spheres in Figure 14.5.

figure are the points corresponding to two compaction pressures. The spherical particles give the lower strength, but the packing density is higher. Thus, irregular particle shapes are associated with higher strengths at a given density [42,44]. For rough particles, the size of the surface asperities is another factor needed to accurately predict strength.

Summary

This chapter has examined the strength characteristics of particle assemblages, both of loose particles and compacted particles. In the loose condition, the effect of surface films is very significant.

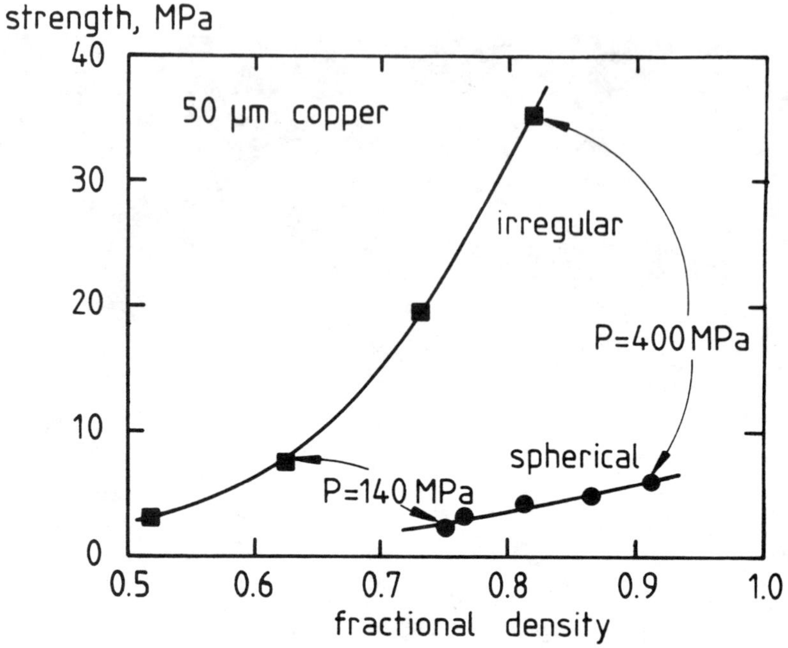

Figure 14.10 Strength versus fractional density for irregular and spherical copper particles. The irregular particles exhibit a higher strength [24].

The simple case is a dry, loose powder. In this condition, the interparticle bond strength is determined by the weak forces such as the van der Waals force, which is only significant for very small particles. In most realistic situations, adsorbed surface films on the particles contribute capillary bonding forces. The capillary force associated with a wetting liquid is a significant contributor to interparticle strength in many situations.

When particles are compacted, the number of contacts and the interfacial bond area increase, thereby improving strength [48]. Furthermore, the shear deformation at the interface disrupts surface films, allowing strong interparticle bonding. Thus, compacted particles are much stronger than loose particles. Various models linking strength to density have been proposed. Unfortunately, most of these models are incomplete since pore characteristics have a large effect on the strength of particle compacts. As demonstrated in Chapter Eleven, pore size and spacing vary with the particle size distribution and particle shape. Neither of these factors are adequately included in the current models of compact strength. Thus, the models for strength and other mechanical properties need to be extended to include particle size distribution and particle shape factors. Until these modifications are attained the present models based on fractional density are only nominally useful, because of the large number of unknown or adjustable parameters. It is established that a high compact strength requires a high density, small particle size, and irregular particle shape.

References

1. K. S. Hwang, R. M. German and F. V. Lenel, "Capillary Forces Between Spheres During Agglomeration and Liquid Phase Sintering," *Metallurgical Transactions A*, 1987, vol. 18A, pp. 11-17.
2. H. Krupp, "Particle Adhesion Theory and Experiment," *Advanced Colloid Interface Science*, 1967, vol. 1, pp. 111-239.
3. R. L. Brown and J. C. Richards, *Principles of Powder Mechanics*, Pergamon Press, Oxford, UK, 1970.
4. J. T. Carstensen and P. C. Chan, "Relation Between Particle Size and Repose Angle of Powders," *Powder Technology*, 1976, vol. 15, pp. 129-131.
5. G. S. Riley, S. Mann and R. O. Jesse, "Angles of Repose of Cohesive Powders," *Journal of Powder and Bulk Solids Technology*, 1978, vol. 2, pp. 15-18.
6. K. Shinohara and T. Tanaka, "Ultimate Tensile Strength of Cohesive Powders Based on Non-Uniform Packing Model," *Journal of Chemical Engineering of Japan*, 1975, vol. 8, pp. 50-56.
7. J. F. Carr, "Tensile Strength of Granular Materials," *Nature*, 1967, vol. 213, pp. 1158-1159.
8. A. G. Douglas, "Tensile Strength of Granular Materials," *Nature*, 1967, vol. 215, pp. 952-953.
9. D. M. Newitt and J. M. Conway-Jones, "A Contribution to the Theory and Practice of Granulation," *Transactions of the Institution of Chemical Engineers*, 1958, vol. 36, pp. 422-442.
10. H. Rumpf, "The Strength of Granules and Agglomerates," *Agglomeration*, W. A. Knepper (ed.), Interscience, New York,

NY, 1962, pp. 379-414.
11. K. Umeya, R. Hara and J. Kikuta, "On Two-Dimensional Shear Tests by Model Powders," *Journal of Chemical Engineering of Japan*, 1975, vol. 8, pp. 56-62.
12. M. J. Adams, "The Strength of Particulate Solids," *Journal of Powder and Bulk Solids Technology*, 1985, vol. 9, pp. 15-20.
13. G. Y. Onoda and M. A. Janney, "Application of Soil Mechanics Concepts to Ceramic Particulate Processing," *Advances in Powder Technology*, G. Y. Chin (ed.), American Society for Metals, Metals Park, OH, 1982, pp. 53-74.
14. M. D. Ashton, D. C. H. Cheng, R. Farley and F. H. H. Valentin, "Some Investigations into the Strength and Flow Properties of Powders," *Rheological Acta*, 1965, vol. 4, pp. 206-218.
15. I. Nikolakakis and N. Pilpel, "Effects of Particle Size and Particle Shape on the Tensile Strengths of Powders," *Powder Technology*, 1985, vol. 45, pp. 79-82.
16. V. Smalley and I. J. Smalley, "Tensile Strength of Granular Materials," *Nature*, 1964, vol. 202, pp. 168-169,
17. R. Farley and F. H. H. Valentin, "Effect of Particle Size upon the Strength of Powders," *Powder Technology*, 1967, vol. 1, pp. 344-354.
18. D. C. H. Cheng, "The Tensile Strength of Powders," *Chemical Engineering Science*, 1968, vol. 23, pp. 1405-1420.
19. N. G. Stanley-Wood and A. M. Abel Karim, "Relationship Between Compact Strength and Functions of the Shear-Normal Stress Curve Obtained from Uniaxial Compression," *Powder Metallurgy International*, 1982, vol. 14, pp. 135-138.
20. R. Kusy and D. T. Turner, "Effects of Nickel on Mechanical Properties of PVC," *Society of Plastics Engineers Journal*, 1973, vol. 29, pp. 56-61.
21. I. G. Shatalova, N. S. Gorbunov and V. I. Likhtman, "Physiochemical Principles of Vibratory Compacting," *Vibratory Compacting Principles and Methods*, H. H. Hausner, K. H. Roll and P. K. Johnson (eds.), Plenum Press, New York, NY, 1967, pp. 1-206.
22. H. Mirota and T. Oshima, "Shear Properties of Uniaxially Preconsolidated Powder Bed - Nonuniformity of the Shear Stress Acting on the Shear Plane," *Powder Technology*, 1987, vol. 53, pp. 49-53.
23. I. M. Fedorchenko, A. E. Kushchevskii, T. F. Mozol and V. F. Chudovskii, "Densification Behavior of Metal Powders During Pressing," *Soviet Powder Metallurgy and Metal Ceramics*, 1987,

vol. 26, pp. 196-199.
24. E. Klar and W. M. Shafer, "On the Nature of Green Strength," *International Journal of Powder Metallurgy*, 1969, vol. 5, no. 2, pp. 5-12.
25. G. W. Halldin, K. T. Yung and T. H. Tsai, "A New Analysis of Compressibility and Green Strength of Metal Powders," *Progress in Powder Metallurgy*, 1981, vol. 37, pp. 383-391.
26. R. M. German, "Strength Dependence on Porosity for P/M Compacts," *International Journal of Powder Metallurgy and Powder Technology*, 1977, vol. 13, pp. 259-271.
27. G. F. Bocchini, "Influences of Porosity on the Characteristics of Sintered Materials," *Metal Powder Report*, 1986, vol. 41, pp. 829-832.
28. A. Salak, V. Miskovic, E. Dudrova and E. Rudnayova, "The Dependence of Mechanical Properties of Sintered Iron Compacts upon Porosity," *Powder Metallurgy International*, 1974, vol. 6, pp. 128-132.
29. R. Haynes, *The Mechanical Behaviour of Sintered Metals*, Freund Publishing House, London, UK, 1981.
30. T. Sheppard and H. B. McShane, "Strength of Cold Pressed Compacts," *Powder Metallurgy*, 1980, vol. 23, pp. 120-125.
31. R. M. German, "Particle Size Influence on the Strength Relation for Air Sintered Aluminum," *Metallurgical Transactions A*, 1975, vol. 6A, pp. 1964-1965.
32. M. Hamiuddin, "Prediction of Mechanical Properties of Porous Materials and Cermets - A Critical Analysis," *Powder Metallurgy International*, 1987, vol. 19, no. 6, pp. 25-27.
33. R. A. Thompson, "Mechanics of Powder Pressing: III, Model for the Green Strength of Pressed Powders," *Ceramic Bulletin*, 1981, vol. 60, pp. 248-251.
34. C. Ludwig, "Principles of Agglomerating Processes," *Transactions of the American Foundrymen's Society*, 1955, vol. 63, pp. 466-469.
35. E. Klar and W. M. Shafer, "On Green Strength and Compressibility in Metal Powder Compaction," *Modern Developments in Powder Metallurgy*, vol. 9, H. H. Hausner and P. W. Taubenblat (eds.), Metal Powder Industries Federation, Princeton, NJ, 1977, pp. 91-113.
36. A. E. Butcher and J. M. Newton, "Tensile Failure Planes of Powder Compacts," *Powder Technology*, 1974, vol. 9, pp. 57-59.
37. T. H. Chilton and A. P. Colburn, "Heat Transfer and Pressure

Drop in Empty, Baffled and Packed Tubes II - Pressure Drop in Packed Tubes," *Industrial and Engineering Chemistry*, 1931, vol. 23, pp. 913-919.

38. A. Stewart, "The Pelleting of Granular Materials," *Engineering*, 1950, vol. 169, pp. 175-177 and 203-204.
39. E. G. Schwartz and A. S. Weinstein, "Model for Compaction of Ceramic Powders," *Journal of the American Ceramic Society*, 1965, vol. 48, pp. 346-350.
40. R. Haynes, "A Study of the Effect of Porosity Content on the Ductility of Sintered Metals," *Powder Metallurgy*, 1977, vol. 20, pp. 17-20.
41. R. W. Rice, "Extension of the Exponential Porosity Dependence of Strength and Elastic Moduli," *Journal of the American Ceramic Society*, 1976, vol. 59, pp. 536-537.
42. R. M. German, *Powder Metallurgy Science*, Metal Powder Industries Federation, Princeton, NJ, 1984, pp. 126-132.
43. A. Buch and S. Goldschmidt, "Influence of Porosity on Elastic Moduli of Sintered Materials," *Materials Science and Engineering*, 1969, vol. 5, pp. 111-118.
44. B. B. Morgan, "Some Measurements of the Tensile Strength of Dry Powders of Irregular Particle Shape," *Nature*, 1965, vol. 205, pp. 36-37.
45. T. W. Seaton, "Density - Sand Grain Distribution Effect on Physical Properties," *Transactions of the American Foundrymen's Society*, 1960, vol. 68, pp. 45-48.
46. A. Bonissent and B. Mutaftschiev, "A Computer Built Random Model for Simulation of the Crystal-Melt Interface," *Philosophical Magazine*, 1977, vol. 35, pp. 65-73.
47. E. Klar and W. M. Shafer, "On the Nature of Green Strength II. Correlations Between Pore Structure of Powder Particles and Their Electrical Conductivity and Green Strength," *International Journal of Powder Metallurgy*, 1969, vol. 5, no. 4, pp. 5-16.
48. J. A. Dodds, "The Porosity and Contact Points in Multicomponent Random Packings Calculated by a Simple Statistical Geometric Model," *Journal of Colloid and Interface Science*, 1980, vol. 77, pp. 317-327.

Chapter Fifteen

Summary of Particle Packing Concepts

Chapter Outline

Review of Important Concepts

The study of particle packing involves several simplifying assumptions concerning the nature of the particles. Unfortunately, the characteristics of most practical powders are highly variable as illustrated by the scanning electron micrographs in Figure 15.1. Clear differences are evident in the particle size, particle shape, surface texture, and agglomeration. Thus, in dealing with particulate systems it is necessary to define these attributes. The usual assumption is that the particles are dense and discrete, and that they have a well defined size, often given in terms of an equivalent spherical diameter. A full description of a particle involves information about its size, shape, hardness, surface chemistry, surface texture, agglomeration, friction, and internal structure [1,2]. In

Figure 15.1 Scanning electron micrographs of three powders that illustrate the diversity of characteristics in real powder systems. Pictured are a) iron-titanium intermetallic, b) tungsten alloy, and c) boron.

considering a group of assembled particles, several questions relating to such characteristics as the packing density, coordination number, pore size, connectivity, permeability, capillary pressure, surface area, strength, and compactability arise. For naturally occurring particle systems, these characteristics are variable according to the details of the packing structure. Furthermore, the packing can be influenced by secondary factors such as the container shape and size, or the procedure used in assembling the packing.

Although spheres do not fill space efficiently, they do constitute the basis for most packing studies [1,3]. An advantage of the sphere is that it can be specified by a single dimension, the diameter. Also, the sphere has a relatively low interparticle friction that aids packing to a high density in spite of the poor inherent space filling character. There are regular polygonal shapes that can give higher packing densities when placed in an ordered array.

Clear differences exist in the structures of ordered and random packings. Ordered structures of monosized spheres, such as shown in Figure 15.2, are easily visualized and treated mathematically; thus, they are widely used in studies of particle packing characteristics. Additionally, ordered packings are encountered in areas such as crystallography. The close-packed ordered packing of spheres shown in Figure 15.2, with a coordination number of twelve, gives a higher packing density than is possible with a random packing of monosized spheres.

There is an important similarity between ordered close-packed structures of different dimensional levels. In one dimension, the highest packing density corresponds to segments (spheres) in point-to-point contact along a straight line, giving a density of 1.000 with a coordination number of two. In two dimensions, the highest density corresponds to three of these one-dimensional structures aligned in a plane, corresponding to a close-packed plane with six contacts per disk. The fractional packing density is 0.9069. For the three-dimensional case, the maximum density occurs with the two-dimensional planes stacked one upon another, giving a coordination number of 12 and a fractional density is 0.7405. This progression is not paralleled by a similar systematic behavior for the fractional density in random packing cases, which is 0.75 for

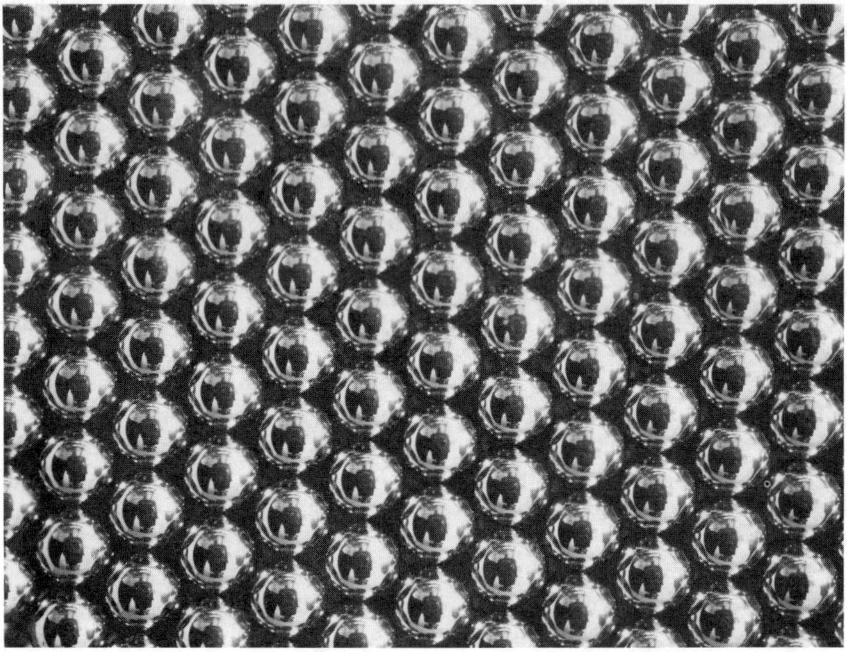

Figure 15.2 The classic model for packed particles is a close-packed array of monosized steel spheres. Contrast this with the particles shown in Figure 15.1.

one-dimensional, 0.82 for two-dimensional, and 0.64 for three-dimensional packings. Note the latter two packings include a gravitational contribution to the stabilization of the structure.

Predicting the density for a random packing is difficult because the construction procedure influences the density. The packing density increases as the energy of deposition and the particle hardness increase. Likewise, vibration after assembling particles will increase the density up to the random dense packing limit [4]. In such structures, there is a distribution to the coordination number, which provides one of the most sensitive measures of randomness.

In model packings of large spheres, several principles relating to particle shape, surface texture, surface films, agglomeration, and vibration have been isolated with respect to the packing density. However, concerns with adhesion and capillary forces arise as the particle size decreases [5]. The surface area to volume ratio increases in inverse proportion to the particle size. Thus, surface frictional forces from oxides, adsorbed films, and atomic scale asperities can influence the density, bridging, strength, pore size, pore shape, coordination number, and other packing attributes. This is evident from the low packing densities of very small particles. With some submicrometer sized particles fractional densities as low as 0.05 are observed, especially with chain-like clusters of particles. Obviously, the attributes of a packing of large monosized spheres (with a fractional density of 0.64) will not correlate well with the packing attributes of these small particles.

The packing behavior of irregular particles is important because of the fragmented character of many particulate systems, such as coal, sand, gravel, and minerals. In spite of the importance of these materials, we have a relatively poor quantitative understanding of the effects of particle shape on packing. There are some similarities in the packing behavior of spheres and nonspherical particles, but irregular particles usually have an inherently lower packing density. Irregular particles increase in density with vibration more than do spherical ones, but the starting density for irregular particles is lower. An important application of mixtures of spheres and fibers is in fabrication of composites, where the fibers are used to strengthen the consolidated structure. Long fibers with large aspect ratios are desired for the best properties in the final composite. Such composite materials typically attain a high packing density with relatively small particles that fill the voids between the fibers.

A large difference in particle sizes can improve the packing density, both for the ordered and random situations and for spherical as well as nonspherical particles. The small objects fit into the interstitial voids between the large objects. Although the relative gain in density is similar for spherical and nonspherical particles, the starting density for the latter is lower. Hence, the final density cannot equal that for mixed spheres. Interestingly, the mean coordination number shows little variance with composition in binary mixtures. The optimal packing occurs at an intermediate composition where the small objects fill all of the voids between the large objects without dilating the structure. An excess of either the large or the small particles will decrease the packing density. Optimization occurs with a majority of small objects when measured on a population basis, but with a majority of large objects by volume (approximately 70%). As the number of component powders increases, the relative increase in density is continually smaller. Hence, it is rare for a packing to have more than three mode sizes.

From such concepts, it is possible to create continuous size distributions that give high packing efficiencies. The continuous distribution for a high packing density is skewed to give an abundance of large particles (by weight) with a long tail to the distribution. Figure 15.3 plots the Andreasen particle size distribution with a high packing density. Shown are both the differential and cumulative particle size distribution curves versus the relative size on a linear scale. The broad character of this distribution and the high content of small particles contribute to the optimized packing density.

When mixed particles are dissimilar in properties the connectivity becomes important [6,7]. Investigation of connectivity problems in packed particle systems is dominated by studies of the electrical conductivity in metal-filled polymers. When the particle shape is that of a fiber or flake, low concentrations of metallic particles are sufficient to form a conducting composite. For the general case of mixed spheres of equal size, the minor phase will be connected when it constitutes roughly 30% of the particles, depending on the packing structure. The critical point at which the minor phase becomes connected throughout a packed structure is highly dependent on the randomness or segregation of the structure. A segregated structure occurs as the particle size of the minor phase becomes smaller than the particle size of the major phase. In this case, the critical volume fraction decreases. In practical experiences with mixed spheres of unequal size, volume fractions as low as 0.05 have provided a conductive composite of conductive particles mixed with nonconductive particles. This allows one to use low concentrations of expensive materials in fabricating electromagnetic shields and conducting pastes. The lower critical volume fractions are possible using nonspherical filler particle shapes. Connectivity depends on attaining a critical mean number of contacts per

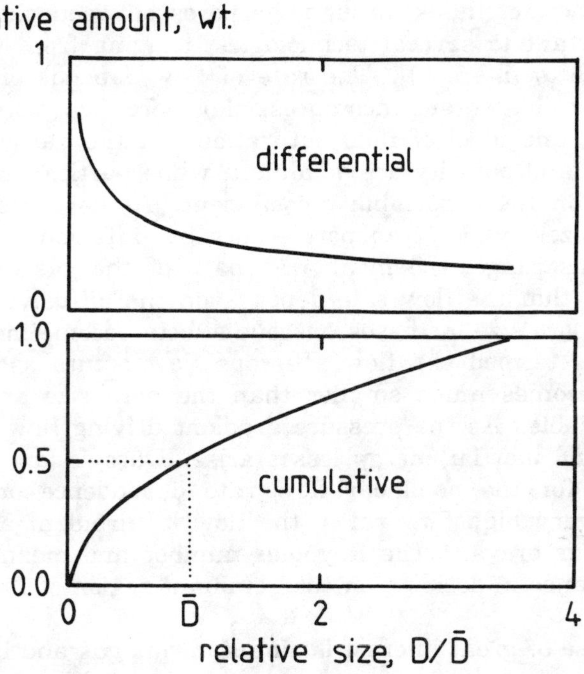

Figure 15.3 A high packing density continuous particle size distribution has a nonsymmetric character that provides a high proportion of small particles and a wide size range.

particle with similar composition particles.

Various relations linking the pore size to the particle size and packing density have been proposed. These models are most applicable to spherical particles, and are inadequate for random packings, especially of irregular particles. Thus, efforts to relate the particle packing features to the particle size are generally prone to error. Combinations of techniques can be potent in characterizing packing structures: for example, mercury porosimetry and gas adsorption measurements can be cross-calibrated to pore volume, surface area, and pore size [8]. These provide an excellent basis for assessing the packing characteristics of very small particles.

The flow of fluids through the pores between packed particles is important to several technologies, ranging from the design of heat pipes to filters [9]. The rate of flow depends on the fluid type (compressible versus incompressible), pore size, porosity, surface texture, and level of fluid saturation. Of the various factors, the pore size and porosity are dominant, while particle shape has a small effect on the permeability coefficient. At low pressures and small pore sizes, vapor transport occurs by diffusion through the pores. In this range, the mean free path of the gas exceeds the pore size, so that the flow rate depends on the diffusivity. As the pressure or pore size increases, viscous effects become more significant. This is termed slip flow. Viscous flow occurs as the mean free path becomes much smaller than the pore size and Darcy's law is applicable. As the pressure gradient driving flow increases, turbulence and inertial energy losses arise. Darcy's law must then be modified for the nonlinear flow rate dependence on velocity. Finally, at very high flow rates, the flow is turbulent and choked flow conditions prevail. The Reynolds number and mean free path provide guidance in deciding on the dominant region of flow.

A case of great interest is simultaneous gas and liquid flow. The relative permeability of each phase is reduced by the presence of the other phase. Because of the disruption in the connectivity of each phase, the sum of the relative permeabilities is less than unity for all intermediate liquid saturations. More extensive treatment of this and several other complex problems in fluid flow through porous materials is given in the classic books [10,11].

A wetting fluid provides an attractive force on particles which explains the various aspects of liquid interactions with packed particles [12]. First, at low saturations, the wetting liquid will preferentially flow into the smallest pores (those with the largest surface area-to-volume ratio). This is part of general capillary behavior. A pendular liquid bond between particles exerts a force which pulls the particles into contact. This compressive force explains agglomeration of powders in the presence of humidity. The vapor pressure of a fluid in a porous structure depends on the curvature of the liquid surface, which in turn depends on the particle size, contact angle, and saturation.

It is characteristic of the liquid-pore interaction that a wetting liquid flows to reduce the energy of the system. Initial flow is by condensation or wicking of liquid into the pores. During filling there is preferential flow into the smaller pores and the formation of pendular bonds. During withdrawal of the liquid, these same small pores and pendular bonds are most resistant to drainage and evaporation. Efforts to enhance drainage by use of a displacing fluid are often frustrated by fingering. Once a liquid penetrates into a packing, evaporation is the only way to attain complete removal, since only partial removal is obtainable by wicking or displacement. In all aspects of liquid-pore interactions there is a hysteresis between flow into and out of the pores.

Pressure is routinely used to form loose particles into controlled bulk shapes with prescribed dimensions and increased packing density [8,13-15]. One major difference between two-dimensional and three-dimensional packings is the effect of contraction. In a two-dimensional packing of disks, the structure can be ordered by a contraction force, while in three dimensions, spheres resist ordering from an external force. Indeed, extremely high pressures are needed to densify spheres beyond the random packing density, because a shape change is produced by the pressurization. Compaction of particles creates both rearrangement and deformation of the particles, with fracture occurring in some cases. The generalized sequence of stages involves particle rearrangement followed by localized deformation. As the compaction pressure is increased, the zone of deformation expands to eventually encompass the entire particle. Because of shear during compaction strong interparticle bonds are formed. The mechanism of densification depends on the specific powder and material characteristics, including the material hardness, yield strength, ductility, fracture toughness, strain hardening, particle size, and particle shape.

Compaction is most typically performed using rigid tooling and uniaxial pressurization. The particles do not uniformly respond to such compaction, and the particles and pores both become squat as the pressure increases. Furthermore, die wall friction induces density gradients in uniaxial die compaction. Lubricants reduce die wall friction, thereby minimizing density gradients and tool wear. As the compaction pressure is increased the optimal amount of lubricant decreases.

The predominant goal in particle compaction is to achieve an increased density by the application of an external pressure. The compact height-to-diameter ratio is important to attaining a uniform compact. Generally, compacts with large height-to-diameter ratios cannot be uniaxially pressed in rigid tooling. Alternative hydrostatic compaction technologies provide more homogeneous pressurization.

In the dry condition, the bond strength between particles is determined by weak forces such as the van der Waals force, which is only significant (and measurable) for very small particles. In most realistic situations, adsorbed surface films on the particles contribute capillary bonding forces. The capillary force associated with a wetting liquid is a significant contributor to interparticle strength in many situations.

When particles are compacted, the number of contacts and the interfacial bond area increase. Both factors contribute to improved strength and to changes in several other attributes, including elastic modulus, toughness, electrical conductivity, and the magnetic response. Shear deformation at the interface disrupts the surface films common to particles, allowing strong interparticle bonding. Thus, the strength of compacted particles is much larger than loose particles, especially if they can undergo plastic deformation during compaction. Various models have been proposed linking the strength to the fractional density [16], but most are incomplete. Pore characteristics have large effects on strength. Pore size and pore spacing vary with the particle size distribution and the particle shape, but neither of these factors are adequately included in the current models for compact strength. It is established that a high compact strength requires a high density, small particle size, and irregular particle shape.

Interest in particle packing characteristics extends over a wide range of technologies. Examples that have been cited in this book touch on such diverse materials as asphalt, sands, metal powders, gravel, catalysts, amorphous solids, nuclear fuels, ceramics, composites, pharmaceuticals, and metal-filled plastics. Because of this range of interests, there has been considerable investigation into the characteristics of packed particles. Accordingly, much is known about this field. However, because this knowledge has not been organized, there has been considerable rediscovery of basic principles in these widely separated technologies. It is anticipated that through this treatment this element of rediscovery can be avoided so that future efforts will make meaningful contributions and advances.

Comments on Computer Simulations

Clearly, the experimental determination of the structure of a particle packing is a tedious task. A desirable alternative has been to assemble spheres in a computer simulation [17-61]. This is particularly attractive in modeling the structure and properties of amorphous liquids or solids. Unfortunately, such an approach often replaces tedium with inaccuracy.

Initially, simple techniques of model construction were employed using minimum energy schemes, such as adding each new sphere to an existing cluster at the site closest to the cluster center. These simple sequential building models give a density significantly below the experimental value. Furthermore, these central growth models give densities that decrease as the computational cell increases in size. Some of the early attempts gave limiting densities below 0.40, far from the experimental random loose packing fractional density of 0.60. Since these approaches failed to give realistic densities, novel assembly techniques have evolved. Eventually, the use of randomly placed sphere centers and subsequent dilation of centers to eliminate sphere overlap produced a realistic simulation of packing. Accordingly, the computer simulated three-dimensional hard sphere packings have converged to the experimental packing density result [19,22,39]. However, in spite of the density match, the radial distribution function, showing the location of sphere centers, and the coordination number distribution do not satisfactorily match with experiment. This indicates lack of homogeneity in the structure. The errors in the long-range positioning of spheres with respect to one another create doubts as to the validity of many of the computer simulations.

With the more recent advances in computer modeling, some important questions have been addressed. For example, the statistics associated with particle mixtures have been attacked using computer models. This has included calculation of parameters such as density, radial distribution function, coordination number distribution, connectivity, pore size, and Voroni polyhedra for random dense packings. Additionally, the simulated packings can be compressed in the computer to observe changes in the coordination number and local order with deformation [26]. It is reassuring that these experiments converge to a packing coordination near 14, in agreement with experiments on lead spheres. However, generation of a realistic random packing in the computer is not a simple task and is still more difficult than the production of an error-free random experimental packing.

Practical Approaches to High Packing Densities

Vibration in the presence of a wetting liquid is one approach to attaining a high packing density. The liquid aids packing and increases the packing homogeneity. Packing density is also higher in bimodal particle mixtures. There is less benefit from mixing beyond the bimodal system to a trimodal or more complex system. Figure 15.4 shows the fractional packing density expected from binary, ternary, and quaternary particle mixtures versus the size ratio for the particles. For these plots the inherent packing density of each individual component is assumed to be 0.6. This figure shows that there are gains in density from multiple mode systems composed of particles with large size differences, but that these gains must be balanced against the practical difficulties in obtaining and handling such different particle sizes. If the particle size ratio is not too large, then the bimodal packing is best, in terms of balancing the density gain against the number of components. Only as the particle size ratio becomes larger than approximately 10 is a trimodal packing attractive. This requires that the largest and smallest particles differ by a factor of 100 in size.

Furnas [62] considered the problem of the number of component powders needed for optimized packing. Figure 15.5, which is taken from his work, shows the optimal number of component powders versus the particle size ratio for four inherent packing densities. For this plot, the ordinate is based on the ratio of the largest to smallest particles, rather than the interval ratio as used in Figure 15.4. This plot indicates that large size differences are necessary for a benefit from even a trimodal packing.

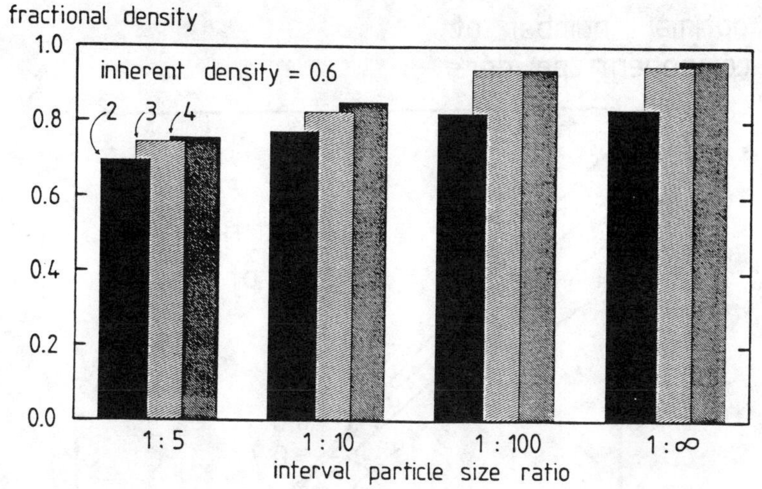

Figure 15.4 The effect of particle size and number of modes on the packing density of spheres; the number of component powders are designated by 2, 3, and 4.

A wide particle size distribution is helpful in attaining a high density. This is a practical option because it minimizes some of the handling difficulties. Figure 15.3 illustrates the character of an optimized particle size distribution for spherical particles. Optimal packing occurs when the largest particles form a skeleton network with the smaller particles filling the voids between the large particles. Accordingly, on a weight or volume basis the distribution is skewed toward the large particles, with an extended tail toward the smallest particles. Using this distribution, the optimal packing density for spheres is typically in the range of 0.82 to 0.87, depending on the particle characteristics. For comparison, a trimodal packing with a particle size ratio greater than 7 will give a comparable packing density.

Generally, to obtain a high packing density the first approach would be to select a base particle size. Subsequently, a smaller particle size is selected and a composition of large and small particles is mixed. The optimal composition depends on the

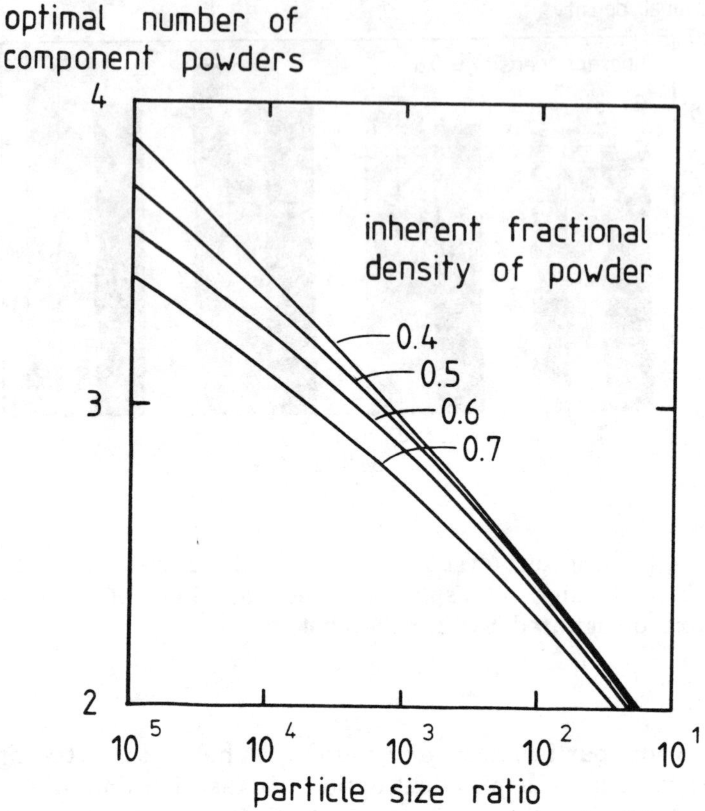

Figure 15.5 The optimal number of components in a multiple modal packing as a function of the particle size ratio (largest to smallest) for four inherent packing densities [62].

particle size ratio and inherent particle packing density, with the main goal of filling the interparticle voids between the large particles with the small particles [63]. This procedure can be repeated with a third size class, but the gain in packing density decreases as the number of classes increases. The final density is highest if the particles are spherical. For small particles, surface treatments with polarizing molecules will further aid packing by reducing the interparticle friction.

Opportunities for the Future

Several questions remain unanswered in spite of many prior investigations of particle packing. It took over 100 years to mathematically prove that a packing coordination of 13 was not possible for monosized spheres. Only recently has it been mathematically established that a density of 0.7405 is the maximum possible in an ordered packing of spheres. It is unclear why the fractional density for a vibrated random monosize sphere packing is 0.637.

It is clear that computer techniques are largely inaccurate when applied to simulation of random particle packings. Recent research has established that a combination of dilation and relaxation approaches can give realistic random packing structures. However, the structure may not be representative of liquids and glasses, contrary to the speculation associated with many computer simulations. The several results based on computer generated packings, especially of multimodal systems, must be treated with caution.

These comments set the stage for some suggestions of opportunities for the future. Clearly, insufficient research has been performed on elementary packing structures. For example, in one dimension what is the size distribution of segments that gives the highest parking fraction? The assembly procedure for placing the segments on a line influences the parking density. There is an interaction between the procedure for filling a line and the optimal segment size distribution. If the segments are added largest first, then the optimal packing may be different from that for randomly assembled segments. Likewise, for two-dimensional packings there are unanswered questions relating to the size, shape, and distribution of objects that lead to optimized coverage. Undoubtedly, similar concepts are applicable to three-dimensional random mixtures.

In many aspects of particle technology there is a need for better definitions of size and shape, especially in quantitative terms. This is apparent in discussions of both the particles and the pores between particles. An effort to link the particle and pore characteristics together via predictive models is needed. One difficulty is that pore size is often measured by mercury porosimetry, most typically under assumptions that result in highly distorted pore size information. The inhomogeneities in packed particles greatly increase the error in mercury porosimetry (this is evident in the hysteresis). Needed are better models for pore shape which provide accurate information on the packing structure. A statistical model is needed that includes pore size, packing coordination number, packing homogeneity, and possibly the radial distribution function for the particles.

There is considerable merit to multimodal powder systems and various continuous particle size distributions for attaining high packing densities. Little attention has been directed to linking the packing density to the other important attributes. Both mathematical analysis and characterization experiments are needed to identify the optimal particle size distribution for high packing density, and to predict the details of the packing such as pore size and coordination number. Furthermore, these concepts must be extended to particle shapes that are not spherical.

Much of the knowledge concerning particle packing can be put into working models for predicting many of the attributes of concern. There is definite need for a reliable computer simulation that would allow initial optimization or exploration of packing characteristics based on input parameters like particle size, container size, particle size distribution, particle shape, vibration, homogeneity, and interparticle friction. Sufficient knowledge on the role of each factor exists such that a reasonable predictive model could be constructed.

Percolation concepts have considerable importance to both the pore and particle structure. In mixed particle systems, there is an opportunity to optimize behavior using proper particle size ratios and compositions to maintain connectivity. One example is reactions involving particle systems, in which the connectivity of each phase should have an influence on the rate of reaction and the character of the final microstructure. Additionally, investigations of particle shape effects on connectivity are needed. Many composite materials can be formed from mixtures of particles with very different sizes and shapes. With an understanding of connectivity, it should be possible to control the consolidated microstructure to tailor the final properties to the performance goals. For example, the thermal expansion of a composite material does not follow the rule of mixtures; there is a role of interfacial stress. Thus, it is possible to adjust the thermal expansion of a composite through the connectivity of the two phases. This is just one opportunity evident from application of the concepts organized in this book.

Some industrial processes have capitalized on particle packing concepts to deliver some innovative new products. For example, novel ceramics are being fabricated using spherical particles packed to high densities with transient surface polarization techniques. Not only is the dimensional change in processing to full density reduced by the high initial packing density, but the quality of the final product is improved because of a more homogeneous initial packing. In another case, high performance composites are being formed using optimized packing of ceramic fibers and particles. The future should provide many new opportunities for use of these basic concepts in many fields of technology. The fundamental principles are well established.

References

1. D. J. Cumberland and R. J. Crawford, *The Packing of Particles*, Elsevier Science, Amsterdam, The Netherlands, 1987.
2. R. M. German, *Powder Metallurgy Science*, Metal Powder Industries Federation, Princeton, NJ, 1984, pp. 9-54.
3. E. M. Tory and W. S. Jodrey, "Comments on Some Types of Random Packing," *Advances in the Mechanics and the Flow of Granular Materials*, vol. 1, M. Shahipoor (ed.), Trans Tech Publ., Clausthal, Federal Republic of Germany, 1983, pp. 75-106.
4. E. V. Degtyareva and Y. M. Grinberg, "Vibrovacuum Packing of Powder Masses," *Refractories*, 1982, vol. 23, pp. 331-339.
5. E. A. Boucher, "Porous Materials: Structure, Properties and Capillary Phenomena," *Journal of Materials Science*, 1976, vol. 11, pp. 1734-1750.
6. V. K. S. Shante and S. Kirkpatrick, "An Introduction to Percolation Theory," *Advances in Physics*, 1970, vol. 20, pp. 325-356.
7. T. J. Coutts, "Percolation Conduction in Mixed Insulator-Conductor Systems," *Thin Solid Films*, 1976, vol. 38, pp. 313-321.

8. S. Modry, M. Svata and J. van Brakel, "Thematic Bibliography of Mercury Porosimetry," *Powder Technology*, 1981, vol. 29, pp. 13-43.
9. I. F. Macdonald, M. S. El-Sayed, K. Mow and F. A. L. Dullien, "Flow Through Porous Media - The Ergun Equation Revisited," *Industrial and Engineering Chemistry*, 1979, vol. 18, pp. 199-208.
10. A. E. Scheidegger, *The Physics of Flow Through Porous Media*, revised edition, University of Toronto Press, Toronto, Ontario, Canada, 1960.
11. R. E. Collins, *Flow of Fluids Through Porous Media*, Reinhold Publishing Corp., New York, NY, 1961.
12. K. S. Hwang, R. M. German and F. V. Lenel, "Capillary Forces in the Early Stage of Liquid Phase Sintering," *Reviews in Powder Metallurgy and Physical Ceramics*, 1986, vol. 3, pp. 113-164.
13. H. H. Hausner, *Bibliography on the Compaction of Metal Powders*, Hoeganaes Corp., Riverton, NJ, 1967.
14. C. G. Goetzel, *Treatise on Powder Metallurgy*, vol. 1, Interscience Publishers, New York, NY, 1949, pp. 259-312.
15. R. P. Seelig, "Introduction to Seminar - Review of Literature on Pressing of Metal Powders," *Transactions of AIME*, 1947, vol. 171, pp. 506-518.
16. G. F. Bocchini, "Influences of Porosity on the Characteristics of Sintered Materials," *Metal Powder Report*, 1986, vol. 41, pp. 829-832.
17. T. J. Coutts and B. Hopewell, "Monte-Carlo Studies of Thin Film Growth," *Thin Solid Films*, 1972, vol. 9, pp. 37-55.
18. M. T. Kamel, E. M. Tory and W. S. Jodrey, "The Distribution of kth Nearest Neighbors and its Application to Cluster Settling in Dispersions of Equal Spheres," *Powder Technology*, 1979, vol. 24, pp. 19-34.
19. W. S. Jodrey and E. M. Tory, "Computer Simulation of Isotropic, Homogeneous, Dense Random Packing of Equal Spheres," *Powder Technology*, 1981, vol. 30, pp. 111-118.
20. K. Gotoh, W. S. Jodrey and E. M. Tory, "Variation in the Local Packing Density Near the Wall of a Randomly Packed Bed of Equal Spheres," *Powder Technology*, 1978, vol. 20, pp. 257-260.
21. M. Suzuki, K. Makino, M. Yamada and K. Iinoya, "A Study on the Coordination Number in a System of Randomly Packed, Uniform-Sized Spherical Particles," *International Chemical*

Engineering, 1981, vol. 21, pp. 482-488.

22. D. C. Koskenmaki, "A Model for the Structure of Amorphous Metals," *Materials Science and Engineering*, 1976, vol. 23, pp. 207-210.
23. P. Mrafko and P. Duhaj, "Analysis of an Aggregate of Hard Spheres," *Physica Status Solidi*, 1974, vol. 23, pp. 583-589.
24. J. P. Straley, "Critical Experiments for the Conductivity of Random Resistor Lattices," *Physical Review B*, 1977, vol. 15, pp. 5733-5737.
25. P. Mrafko and P. Duhaj, "A Compromise Hard-Sphere Model for Amorphous Metallic Alloys," *Journal of Non-Crystalline Solids*, 1975, vol. 17, pp. 143-145.
26. J. L. Finney, "Fine Structure in Randomly Packed, Dense Clusters of Hard Spheres," *Materials Science and Engineering*, 1976, vol. 23, pp. 199-205.
27. A. Sur, J. L. Lebowitz, J. Marro, M. H. Kalos and S. Kirkpatrick, "Monte Carlo Studies of Percolation Phenomena for a Simple Cubic Lattice," *Journal of Statistical Physics*, 1976, vol. 15, pp. 345-353.
28. C. H. Seager and G. E. Pike, "Percolation and Conductivity: A Computer Study. II," *Physical Review B*, 1974, vol. 10, pp. 1435-1446.
29. anonymous, "What is Random Packing?," *Nature*, 1972, vol. 239, pp. 488-489.
30. J. A. Barker, M. R. Hoare and J. L. Finney, "Relaxation of the Bernal Model," *Nature*, 1975, vol. 257, pp. 120-122.
31. M. Rubinstein and D. R. Nelson, "Order and Deterministic Chaos in Hard-Disk Arrays," *Physical Review B*, 1982, vol. 26, pp. 6254-6275.
32. B. J. Alder, S. P. Frankel and V. A. Lewinson, "Radial Distribution Function Calculated by the Monte-Carlo Method for a Hard Spheres Fluid," *Journal of Chemical Physics*, 1955, vol. 23, pp. 417-419.
33. H. H. Kausch, D. G. Fesko and N. W. Tschoegl, "The Random Packing of Circles in a Plane," *Journal of Colloid and Interface Science*, 1971, vol. 37, pp. 603-611.
34. D. S. Boudreaux and J. M. Gregor, "Structure Simulation of Transition-Metal-Metalloid Glasses," *Journal of Applied Physics*, 1977, vol. 48, pp. 152-158.
35. L. V. Woodcock and C. A. Angell, "Diffusivity of the Hard Sphere Model in the Region of Fluid Metastability," *Physical Review Letters*, 1981, vol. 47, pp. 1129-1132.

36. M. J. Powell, "Site Percolation in Randomly Packed Spheres," *Physical Review B*, 1979, vol. 20, pp. 4194-4198.
37. G. F. Round and R. Newton, "Random Packing of Equal Spheres on a Plane Surface," *Nature*, 1963, vol. 198, pp. 747-749.
38. J. L. Finney, "Modelling the Structures of Amorphous Metals and Alloys," *Nature*, 1977, vol. 266, pp. 309-314.
39. E. J. LeFevre, "Equation of State for Hard Sphere Fluid," *Nature Physical Science*, 1972, vol. 235, p.20.
40. A. Bonissent and B. Mutaftschiev, "A Computer Built Random Model for Simulation of the Crystal-Melt Interface," *Philosophical Magazine*, 1977, vol. 35, pp. 65-73.
41. A. J. Matheson, "Computation of a Random Packing of Hard Spheres," *Journal of Physics C: Solid State Physics*, 1974, vol. 7, pp. 2569-2576.
42. T. Ichikawa, "The Assembly of Hard Spheres as a Structure Model of Amorphous Iron," *Physica Status Solidi*, 1975, vol. 29, pp. 293-302.
43. A. S. Goldman, H. D. Lewis and W. M. Visscher, "Random Packing of Particles: Simulation With a One Dimensional Parking Model," *Technometrics*, 1974, vol. 16, pp. 301-309.
44. J. Rodriguez, C. H. Allibert and J. M. Chaix, "A Computer Method for Random Packing of Spheres of Unequal Size," *Powder Technology*, 1986, vol. 47, pp. 25-33.
45. N. R. Schott, "Precision Injection Molding of Thermoplastics," *Innovations in Materials Processing*, G. Bruggeman and V. Weiss (eds.), Plenum Press, New York, NY, 1985, pp. 219-230.
46. T. Itoh, Y. Wanibe and H. Sakao, "Relation Between Packing Density and Particle Size Distribution in Random Packing Models of Spheres," *Journal of the Japan Institute of Metals*, 1986, vol. 50, pp. 740-746.
47. T. M. Hare, "Statistics of Early Sintering and Rearrangement by Computer Simulation," *Sintering Processes*, G. C. Kuczynski (ed.), Plenum Press, New York, NY, 1980, pp. 77-93.
48. R. C. Burk and P. S. Apte, "A Packing Scheme for Real Size Distributions," *Ceramic Bulletin*, 1987, vol. 66, pp. 1389-1392.
49. M. M. Levine and J. Chernick, "A Numerical Model of Random Packing of Spheres," *Nature*, 1965, vol. 208, pp. 68-69.
50. F. A. L. Dullien, E. Rhodes and S. R. Schroeter, "Comparative Testing of Statistical Methods for Obtaining Particle Size

Distributions," *Powder Technology*, 1969, vol. 3, pp. 124-135.

51. M. Suzuki and T. Oshima, "Estimation of the Coordination Number in a Two Component Mixture of Cohesive Spheres," *Powder Technology*, 1983, vol. 36, pp. 181-188.
52. M. Suzuki and T. Oshima, "Comparison Between the Computer Simulated Results and the Model for Estimating the Coordination Number in a Three Component Random Mixture of Sphere," *Powder Technology*, 1985, vol. 43, pp. 19-25.
53. E. M. Tory, B. H. Church, M. K. Tam and M. Ratner, "Simulated Random Packing of Equal Spheres," *The Canadian Journal of Chemical Engineering*, 1973, vol. 51, pp. 484-493.
54. E. M. Tory, N. A. Cochrane and S. R. Waddell, "Anisotropy in Simulated Random Packing of Equal Spheres," *Nature*, 1969, vol. 220, pp. 1023-1024.
55. W. M. Visscher and M. Bolsterli, "Random Packing of Equal and Unequal Spheres in Two and Three Dimensions," *Nature*, 1972, vol. 239, pp. 504-507.
56. M. J. Powell, "Distribution of Near Neighbours in Randomly Packed Hard Spheres," *Powder Technology*, 1980, vol. 26, pp. 221-223.
57. M. J. Powell, "Computer Simulated Random Packing of Spheres," *Powder Technology*, 1980, vol. 25, pp. 45-52.
58. H. J. Frost, "Cavities in Dense Random Packings," *Acta Metallurgica*, 1982, vol. 30, pp. 889-904.
59. T. Tanaka and N. Ouchiyama, "Coordination Numbers and Porosity of Packing Calculated from Particle Size Distribution," *Proceedings of the Fourth International Symposium on Agglomeration*, C. E. Capes (ed.), Iron and Steel Society, Warrendale, PA, 1985, pp. 9-16.
60. D. J. Adams and A. J. Matheson, "Computation of Dense Random Packings of Hard Spheres," *Journal of Chemical Physics*, 1972, vol. 56, pp. 1989-1994.
61. R. H. Wang and L. T. Fan, "Critical Contact Number for Percolation Processes," *Powder Technology*, 1979, vol. 23, pp. 139-142.
62. C. C. Furnas, "Grading Aggregates I - Mathematical Relations for Beds of Broken Solids of Maximum Density," *Industrial and Engineering Chemistry*, 1931, vol. 23, pp. 1052-1058.
63. T. H. Ferrigno, "Principles of Filler Selection and Use," *Handbook of Fillers and Reinforcements for Plastics*, H. S. Katz and J. V. Milewski (eds.), Van Nostrand Reinhold, New York, NY, 1978, pp. 11-58.

Index

A

B

C

D

E

F

G

M

N

Q

R

S

T

U

V

W

X

Y

Z

—NOTES—

—NOTES—

—NOTES—